Economics of the Energy Industries

WILLIAM SPANGAR PEIRCE
Case Western Reserve University

Wadsworth Publishing Company
Belmont, California
A Division of Wadsworth, Inc.

For Nynke, who bore the costs,
and Arjen, Charles, and Hester, who expect some benefits

Economics Editor: Stephanie Surfus
Production Editor Deborah O. McDaniel
Managing Designer: Paula Shuhert
Designer: Salinda Tyson
Print Buyer: Barbara Britton
Copy Editor: Elaine Linden
Cover Designer: Salinda Tyson
Cover Photographer: Ellis Herwig/The Image Bank
Signing Representative: Art Minsberg

Cover: Windmills at sunset suggest the interplay of technology, economics, aesthetics, policy, and chance in the choice of energy sources.

Printed in the United States of America
1 2 3 4 5 6 7 8 9 10—90 89 88 87 86
ISBN 0-534-05286-X

Library of Congress Cataloging in Publication Data

Peirce, William Spangar.
Economics of the energy industries.

Includes index.
1. Energy industries—United States. I. Title.
HD9502.U52P43 1986 338.4'7621042'0973 85-8998
ISBN 0-534-05286-X

C O N T E N T S

FIGURES

TABLES

PREFACE

This book is an introduction to the way economists analyze energy and to the central economic questions of each of the energy industries. The objective is to provide the reader with the background information, analysis, and vocabulary necessary for understanding the mass of specialized literature currently being published and the policy issues discussed almost daily in the news media.

The intended audience is the person who knows nothing about energy and knows only the basics of microeconomics as recalled from an economic principles course. Those who have more specialized knowledge of economics or some aspect of energy may also gain from the broader perspective offered here.

Part I provides the framework of economic concepts and gives an overview of the changes in energy markets since the turbulent 1970s. An underlying theme is the need to rely on the data that markets provide about human preferences when trying to analyze energy. Part II examines the extractive industries, beginning with the concepts of reserves and resources and the theory of optimal exploitation, and then moving on to consider the coal, oil, and natural gas industries. Part III deals with electricity, nuclear power, and various other sources of energy that have recently attracted attention. Finally, Part IV summarizes a number of the public policy questions raised by the energy industries. The objective throughout is to provide just enough detail so that the beginner can read more specialized studies and grasp their relationship to the larger picture.

Many people have helped me with this project. The East Ohio Gas Company provided funding to the Department of Economics of

Case Western Reserve University to permit me to initiate an Economics of Energy course. Case Western Reserve provided me with the sabbatical during which the first draft was written. My wife and children were very indulgent of my eccentricity in starting a third book while the first two were still in process. My colleagues were helpful and supportive; and Bela Gold (now of Claremont Graduate School) and Gerhard Rosegger shaped my views and approach during nearly two decades of collaboration. The enthusiasm of John Wierzbicki enabled me to deliver an early draft when my promises outran my energy. The secretarial staff of the economics department surmounted all obstacles to ensure that students received the text on time. I am indebted to David A. Huettner, James B. Lindberg, and John M. Peterson for many useful suggestions. Most of all, however, I want to thank the students who endured early versions of the course, improved two drafts of the text, and encouraged me to complete it.

P A R T O N E

An Economic Approach: Measures and Analysis

CHAPTER ONE

Energy: An Economist's View

THE GREAT ENERGY CRISIS

Shortages, rapidly rising prices, and talk about an energy crisis focused public attention on energy issues during the 1970s. The shortages during the fall of 1973 were the most dramatic symptoms of trouble in the petroleum markets; Americans are not used to waiting in line for an hour to buy *any* commodity. Not many months earlier, dealers had tried to attract customers by giving away maps, glassware, or other gifts. In 1973, however, the daily news carried pictures of lines stretching around the block, reports of shouting, fighting, and even killings in the waiting lines. Interviews featured people whose daily lives were completely disrupted by the inability to buy the gasoline they needed to reach work, home, school, or play.

The shortages were short-lived. The gasoline lines ended even before the Arab nations resumed shipments to the United States, although they recurred in 1979 when production fell in Iran. Nevertheless, much of the decade seemed to be characterized by reports, fears, memories, or forecasts of shortages of natural gas, fuel oil, gasoline, or electricity.

Although fear of shortage was more common than shortage itself, prices certainly showed conclusively that the markets for energy had changed in the 1970s. Figure 1.1 shows the behavior of some of the energy prices confronted by ordinary consumers. The price of natural gas increased at about the same rate as the Consumer Price Index (CPI) from 1950 to 1967, and the prices of fuel oil, gasoline, and electricity were even more stable. After 1973 all four of the energy prices increased rapidly. By 1981, when the CPI stood at 272 percent of the 1967 level, electricity came close to matching that at 267 percent, gas stood at 385 percent, fuel oil was 656 percent, and gasoline was 395 percent. These increases in the prices paid by a typical household were shocking because of both their magnitude and the publicity that OPEC (the Organization of Petroleum Exporting Countries) received. The amount and the timing of the price increases for various fuels differed because of such special characteristics as taxes, price controls, and the importance of fixed costs

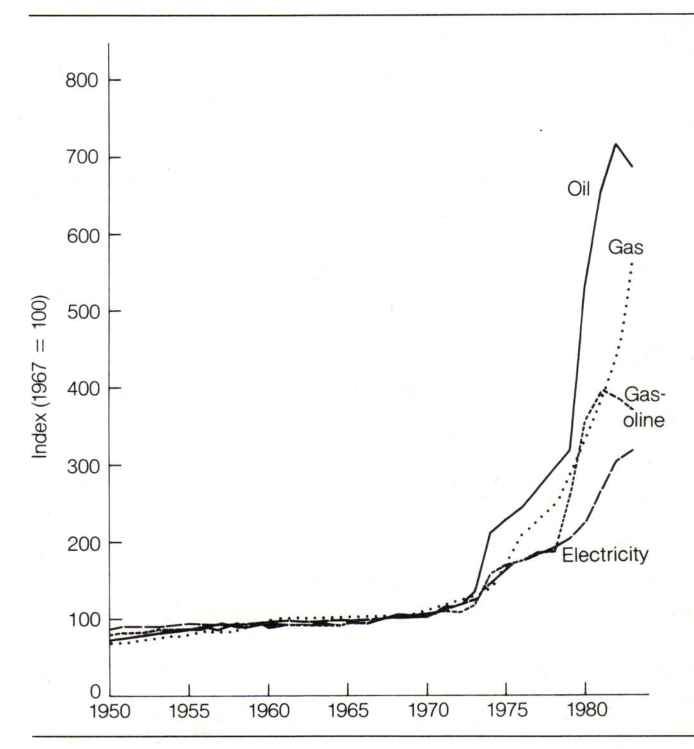

Figure 1.1 *U.S. Retail Price Indexes of Electricity, Gas, Oil, and Gasoline, 1950–1983*

Sources: U.S. Historical Statistics *(1975), p. 214; U.S. Department of Energy,* Annual Report *(1981), vol. 2, p. 95;* Monthly Labor Review *(various issues).*

and labor costs to the total cost of the industry. Despite such differences, however, wherever consumers turned they saw energy prices rising fast in the 1970s.

Before 1970 most people complacently assumed that electricity, gasoline, heating oil, and natural gas would be readily available at prices that yearly became less burdensome. Occasional power failures and shortages aroused irate disbelief from a public thoroughly conditioned to expect utmost reliability. Indeed, the substitution of mechanical and electrical devices for human effort was considered a symbol of economic progress. Furthermore, the assumption that energy prices would remain low was reflected in the design of cars and buildings.

Events of the 1970s popularized the gloomy forecasts that had once been restricted to an eccentric minority.[1] People waiting in line for gasoline, out of work because industrial plants were ordered to stop using natural gas, or shivering at home because oil was so expensive found it easy to believe that the age of plenty had ended. The eccentrics now seemed to be that small minority who argued that the shortages resulted from stupid government policy, not the niggardliness of nature.

Yet gradually the shortages eased. Even the coldest winter in a century in the Northeast produced only minor disruptions in 1980–81. Had the crisis that was supposed to herald the end of our civilization vanished? Did government policy suddenly become effective and intelligent? Had nature provided us with additional resources?

The most striking change after the shortages and fears ended was the new alignment of prices. Energy had become more expensive during the decade, dashing the dreams of unbounded supplies of cheap energy that had been inspired by vast reserves of cheap oil in the Middle East and by optimistic forecasts of the future of nuclear power. People are still adjusting to the new energy prices. Smaller cars and tighter houses are only the most obvious changes by consumers. Businesses are making equivalent improvements in insulation and motors. The more subtle and complex adjustments in industrial processes, location of economic activity, and choice of products are just beginning.

People have responded to high energy prices by trying to produce more energy, as well as by dimming the lights and turning down the heat. The number of crews exploring for oil in the United States and the rest of the world outside OPEC increased dramatically after prices rose.[2] The effort to increase non-OPEC production was successful, as the data in Chapter 9 show. Some users of oil responded to the high OPEC prices by shifting to cheaper fuels. The increased demand raised the prices of such fuels as well. Suppliers responded by searching for natural gas and developing more coal mines. Other people have turned their attention to using sources of energy as traditional as burning wood or angling a house toward the sun or to studying ideas as esoteric as the direct production of electricity from sunlight or from nuclear fusion.

All these responses are economic, although the basic problem was perceived as either political (OPEC power) or physical (exhaustion of natural resources). The decisions of individuals and businesses to change consumption and production patterns were motivated primarily by the desire to profit from the new prices and conditions or to minimize the cost and inconvenience stemming from them. It is notable, however, that such purely private actions succeeded in solving the public problems of shortage and waiting lines.

Public policy during the 1970s was not able to come to grips with energy problems.[3] The political discussion centered on the question of the distribution of income and wealth, rather than the production and consumption of energy. The initial political reaction to scarcity and rising prices was to impose price controls. As shortages worsened, the next political response was to control the use of oil and gas as well. Controls were finally lifted when those who wanted free markets were able to combine with those who wanted high prices to encourage conservation. Nevertheless, government still chose to tax away a substantial part of the higher profits that the new prices implied. The particular policy problems of each energy industry are discussed in Part II.

ARE WE RUNNING OUT OF ENERGY?

Although the market rescued us from shortages in the 1970s, perhaps we really are running out of energy, as the pessimists feared. Were the problems of that decade a completely false alarm or just an early warning? The old quip states that an economist is someone who believes that you get oil by drilling a deep hole and stuffing enough money down it. If you drill enough holes to the right depth in the right places the technique will work for a while. But surely the amount of oil or any other resource underground really is finite, which means that we should think seriously about the prospect of exhaustion.

The most direct approach to the question of exhaustion of fuels is to obtain detailed geologic information about the quantity, cost, and quality of various mineral deposits. Techniques for obtaining such information are discussed in Chapter 6. Information based on exploration in the field is expensive, however, so some people have attempted to grapple with the issue of scarcity at a more theoretical level.

In particular, some have tried to use insights from thermodynamics (the physicist's study of heat) to answer economic questions.[4] The first law of thermodynamics holds that the quantity of energy in the universe remains constant. This is not very comforting, because according to the second law the energy available to do work continually decreases, whereas the unavailable energy (entropy) continually increases. For example, if we have a source of intense heat (a furnace burning coal) and relatively cool surroundings (the outside air), heat will flow from the furnace to the air. By building some device, such as a steam turbine and electric generator, we can use some of the energy to generate electricity. At the end of the day the coal is gone and the air is a little warmer around the power plant and around all the electric motors, lights, and heaters. The energy still exists, but not in a form in which we can use it. By more efficient engineering we can make more use of the energy from each pound of coal before it is dissipated, but that only slows the gradual running down of the stocks of available energy; it cannot end the process or reverse it.

Despite its usefulness for other purposes, thermodynamics does not provide a shortcut to economic information because the assumptions of the two fields are different. In particular, this simple formulation of the laws of thermodynamics applies to a closed system (the universe), whereas economists restrict their analysis to the earth. Furthermore, since economics is a study of human behavior, economic analysis is concerned only with that time period during which human beings will live on the earth. These differences are significant because the economist can treat sunlight as a net energy input to the economic system without noting that the rest of the universe is losing that amount of energy. Moreover, astronomers predict that someday the sun will either flare up and vaporize everyone or just cool off and freeze the earth. Any resource that will last that long can be treated as inexhaustible, even if the quantity available is finite. The important economic questions about prices and quantities of various fuels can be answered only by the hard work of examining the various energy industries and the behavior of consumers.

Let us rephrase the question from "Are we running out of energy?" to "Will you encounter shortages of energy this year?" This is a meaningful question that falls within the realm of the economist. The answer, however, will depend on the behavior of government. There is no physical reason for a shortage to occur during the next year, but as every text in the principles of economics stresses, government can always create a shortage by setting a low enough price ceiling. It is difficult to imagine any other event that could cause a widespread shortage this year. It is easy, however, to think of events that could cause spot shortages (for example, extremely cold weather in an area with inadequate natural gas storage and distribution facilities), and it is easy to imagine incidents that could cause prices to rise so rapidly that some people would suffer.

The most meaningful economic time period is something between a year and a millennium. Ideally, a firm or an individual wants to know about the availability and prices of various inputs during the entire life of major investments. An electric utility commits itself to a particular fuel for the twenty-year economic life of a plant, but if the planning, licensing, and construction require ten years, the decisions are really based on guesses about what will happen to fuel prices and electricity markets for the next generation. Individuals also try to look a generation ahead for major decisions on occupation and location. People may be interested in the shape of the world a half century from now, when they plan to retire, or even of the world that their children and grandchildren will inherit. Such musings become economically relevant if they influence overt behavior in the markets. If I mumble to myself that coal will be very valuable in another century, that mumbling has no economic implications. If, however, I buy coal reserves, perhaps to provide for my grandchildren, my behavior affects the market.

Some who try to look several hundred years ahead suggest that the availability of energy is the central technical issue.[5] If the traditional fuels can be replaced by plentiful supplies of something like solar or fusion energy at reasonable cost, then scarcity of other minerals will not be a serious problem. Although particular minerals might become very expensive, practically unlimited amounts of such elements as iron, aluminum, and silicon can be used as substitutes if other sources of energy are available to replace the fossil fuels as they approach exhaustion.

Within the economic time horizon of, say, two decades, the problem is not one of running out but rather of relative prices. Only by mismanagement could we actually run short of oil, gas, coal, wood, or electricity during my lifetime or yours. But as people make decisions they want to know about relative prices of various forms of energy, labor, and other inputs and also about the prices of the products they sell. It is not a question of whether energy is scarce—all goods that are sold at a positive price are scarce—but rather of directing our efforts to conserve one scarce input at the expense of another.

SCARCITY

Why speak of scarcity when solar energy is unlimited and supplies of coal and oil shale are great enough for any economically relevant period? Although doomsday will not be brought about by an energy shortage, the fact remains that the resources to capture solar energy and to exploit other energy sources *are* scarce. The energy industries absorb a huge quantity of capital. This quantity may very well increase as less favorable deposits are exploited and the economy comes to depend on such diffuse sources as sunlight, wind, vegetation, or small differences in the temperature of different ocean currents.

The very best is always scarce, of course, and this poses a particular problem in the case of natural resources. It is commonly assumed that the best of the known deposits of oil and gas will be exploited first. This leads to a number of questions: (1) How much scarce labor and expensive equipment should be devoted to searching for superior deposits? (2) How much capital should be devoted to exploiting the known inferior deposits? (3) How much capital and labor should be used in switching to alternate sources of energy or in reducing consumption? (4) How much of the best deposits should be saved for future generations? These are all economic questions. Theory provides one set of answers about what people will do under certain assumptions, but actual behavior may not correspond with the predictions. These matters are examined later.

In the case of coal, the vast quantity of known resources makes the questions somewhat different. Here too we can ask how much of the best should be saved. The more important questions, however, relate to environmental impact. How much capital and labor will we expend to decrease the damage to the environment? The other side of this question is whether the same deposit remains the best when environmental costs are taken into account.

Analyzing solar (and perhaps nuclear) power, we see no meaningful limit to the amount of energy available, but it can be captured only at the cost of scarce capital and labor. Health, safety, and aesthetic costs can also be significant in particular applications. Similarly, growing plants for use as energy—trees for firewood or corn and sugar cane for alcohol to power autos—is certainly possible, but the cost must be measured in alternative uses of the land (food production, for example) and the capital and labor required.

MEASURING ENERGY

Analyzing such problems requires the measurement of energy. This means that various energy sources like electricity, coal, and oil must be expressed in common units. It is customary to use heat content as the weighting device. For example, data are often stated in barrels of oil equivalent (BOE), in calories (the amount of heat necessary to raise one gram of water one degree Celsius), or in British thermal units (the amount of heat necessary to raise one pound of water one degree Fahrenheit, abbreviated as BTU). In this book the BTU will be the standard unit because it is commonly used for reporting national energy statistics. Of course, for weighting various inputs any other consistent units would work just as well. Despite the availability of a common measure, however, the addition of firewood, nuclear power, and gasoline presents some conceptual problems. These are considered in Chapter 3.

In most industry studies the output is described both in physical terms (kilowatt-hours of electricity, tons of steel, number of automobiles) and in terms of the aggregate value of output. Both forms of weighting have their uses and limitations. Adding one pickup with a seat in back plus one truck that hauls 200-ton loads off the highway yields two trucks, but for many purposes it is more meaningful to state the total dollar value of the two. Similarly, different forms of energy have different prices per BTU and different uses. Yet the apparent physical homogeneity of energy sources has been so beguiling that energy statistics are nearly always aggregated on the basis of heat content alone.

This raises two questions. First, are the products of the various energy industries sufficiently interchangeable in use so that we are justified in treating them as one big industry? Second, if so, what is the appropriate weighting to use in the aggregation? The first question is answered in the markets for energy, where consumers reveal their preferences. These are sketched in Chapter 2. In the production of heat, for example, most of the possible energy sources are good substitutes. Even if conditions remained stable for a long time, however, coal and oil would not sell at identical prices per BTU. The commodities are substitutes in an economic sense, but this does not solve the aggregation problem.

Homes can easily be heated with oil, gas, coal, wood, or electricity. But at what price are homeowners indifferent among them? In the short run, of course, the homeowner will use whatever fuel the house is equipped for, unless cost differences become very large and are expected to persist. If conditions remain stable for a long period of time, however, would we expect to see identical prices? No, because the efficiency with which different energy inputs are used differs. In typical installations, about 60 percent of the energy content of oil or gas would go to heat the home, with the other 40 percent lost through incomplete combustion, heat vented through the flue, pilot lights burning when heat is not wanted, or similar losses. These losses can be reduced, but that takes money and effort. Electric resistance heating is 100 percent efficient within the home. All the energy purchased is converted into heat with no thought or effort by the consumer. An electric heat pump is more than 100 percent efficient (perhaps 150 percent in a particular application).

Heat pumps do not violate the laws of thermodynamics; they extract heat from the environment and pump it into the house. Coal and wood, by contrast, may be burned about as efficiently as oil or gas in well-designed stoves and furnaces, but they also require more work by the consumer.

SUBSTITUTION AND THE BOUNDARIES OF THE INDUSTRY

This suggests that the various forms of energy should not be aggregated solely according to BTU content, even as sources of heat where substitution is easiest. Furthermore, defining industries and markets according to the possibilities for substitution is not totally clear-cut because we can substitute insulation (or insolation) for fuel to maintain the same air temperature in the house. If the goal is to maintain the body at a comfortable temperature, then a sweater becomes a substitute for fuel oil; we might even choose to eat more and do a few push-ups to keep warm, in which case peanut butter becomes a substitute for coal. But this way lies madness. The industry must be defined in a conventional way, even though consumers will make a much wider range of substitutions.

At the industrial level, processes vary widely in their adaptability to different fuels. For producing heat, firms have the same range of options as households, but the convenience of electricity and the virtue of being able to contribute our own land and labor to provide wood become insignificant except in specialized situations. Some industrial processes have highly specific input requirements, such as the use of coke in the blast furnace or natural gas as a feedstock in fertilizer manufacturing. Even the most highly specific requirements can be varied in the long run, but sometimes only by rebuilding an entire plant. At the other extreme, electricity can be generated from anything that can be burned, from moving water, wind, sunlight, nuclear fission, or the heat of the earth.

To the extent therefore that final consumers require either heat or electrical power, the various energy sources are technically useful substitutes. The two main areas where substitution is difficult are the uses of particular fuels as chemical or process inputs – about 12 percent of oil and 3 percent of natural gas consumption – and the use of liquid fuels for transportation. Trucks, cars, ships, and airplanes use 58 percent of all petroleum in the United States. The liquid fuels have special advantages for transportation because of the easy handling and storage combined with the high ratio of energy to weight and the small amount of space required for storage. This does not mean that liquid fuels are essential for transportation. Ships are also powered by nuclear reactors, wind, and coal; and electricity and propane are used to power some specialized vehicles. For most transportation uses, however, liquid fuels will be purchased even when they are much more expensive per BTU.

WEIGHTING ENERGY SOURCES BY PRICE

The appropriate premium for the portability of liquid fuels or the convenience of electricity is a subjective matter. It depends on the evaluation by users of the relative benefits of various energy sources. In effect, we would like to ask consumers, "What would an extra gallon of gasoline be worth to you?" and compare that with the value of an extra ton of coal.

If markets were perfect, the prices observed in the market would provide this information. As every introductory economics text stresses, the significance of the intersection of the supply and demand curves is that the point shows both the value of additional production to the consumer and the cost of additional production to the society. Specifically, consumers are constantly faced with the choice of whether an additional unit of the commodity will give them pleasure equal to its cost. In the case of energy, typical consumers in 1981 were paying $18 per million BTU for electricity and $3 per million BTU for coal, yet aggregate residential consumption of electricity was 4.4 quads (1 quad $= 1 \times 10^{15}$ BTU), whereas coal consumed residentially was less than 0.2 quads and most households used no coal at all. Of course, decisions are made by particular consumers facing particular prices and other conditions (if your house has no chimney, you will not buy much coal). Nevertheless, the cost of buying another unit of the commodity is a good measure of its value to consumers. Similarly, under pure competition with perfect markets, the price will also be equal to the cost to society of producing another unit.

Such prices from perfect markets would be the appropriate weights to use in aggregating energy production. From the consumer's view, they take account of all the reasons for which consumers are willing to pay a premium for particular energy sources. These include questions of engineering efficiency (an electric motor will use fewer BTU than a coal-fired steam engine in sporadic operation) and matters of convenience (it is easier to flip a switch than to light a kerosene lamp in the middle of a dark night). On the producer side, the prices resulting from perfect markets include allowances for all the energy and other resources directly and indirectly consumed in the production of the energy source in question.

So much for the prices established in perfect markets. What can we expect of the prices observed in the world we live in? On the demand side, regulation of rates by government causes prices to depart from marginal cost. Often the consumer faces declining block rates (discussed in Chapter 11), which means that successive amounts of gas or electricity cost the consumer less than the first units consumed. The consumer may use one fuel for all purposes instead of the correct one for each purpose if the rate schedule declines sufficiently with use. A more serious problem is the neglect of externalities. Individuals may find coal or wood cheaper than gas or oil because they are not charged for the air pollution that coal and wood cause.

On the production side, the situation is more serious. In the past price controls on oil, electricity, and natural gas have encouraged excessive use. In particular, the tendency of regulators to set prices to cover average cost can mean a long delay in adapting to new conditions when marginal cost is increasing rapidly. Such problems are particularly severe in electric power production, where the costs of adding generating capacity have risen rapidly and the relative prices of fuels have changed as well. If electricity were priced everywhere at marginal cost, prices would be much higher and consumption correspondingly depressed. This topic is examined in Chapter 11.

Market prices can depart from marginal social cost for reasons other than government intervention. Indeed, economists sometimes call for government to correct departures that result from such factors as externalities (discussed in Chapter 14) or monopoly. If a power plant produces air pollution as well as electricity, the marginal social cost of electricity is greater than the marginal private cost to the electric company. The government has intervened in recent years to force producers to pollute less, but only a very brave analyst could argue that price is now equal to marginal social cost. When a firm faces a downward-sloping demand curve, it will maximize profits at a price that exceeds marginal cost. This furnishes the economic rationale for regulating public utilities, which is discussed in Chapter 10.

Although actual prices depart from marginal social cost and lead consumers to decisions that are inefficient for the economy, the bias that the imperfections produce in the statistics is not as great. If the market price is too low, the quantity purchased will be too high. Total spending on the product may vary either way, depending on the elasticity of demand. Ordinarily, the product of price times quantity will be closer to its "correct" value than will either price or quantity alone.

Ideally, data should be weighted by both BTU and by prices in various tables and charts. Neither is correct. It is just a question of what is useful for the purpose at hand. The central point is that commodities are valued for their qualities, not for themselves. Energy inputs have a variety of qualities, of which heat value is only one. For some purposes it is meaningful to aggregate. For others it is more useful to consider the industries separately, especially since production conditions vary so much.

THE IMPORTANCE OF ENERGY

How important are energy costs to the economy? This became a key question during the rapid rise of prices in the 1970s. One indication of the direct spending by an urban household on energy is the weight of 6.6 percent for fuel and other utilities and 5.9 percent for gasoline in the Consumer Price Index. These weights were based on detailed budget studies made in 1978 and contrast with 5.2 percent and 3.0 percent for the same categories in 1964.[6] Expenditures on energy as a proportion of the household budget may understate the visibility of rapidly rising energy prices to consumers. Motorists think of gasoline prices whenever they fill their tanks, and homeowners ponder utility bills monthly as they are assaulted by them. Even more important is the intertwining of energy costs with the major household decisions. Automobile purchases, the type and location of housing, and the search for a new job have all been influenced by the recent changes in energy prices, which gives energy a prominence that exceeds its quantitative importance.

More dramatic arguments about energy as the basis for all civilization or material progress have also been made.[7] The discrete form of the argument takes episodes from the history of technology and relates them to the growth in energy consumption over time. Thus energy use per person increased with the transition from hunting and gathering to agriculture, again in the industrial revolution, again with the age of iron and steam, and again with our modern, urban, automobile-dominated civilization. Further advances in material civilization are ahead of us, according to this view, but only if we can provide the ever-increasing supplies of energy required.

Instead of looking at discrete episodes, we can plot a continuous time series of the ratio of energy consumption to Gross National Product (GNP). This relationship is examined closely in Chapter 4. By looking at the broad sweep of the ratio over time or the differences among countries, the analyst can make a case for the relationship between energy and material progress. Of course, the deviations from the broad sweep are also interesting. In particular, the energy input per dollar of GNP has been declining in the United States since 1920.[8] Similarly, France, Italy, and Japan have far lower ratios of energy to GNP than do the United States and Canada. We discuss these relationships in more detail after examining the conceptual basis.

Before dealing with such refinements, the main point can be reiterated: Historically, cheap and abundant power was broadly associated with economic growth, but there were some exceptions and there are some trends that suggest the declining importance of the relationship. The industrial revolution in England and in the United States was associated with the use of large amounts of water and steam power. Similarly, late nineteenth-century mechanization was associated with the use of huge steam-powered machines to replace human and animal labor. However, Japan, where energy is much more expensive, passed through corresponding stages of industrialization later with much smaller power input. Furthermore, the new technologies involving electronics, automatic controls, and lighter materials give every indication of continuing reductions in the energy-to-GNP ratio. It is the meaning of such reductions that must be considered.

Rosegger's study of energy inputs relative to output in the iron and steel industry suggests some of the problems of analysis.[9] Several energy inputs declined markedly in the period from 1943 to 1970, although prices of fuels were not rising during this period. The largest declines were in steam coal and in fuel oil; coke consumption also continued its long decline. Partly offsetting these were sharply increased consumption of electricity, natural gas, and especially oxygen, which, although not a fuel, requires energy to produce and reduces the requirements for other energy inputs. This raises a number of interesting questions, including the one just analyzed of how to add up electricity, natural gas, coal, and oxygen to determine whether energy input went up or down.

There is an analogous question on the output side. Rosegger's data are expressed in terms of inputs per ton of output, and tonnage is indeed the traditional measure of steel industry output. But for most purposes tonnage is not what people look for when they buy steel. Manufacturers of cans, for example, require a certain area to produce a can of standard size. With the trend toward thinner steel, a ton of steel will produce more cans now than it produced at the turn of the century. Crushing a beer can with one hand is not the feat of strength it once was. For purposes other than testing strength, however, the lighter cans are equally useful. The additional rolling to produce the "thin tin" also absorbs additional energy while making the product lighter. Hence BTU per ton should be expected to rise. But the important question is how the innovation of thin tin influences BTU per can. Theory does not answer that question. A rough empirical estimate suggests that BTU per can is decreased by the additional rolling. Indeed, despite the improved usefulness of a ton of steel, energy input per ton has steadily declined over the years when the various inputs are aggregated in the usual way.

All of this suggests that economic analysis plays about the same role for the energy industries as it does for other industries. Yet the topic is particularly vital for a variety of reasons: (1) Energy is a significant part of the budget for many consumers and firms. (2) Changes in energy prices can cause substantial changes in the distribution of income and wealth among individuals and regions. (3) Energy questions have attracted a vast amount of political rhetoric, many policy proposals, and occasional policy decisions. (4)

Substitution of one form of energy for another, of capital and labor for energy, and the redirection of consumption in less energy-intensive directions open an intriguing variety of possibilities. (5) Energy questions are intertwined with important questions of national security and international relations. (6) Since the timing of exploitation of deposits is so important, energy questions are both interesting from a theoretical viewpoint and significant for the welfare of future generations. Moreover, (7) the energy industries provide numerous empirical counterparts of theoretical economic constructs and therefore interesting cases for economists to study.

NOTES

1. The most publicized vision of doom was Donella H. Meadows et al., *The Limits to Growth* (New York: Universe Books, 1972).
2. The Organization of Petroleum Exporting Countries (OPEC) is discussed in Chapter 9. Member countries in 1980 were Iran, Iraq, Kuwait, Qatar, Saudi Arabia, United Arab Emirates (Abu Dhabi, Dubai, and Sharjah), Algeria, Libya, Gabon, Nigeria, Indonesia, Ecuador, and Venezuela.
3. Several studies of energy policy are available, including Richard B. Mancke, *Squeaking By: U.S. Energy Policy since the Embargo* (New York: Columbia University Press, 1976); Mason Willrich with Phillip M. Marston, David G. Norrell, and Jane K. Wilcox, *Administration of Energy Shortages* (Cambridge, Mass.: Ballinger, 1976); and Joseph P. Kalt, *The Economics and Politics of Oil Price Regulation: Federal Policy in the Post-Embargo Era* (Cambridge, Mass.: MIT Press, 1981).
4. See Nicholas Georgescu-Roegen, *The Entropy Law and the Economic Process* (Cambridge, Mass.: Harvard University Press, 1971), especially pp. 295–306.
5. H. E. Goeller and Alvin M. Weinberg, "The Age of Substitutability," *Science*, February 20, 1976, pp. 683–689.
6. The weights used in the Consumer Price Index are reported periodically in the *Monthly Labor Review*.
7. For an illuminating study of the glorification of machinery and energy by intellectual leaders, see John F. Kasson, *Civilizing the Machine: Technology and Republican Values in America, 1776–1900* (New York: Grossman, 1976). Earl Cook, "The Flow of Energy in an Industrial Society," *Scientific American* 225, no. 3 (September 1971), p. 136, provides estimates of energy consumption per person at different stages of civilization.
8. For a highly informed general overview, see Sam H. Schurr, "Energy Efficiency and Productive Efficiency: Some Thoughts Based on American Experience," *Energy Journal* 3, no. 3 (July 1982), pp. 3–14.
9. Gerhard Rosegger, "Technological Change, Managerial Incentives and Environmental Effects," chap. 2 in *Technological Change: Economics, Management and Environment*, ed. Bela Gold (Oxford: Pergamon Press, 1975).

CHAPTER TWO

Energy Flows

The preceding chapter has stressed the variety of different possibilities open in the energy industry. Uses of particular energy sources by particular sectors are not matters of chance, history, or even technology, but rather the responses of decision makers to the choices presented to them. Yet the choices are not made anew every minute, and the possibilities that are available depend on the efforts that have been made in the past, so it is important to look at actual patterns of consumption and production. In interpreting the data on the present situation, we must avoid two errors—assuming that present patterns can never be changed and assuming that changes will be made quickly.

This chapter gives a quick overview of the sources and uses of energy in the United States and provides some perspectives for the more detailed examination of particular industries and applications of energy in later chapters. Suppose, for example, that production of electricity from wind triples in the next decade. Will that have a big impact on the economy? Suppose that consumers save 20 percent of the energy now used for home heating. What industries will that affect and by how much? In addition to furnishing the basic numbers, the chapter explains how such diverse items as coal, oil, and nuclear power are added to obtain the aggregate data.

AGGREGATE CONSUMPTION

Aggregate energy consumption in the United States fluctuated in the vicinity of 75 quads per year during the 1970s, as shown in Figure 2.1.[1] The interruption of the long uptrend in the series was a striking change that must be attributed to the price increases (shown in Figure 1.1). Not only was overall consumption fairly stable but also the distribution among the major categories of consumers showed little variation. Residential and commercial users consumed about 36 percent of all energy, industrial users accounted for about 39 percent of all energy, and transportation used the other 25 percent. These percentages fluctuate from year to year (and more with the seasons), but the overall stability of the proportions is notable.

In interpreting these (and subsequent) data, the definitions must be borne in mind. The energy consumed by the electric power

Figure 2.1 *Consumption of Energy by Type*

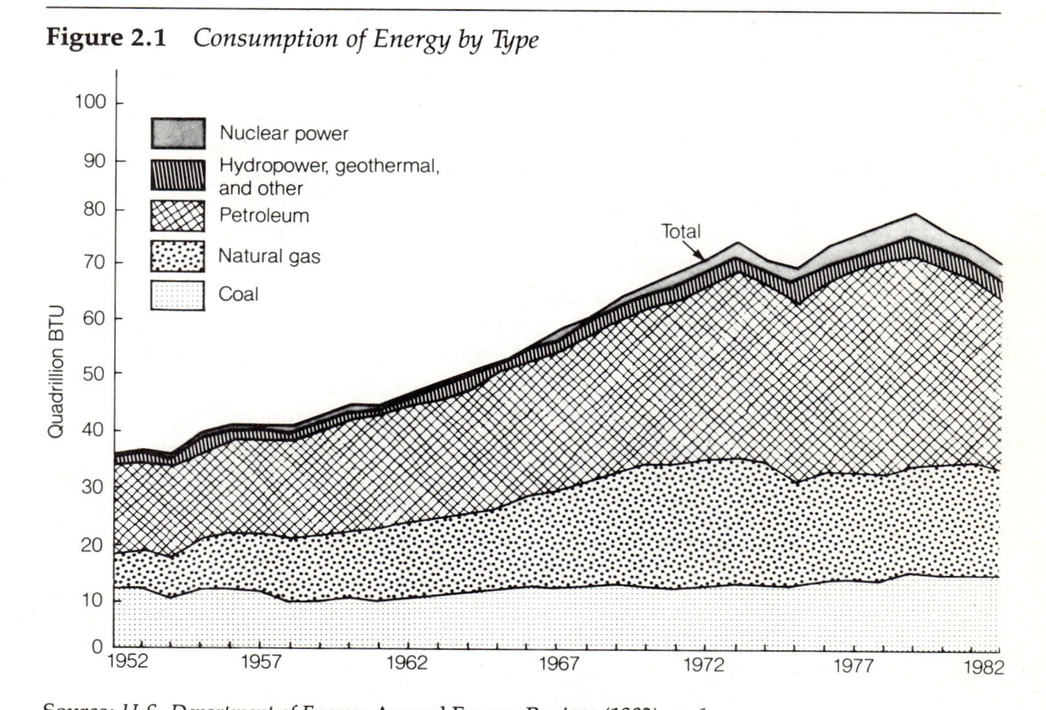

Source: *U.S. Department of Energy,* Annual Energy Review *(1982), p. 6.*

industry is attributed to the residential, industrial, or transportation sector in proportion to the use of electricity by that sector. The residential and commercial sector includes housing, nonmanufacturing business establishments, health and educational institutions, and government office buildings. The industrial sector includes construction, manufacturing, agricultural, and mining establishments. Transportation includes public, private, and government passenger and freight transportation.

In the statistics most electrical energy is counted as though it were generated by burning fossil fuels. If electric power is generated by burning coal, nearly 1 pound of coal is burned per kilowatt-hour (kwh) of electricity produced. One kwh of electricity is exactly equivalent to 3,412 BTU, but burning the coal to generate that kilowatt-hour releases more than 10,000 BTU of heat. (The exact figure varies. It has decreased over time as the efficiency of the generating process has improved. In 1980 the average for all fossil fuel electric plants was 10,435 BTU per kilowatt-hour.) The difference between the coal input and the electrical output is heat that is dissipated into the air and cooling water. In calculating total energy consumption for the economy, it seems appropriate to count the energy content of all the coal that goes into the power plant, not just the heat value of the electricity that comes out. If a household uses electricity, therefore, it will be considered to have used the number of BTUs in the coal that generated that amount of electricity.

This seems straightforward, but the way that the government statistics treat alternatives to coal-fired plants is more complex. The output of a hydroelectric plant is counted as though a fossil fuel plant had generated it—at 10,435 BTU per kilowatt-hour. Nuclear and geothermal power, however, is counted according to actual BTU input. Since nuclear power plants dissipate slightly more heat per kilowatt-hour produced, and geothermal plants considerably more, the data treat a nuclear kilowatt-hour as equivalent to 10,769 BTU and a geothermal kilowatt-hour as 21,611 BTU. Since neither the nuclear fuel nor the geothermal energy would have much of a market in the absence of its use for generating electricity, however, it would seem more rational to use the convention adopted for hydroelectric power; that is, use the fossil fuel equivalent of 10,435 BTU per kilowatt-hour. At present the treatment of nuclear and geothermal does not influence the data noticeably because the nuclear factor is so close to that of fossil-fueled plants, and geothermal is such a small fraction of total power production. The conceptual issues will become more important if substances that are currently wasted (for example, urban rubbish or sunlight) begin to make any significant contribution to energy supply.

Attributing energy consumption to a particular sector involves a conceptual question whenever an energy conversion process is interposed between the raw material and the final consumer. Oil refineries, for example, require energy for their operations just as electrical generating plants do. We could allocate the energy used in refining to residential, industrial, and transport sectors, as is now done with the losses in the electric power industry. Alternatively, losses in generating electricity could be attributed to the manufacturing sector, as is now done with refining. This issue will become serious when oil shale is processed in quantity: Will the data take the BTU contained in the original shale as the base, then allocate all to the consuming sectors or some to the mining activity? Or should the net output after mining and processing be the beginning point? The tendency in the mineral industries generally is to place most emphasis on material as it is shipped from the mine site, because those data are easiest to obtain. In any event, the inherent inconsistencies in the methods of accounting for energy are likely to become more troublesome as we turn increasingly to energy sources that are more like manufactured materials than like the oil that gushed freely from the ground.

Once having decided how to count energy inputs, we are immediately faced with an equivalent problem in counting energy outputs. As Table 2.1 indicates, people in various circumstances are willing to pay different amounts per BTU for different forms of energy. The most obvious implication of Table 2.1 is that the energy used by the residential and transportation sectors is expensive. This is not surprising because convenience, ease of handling, cleanliness, and portability are significant characteristics, along with energy content, for those uses. Table 2.1 also shows that electricity is by far the most expensive form of energy for residential customers, with an average price of $23.59 per million BTU in 1984. In that year natural gas cost $6.26 per million BTU, whereas residential heating oil was $7.48. Industrial users paid substantially less for their energy.

Because the energy purchased by households is so expensive, the residential share of total expenditures for energy is even larger than the share of energy shown in Table 2.2. Households also purchase much of the high-priced gasoline used for transportation. Table 2.2 shows the types of energy used by each sector. Even these data are not adequate indicators of the differences in energy costs, because industrial plants burn cheaper grades of oil and coal and often buy electricity and natural gas at cheaper off-peak or interruptible rates.

Nevertheless, the relatively low prices noted for industrial energy sources are misleading in two respects: The delivery conditions differ from those available to residential customers, and the location where fuel is cheap may determine industrial location. Large industrial users of gas or electricity often contract for interruptible supplies. This means that they must invest in standby equipment to use another fuel or be prepared to shut down when residential and commercial demand is high or generating equipment fails. The industrial plant may also buy electric power at the plant gate, providing its own transformers and lines. The residential customer pays a higher rate per kilowatt-hour, but does not incur those additional capital expenditures.

Location is the most significant determinant of the low average prices of natural gas and coal purchased by electric power plants. In the days when natural gas was flared off at the oil wells, electric power companies and industrial firms in the vicinity could purchase it for not much more than the cost of cleaning and transportation. Thus electricity and industrial production generally in the Southwest came to be based on cheap natural gas. Although these conditions had changed by the 1970s, as we discuss in Chapter 10, a combination of regulations and long-term contracts kept gas prices low for many large users. Even though natural gas prices to power

Table 2.1 *Retail Prices for Energy, 1984*

	Price per unit (dollars)	Energy per unit (million BTU)	Price per million BTU (dollars)
Residential			
Natural gas	6.43/MCF[a]	1.027	6.26
Fuel oil (#2)	1.04/gallon	0.139	7.48
Electricity	0.08/kwh	0.003412	23.59
Coal, anthracite	70/ton	25.	2.80
Wood (hardwood cut and split)	90/cord	16.2	5.56
Transportation			
Gasoline	1.21/gallon	0.125	9.68
Diesel fuel	1.005/gallon	0.139	7.23
Jet fuel	0.969/gallon	0.135	7.18
Industrial			
Natural gas	3.78/MCF	1.025	3.68
Electricity	0.05/kwh	0.003412	15.42
Fuel oil (#6)	28.56/BBL[b]	6.287	4.54
Coal, bituminous	37.58/ton	22.5	1.67

Sources: Monthly Energy Review *(September 1984), pp. 91, 99, 103; wood and anthracite are advertised prices in a local newspaper, not averages; the industrial natural gas price is the average price paid by electric utilities, as is the coal price.*

[a] *MCF means thousand cubic feet at standard temperature and pressure.*

[b] *BBL means barrel. A petroleum barrel equals 42 gallons.*

companies increased tenfold between 1973 and 1984, the price was still less than that of the oil that many eastern utilities and industrial firms relied on.

The price of coal, which was slightly more than that of natural gas in 1973, increased at a much slower rate for the period and so was the bargain by 1984. Again, however, the price in Table 2.1 is exceptionally low because the coal-based electric utilities were those that were favorably located relative to the deposits of coal.

Average prices therefore are not a very good guide to the price that a particular firm or household will pay. Firms consider energy costs or costs for a particular form of energy in their location decisions. Once a firm or household is in a particular location, it will certainly allow the relative prices of energy inputs to influence the choice of fuels (or of capital and labor relative to fuel). Thus the heaviest use of a particular energy source will be concentrated in the location where it is cheapest. Furthermore, electric utilities will pay especially low prices as they are the firms most heavily affected by fuel costs and most flexible in choice of fuel (at the moment when the plant is designed).

The firm or household, moreover, is typically interested not in the cost of fuel per BTU but in the total cost of the process, which includes the cost of the labor and equipment to burn the fuel and to clean up after it. Such costs are heaviest for coal and wood of the energy sources listed. Nevertheless, the data in Table 2.1 give some indication of the choices facing households and firms. A more

Table 2.2 *Sources and Users of Energy, 1983 (quads)*

	Coal	Natural gas	Petroleum	Electric	Hydro	Other	Nuclear	Total
Residential and commercial	0.192	7.244	2.345	4.683	[b]	[b]	[b]	14.464
Industrial	2.458	6.638	7.759	2.648	0.033	[b]	[b]	19.527
Transport	[b]	0.577	18.428	0.010	[b]	[b]	[b]	19.016
Electric utility	13.226	3.011	1.544	−7.341[a]	3.847	0.135	3.235	17.657[c]
Total	15.877	17.477	30.076	0	3.880	0.135	3.235	70.664

Source: Monthly Energy Review *(July 1984), pp. 23–29.*

[a] *The negative sign indicates that the electric utility sector supplies 7.341 quads of electricity to other sectors, while consuming 24.998 quads of coal and other sources of energy.*

[b] *Not significant or not available.*

[c] *Net energy use by electric utilities; that is, consumption minus amount supplied to other users.*

specific conclusion requires detailed study of the prices and availability of energy sources for a particular consumer in a specific location.

ENERGY SOURCES: CHANGES OVER TIME

As firms and households have responded to the specific conditions in the markets, the actual mix of energy inputs has varied substantially over time. Figure 2.2 provides a more systematic view of the variation in the importance of each of the energy sources. At the beginning of the twentieth century, coal was the mainstay of industrial America, furnishing two-thirds of the 10.2 quads of energy used. Wood still retained second place with 2.9 quads, but it was being confined more and more to rural areas or used as an adjunct to coal. The two fluid fuels were of small quantitative importance in 1900: Crude oil and natural gas each totaled about 0.2 quads.

The market for natural gas was restricted by transportation costs, and crude oil had been considered useful mainly as a source for kerosene to fuel lamps. Gasoline was a dangerous waste product of the early oil refineries, but by 1900 it was used to fuel a handful of automobiles. As automobiles developed into a serious form of transportation (by World War I), the oil industry became oriented toward supplying gasoline for the premium transportation market. Any part of the crude oil that could not be refined into gasoline had to be

Figure 2.2 *U.S. Consumption of Energy Inputs, 1900–1982*

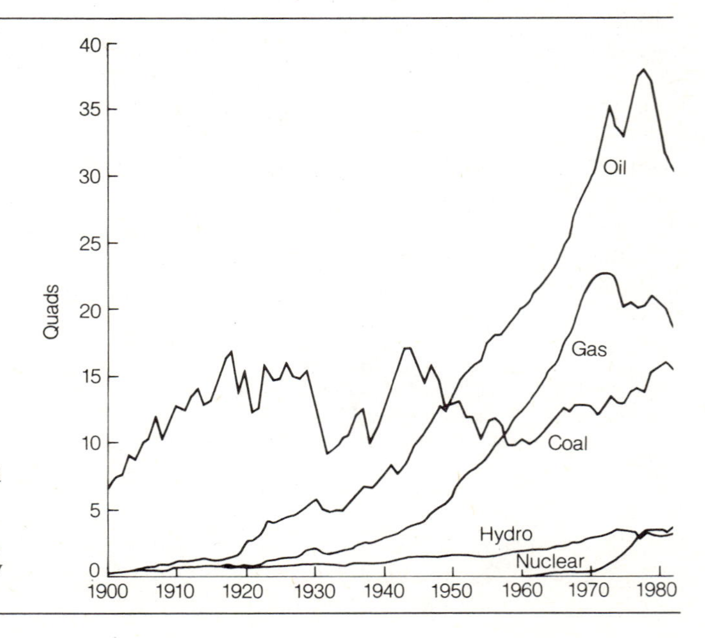

Sources: *1951–1981, U.S. Department of Energy,* Annual Report to Congress *(1981), vol. 2, p. 7; 1900–1950,* U.S. Historical Statistics *(1975), p. 588;* Monthly Energy Review *(December 1983), p. 11.*

sold for whatever price it would bring, so home heating and industrial boilers were gradually converted from coal to the more convenient liquid fuels.

Although the fortunes of the oil industry were tied closely to the automobile, coal was still the dominant industrial fuel. The five big markets were railroads, retail sales (including domestic heating and small commercial and industrial uses), electric utilities, the iron and steel industry, and general industrial use (primarily to produce steam). In 1900 these markets were all growing fast. By the 1920s the five were of roughly equal importance, and the coal industry still contributed about two-thirds of the 25 quads used in the country.

The heavy industrial markets of the coal industry were devastated by the Great Depression of the 1930s. Output recovered again to meet the World War II demands, but even then barely exceeded the World War I peak of 16.2 quads. After the war the industry resumed its decline. Home heating systems were converted to oil and natural gas, the railroads converted to diesel locomotives, and one industrial plant after another converted to oil or natural gas. Even some of the energy requirements of the electric utilities and the iron and steel industry were shifted from coal to the cleaner and more convenient oil and gas. Coal also lost markets whenever bargaining between the coal industry and the United Mine Workers brought strikes or the threat of strikes.

The result is seen in the long decline in coal production from 1944 through 1961, at which time it was only 10.2 quads. The one rapidly growing market for coal during this period was the electric utilities. By 1962 this growth had become important enough to offset losses in other markets, so the latest revival in coal began. By 1980 output had climbed to 15.6 quads—a level first attained in 1917. In the earlier year, however, coal provided 75 percent of the energy used by the United States. In 1980 its share was 20 percent.

The record for oil shows much steadier growth until the 1970s. Even the Great Depression produced only a plateau, not a large decline. Transportation in particular became almost completely dependent on oil with the dominance of the automobile and the truck, the shift of railroads to diesel power, and even the conversion of coal-burning ships to oil-fired steam turbines or diesel engines. By 1980 oil supplied 45 percent of the nation's energy use.

Natural gas, which had been at the same level as oil in 1900, grew much more slowly and fitfully into the 1940s. It was only with the extension of the natural gas pipelines to the industrial and population centers of the Midwest and Northeast after World War II that the era of rapid growth began. By 1951 oil had overtaken coal to become the largest energy source. By 1953 gas had also overtaken coal to move into second place, where it has remained. In 1980 natural gas accounted for 26 percent of energy consumption in the United States and gas and oil jointly accounted for 72 percent, sharing almost the same degree of dominance as coal had held at the beginning of the century.

The minor sources of energy, at the beginning of the century consisting mainly of wood and water power, included geothermal and nuclear by 1980. The total, however, was less than 3 quads in 1900 (mainly wood) and 5.9 quads, or 8 percent of total consumption, in 1980. It is quite likely that the use of wood for domestic heating is underestimated in the official statistics because wood lends itself to informal production. Individuals may cut their own or buy it for cash from an independent supplier who keeps no records. Nevertheless, the possible error in the total amount of energy is negligible even if wood use is two or three times the amount estimated.

The failure of hydropower to grow has been attributed to the lack of additional sites, although the higher prices for fossil fuels are now prompting a reexamination of sites that had been rejected or abandoned previously. Geothermal is here combined with hydro. It is growing but still accounts for very little in the national statistics. Again, the number of potential sites appears to be limited. Nuclear power has fallen farthest short of the forecasts of its early supporters; but this is a story that must await Chapter 12.

THE MATRIX OF ENERGY FLOWS

The patterns of energy use discussed in this chapter are summarized in Table 2.2, which shows the matrix of energy sources and uses for 1980, and Tables 2.3 and 2.4, which are calculated from Table 2.2. The tables give a highly aggregated snapshot of energy sources and uses in 1980. They suggest the general magnitudes of the flows as

Table 2.3 *Distribution of Sources, 1983 (by users of energy; percent)*

	Coal	Gas	Oil	Electric	Hydro	Other	Nuclear	Total
Residential and commercial	a	50	16	32	a	a	a	100
Industrial	12	34	40	14	a	a	a	100
Transportation	a	3	97	a	a	a	a	100
Electric utility	53	12	6	—	16	a	13	100

Source: *Calculated from Table 2.2.*

[a] *Not significant or not available.*

they were in 1980, but they should not be treated either as forecasts of how things will be in another decade or as statements of technological requirements today. Instead, the existing patterns of use reflect the response of consumers to the conditions of the times in which decisions were made.

The residential and commercial sector relies mainly on gas, oil, and electricity. Coal and other sources are negligible in the aggregate data. (Of course, if you live in Vermont chances are that you will not treat the contribution of "other" as negligible, since that includes wood.) In interpreting these three tables, we must keep in mind that electricity is counted at its heat value (3,412 BTU/kwh) to consumers, not at the thermal energy that was required to produce it (more than 10,000 BTU/kwh).

The industrial sector also relies heavily on natural gas, oil, and electricity, but in addition derives about 12 percent of its energy from coal and negligible amounts from hydro and other sources. As in the case of households, we can find particular firms using significant amounts of the minor sources such as water power and wood or waste.

The transportation sector is almost totally dependent on oil, although natural gas has about 3 percent of the market. (Use of natural gas in transportation consists mainly of using natural gas to power the pumps of natural gas pipelines.) Coal use in locomotives and ships has dwindled to negligible amounts. Some electricity is still used for rail transportation, especially in urban areas, but the aggregate amounts are small. Nuclear power is used in military ships, but commercial applications have encountered public opposition. Some minor use of other fuels does occur. A few electric vehicles are used by the residential and industrial sectors. Some gas-powered vehicles are used in special applications, but most of these

Table 2.4 *Distribution of Uses, 1983 (by energy sources; percent)*

	Coal	Gas	Oil	Electric	Hydro	Other	Nuclear
Residential and commercial	1	41	8	64	[a]	[a]	[a]
Industrial	15	38	26	36	1	[a]	[a]
Transportation	[a]	3	61	[a]	[a]	[a]	[a]
Electric utility	83	17	5	—	99	100	100
Total	100	100	100	100	100	100	100

Source: *Calculated from Table 2.2.*

[a] *Not available or not significant.*

are powered by liquified petroleum gas (LPG), which is a product of the petroleum industry. A few vehicles run on compressed natural gas (CNG).

The electricity used by the other sectors of the economy is produced from the various other energy sources, as indicated in the tables. About 53 percent of electric power was generated from coal, 16 percent from water power, and 13 percent from nuclear reactors. Minor amounts were also generated using geothermal heat, various wastes or wood, and wind. The marginal summations of Table 2.2 indicate the fundamental ambiguity in accounting for the energy conversion industries generally, but especially electric power. If all electricity had been generated in fossil fuel plants, it would have required 25 quads of energy, as shown in the horizontal row for electricity. The total electric power consumed in the other sectors, shown in the vertical column, sums to 7.3 quads. The remaining 17.7 quads were dissipated into the air and cooling water. Since all other data in Table 2.2 are inputs, the 7.3 quads of output by the electric utility industry are indicated by a negative sign.

The same flows can be analyzed in the other direction, as shown in Table. 2.4. The electric power industry consumed more than four-fifths of the coal, and the industrial sector consumed nearly all the rest. The industrial sector used 38 percent of the natural gas, residential and commercial used 41 percent, electrical generation used 17 percent, and transportation used 3 percent. Transportation was the biggest user of petroleum, taking more than half of it. Industry used one-quarter and residential and commercial shared the rest with electricity. Nuclear and hydro were used almost entirely to generate electricity, as was the miscellaneous energy. Finally, of the electricity produced, almost two-thirds was used by the residential and commercial sector and the rest by the industrial sector.

THE CHANGING GEOGRAPHY OF ENERGY

The changes in energy supply detailed in Figure 2.2 imply shifts in the geographic origin of energy supplies. The three topics that must be examined are (1) the degree of concentration of geographic origin and of consumption and hence the reliance on a linking network of transportation facilities; (2) the extent of the shifts in the origin of energy within the country; and (3) the reliance on imported energy.

Energy, like many other commodities, has been characterized by an increasing degree of geographic specialization and concentration. This required improvements in transportation to permit superior sources of energy to extend their market areas at the expense of inferior local products. At the same time, energy-intensive industries have clustered around the best energy deposits, substituting transportation of the product for transportation of energy. Both of these trends were evident in the early twentieth century as coal continued its displacement of wood and then the better coal deposits displaced inferior deposits. Some high-quality coal is transported as is, but an increasing share of low-cost coal is converted into electricity for use at remote sites. Except for electrification, these trends were well under way before 1900, as was the movement of energy-intensive industries such as iron and steel into areas served by rich coal deposits.

As the century wore on, the corresponding phenomenon was the shift of industry to the Southwest to take advantage of cheap natural gas or to the Northwest to take advantage of cheap electrical power. The implication is that the changing geography of energy production and consumption is closely intertwined with the development of transportation systems both for energy and for other commodities.

Specific information about geographic shifts in the production of each commodity is reserved for the chapters discussing the particular industries, but two major themes must be noted. The first is the concentration of high-valued hydrocarbon production in a relatively few locations of the South (and now Alaska). The second is the growth in oil imports after World War II, which initially provided cheap energy for the East Coast of the United States but eventually raised questions about the consequences for national security, as well as the economy, of that degree of dependence on a few foreign countries.

THE REAL COST OF ENERGY

The cost of energy in terms of human labor changed enough over the past generation to influence many decisions. Figure 2.3 shows the number of hours a person had to work to buy a fixed amount of energy each year from 1951 to 1982. The chart was constructed by adding the cost of 500 kwh of electricity to the cost of 100 gallons of gasoline and dividing the total by the average hourly wage of a production worker in manufacturing. All data are national averages, so particular individuals may have encountered very different conditions, but the figure does convey a general feeling of how much the average person had to work to pay for gasoline and electricity.[2]

The energy that cost 23.86 hours of work in 1951 was available for only 12.56 hours in 1973. It was this type of calculation that made many forecasters so optimistic in the 1960s. It really seemed that energy would not be a serious constraint on the U.S. economy. Similarly, most individuals, faced with utility and gasoline bills that became steadily less burdensome, did not pay much attention to energy conservation when they purchased homes or cars. The reversal after 1973 was sharp, but even the second jump in 1979 left energy bills less burdensome than they had been in the early 1950s. As 1983 began most factory workers needed about as many hours to pay for 100 gallons of gasoline as they did in the era of tail fins and V-8 engines twenty-five years earlier!

The other side of this same relationship between wage rates and energy prices is the choice it presents to employers. When energy is cheap relative to labor, we would expect managers to seek out ways to substitute energy for labor. At a time when energy is expensive relative to labor and rapidly becoming more so, as in 1980, we would expect managers to be less interested in saving labor by using more energy. The observed relationship is more complex than the simple theory, however, as we see in Chapter 4.

The long-term nature of many energy decisions must be borne in mind, moreover. Many of the decisions that determine today's energy flows were made many years ago when energy was cheap and the price of one form of energy relative to another differed from that which prevails today. When gasoline was cheap, many drivers were happy to pay the small extra cost of having a very large car. When fuel oil was cheap, it made no sense to spend a lot of money for insulation. Similarly, industrial processes were chosen to minimize total costs of production—not energy costs—or in response to other goals of management such as high quality of output and

Figure 2.3 *Real Cost of Energy, 1951–1982*

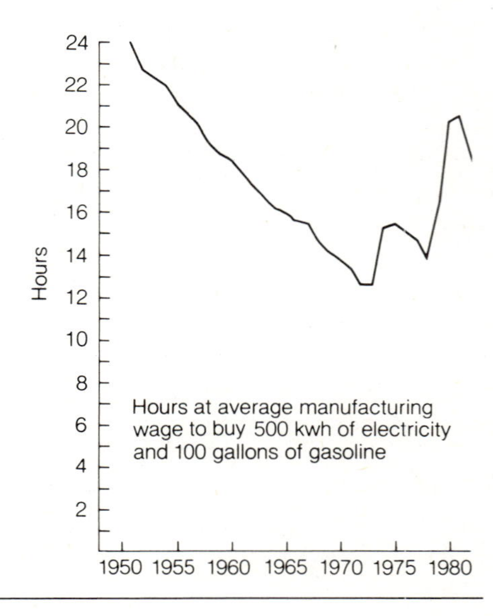

Sources: *U.S. Department of Energy,* Annual Report *(1981), vol. 2, pp. 95, 163;* Monthly Labor Review *(various issues).*

freedom from unplanned interruptions. Much of the capital equipment that was adopted when energy was cheap is still in use.

During the decade of the 1970s, the pressures for change made themselves felt with a vengeance. Some of the adjustments are already under way, as is apparent in the decline in the ratio of BTU to GNP, discussed in Chapter 4. But many of the changes had not yet been reflected in the data for 1980. This suggests that the energy flows for 1990 will look quite different from those in Tables 2.2 and 2.4. Furthermore the period of transition will be of great interest to students of energy or of microeconomics.

NOTES

1. The quad is a convenient unit for national and world data. It is defined as one quadrillion (10^{15}) British thermal units (BTU). One BTU is defined as the heat required to raise one pound of water one degree Fahrenheit. The equivalent metric unit is the kilocalorie; one kilocalorie equals about 3.97 BTU.
2. Often the term *real* is used to describe data that have been adjusted for changes in the price index. In effect, that sort of adjustment would compare energy prices with the prices of goods included in the price index.

CHAPTER THREE

Markets, Prices, and Efficiency

The discussion in the preceding chapters has generally used energy content as the weight for the various fuels because the dollar-weighted data are not collected in as regular and consistent form. There are some who prefer the physical measure, however, and would argue its superiority to a value measure. As in similar disputes, it will be apparent that both the economic measure and the physical one have their uses. The important consideration is to understand the uses of each.

This chapter begins with a rough sketch of the concept of economic efficiency and the role of prices from perfect markets in moving the economy toward efficient allocation of resources. The next section discusses the functioning of actual energy markets and some of the reasons why actual prices are imperfect guides to the pursuit of efficient allocation. An alternative technique, net energy analysis (NEA), is then discussed. A final section gives some illustrations of the necessity for relying on economic analysis.

ECONOMIC EFFICIENCY: A BRIEF REVIEW OF ECONOMIC PRINCIPLES

Economic efficiency is a central concept in a study such as this. It is not meaningful to criticize a government policy as inefficient or to suggest improvements in particular markets without having some standard of efficiency from which the existing situation falls short. Most of microeconomic theory and welfare economics deals with the topic of economic efficiency. It is obviously not possible to condense that information into a few paragraphs.[1] This discussion merely sketches a few of the fundamentals without any attempt at rigor. Aspects of efficiency concerned with production and time are discussed in Chapter 4 and 7, respectively.

Beneath the symbols and the diagrams, the subject matter of economic efficiency is human behavior. The fundamental input is human effort measured not in hours but in the opportunity cost of the effort to the individuals who exert it. Labor is combined with capital and natural resources, both of which have market prices that depend on the preferences of individuals.[2] The fundamental output of the economy is equally subjective because it is the utility or pleasure that individuals derive from the goods and services produced. The perfectly functioning economy can be thought of as a giant machine that maximizes welfare, which is the difference between the pleasure consumers derive from output and the disutility of contributing the inputs. A discussion of economic efficiency therefore is a discussion of how close the actual economy comes to attaining that maximum welfare.

Unfortunately, the data that are available are costs and revenues, or prices and quantities of inputs and outputs, not utilities and disutilities. Investors, managers, and consumers react to prices as they make decisions that allocate resources to produce a particular mix of goods, services, and leisure. The analyst must try to judge whether the actual prices are the prices that would give decision makers the signals that would lead the economic system to maximize welfare.

This is a tall order, but an even taller one follows. If there are problems that interfere with maximizing welfare, the analyst would ideally be able to recommend government action to improve the situation. As the following chapters indicate, however, government behavior is subject to its own failures.[3] When we observe a discrepancy between the ideal economic model that maximizes welfare and the actual world, we cannot assume that government would actually improve welfare unless we know both the magnitude of the existing welfare loss and the likely behavior of politicians and bureaucrats when they set out to correct the problem.

If prices and quantities are what we observe, what do they tell us? Figure 3.1 shows the demand curve for gasoline by a single consumer. The consumer is assumed to be a *price taker*; that is, she observes the price posted on the pump and buys as much as she chooses at that price without exerting any direct influence on it. We can see that this particular consumer buys 50 gallons of gasoline per month when the price is $1.25 per gallon. If we are also willing to accept the theoretical construct of the downward-sloping demand curve, then we can also surmise that the last gallon of gasoline that the consumer bought was worth just $1.25 to her. Also, if she had been rationed to only 40 gallons, the last gallon would have been worth more than $1.25. Conversely, the sixtieth gallon would be worth less.

More generally, for each consumer of a good, the price indicates the marginal utility of the good. Different people buy different amounts of gasoline each month and some may derive much more total utility or pleasure from using it than others, but the last gallon that each person buys is worth just the $1.25 that it costs her.

So far we have said nothing about the source of the price to which the consumer reacts. It is easy to draw market supply and demand curves (Figure 3.2), where the intersection determines price and quantity. Whereas the demand curve for the entire market can be obtained by adding the quantities that individual consumers would buy at each price, obtaining a supply curve is sometimes not so straightforward.

In the simplest case, the firm can be assumed to be a price taker, just like the consumer. The purely competitive firm is by definition too small relative to the entire market to worry about either the direct impact of changes in its output on price or the reactions of its rivals to its behavior. The situation is depicted in Figure 3.3.

The firm maximizes profit by increasing output until the cost of adding another unit of output (marginal cost) is exactly equal to the revenue obtained by selling one more unit (marginal revenue).

Figure 3.1 *Demand Curve of One Consumer*

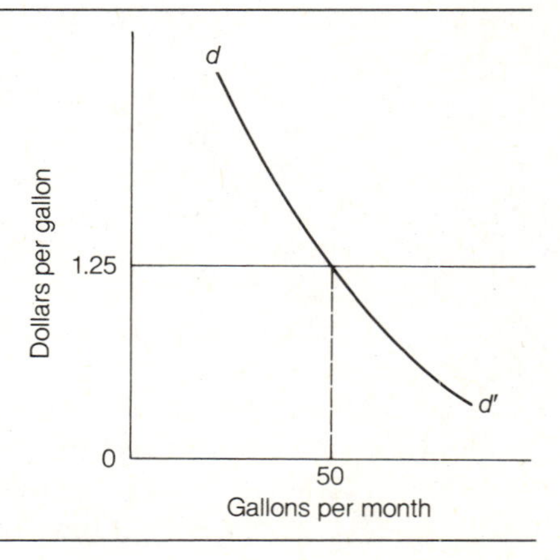

When the firm is a price taker, marginal revenue equals price. This implies that the supply curve for one firm is its marginal cost curve. Under some conditions the quantity that each firm would supply at a particular price can be added horizontally, so that the industry supply curve is the sum of all the individual firm supply curves.

If the costs that the firm counts are the full costs to society for producing another unit of the good, then this simple model yields a very interesting result. The marginal social cost equals marginal private cost, which equals price, which equals marginal utility. Thus the resources used to make the last unit of product produce exactly as much benefit to consumers as they cost. There is no way the resources of society can be reallocated to generate more utility.

It is particularly significant that price is a measure both of marginal benefit to each individual and of marginal cost to society. If pure competition and perfect markets prevail throughout the economy, then the monetary cost of any good, service, or project measures its social cost, and the price at which it can be sold measures its value to society.

Prices of all other commodities in the economy can be interpreted in the same way. Consumers react to relative prices of oil and electricity by choosing one or some mixture of both for their heating systems. They choose between fuel and insulation as ways to keep the house warm and between a warm sweater and a warm house as ways to keep themselves warm. On the producers' side of the market the same sorts of trade-offs are being made: What mix of products is most profitable to produce? What mix of inputs is least costly to buy?

The result of this series of adjustments is to set values for every commodity and factor in the economy. These values reflect the preferences of consumers and the constraints placed on the economy by the available resources, the terms on which commodities may be imported, and our limited technical knowledge. All of these

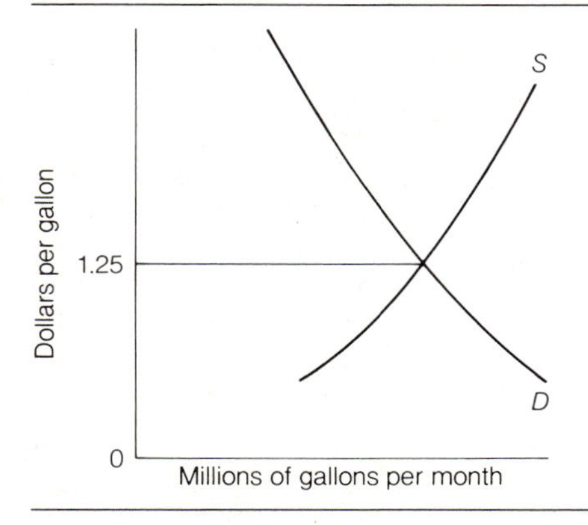

Figure 3.2 *Market Supply and Demand*

items can change, of course. People's tastes may change as may the known stocks of labor, capital, and raw materials, as well as the prices of imports and the available technology.

If we ignore market imperfections for the moment and treat actual prices as though they had the characteristics described, what can we say about the use of prices as weights to attach to energy inputs? In equilibrium the marginal consumer would be indifferent between two types of energy at the prevailing market price. Suppose, for example, I can buy coal at $2 per million BTU (MMBTU), electricity at $16 per MMBTU, and oil at $8 per MMBTU. If I choose to heat mainly with oil, supplementing that with electricity in some rooms, and buy no coal, what have I shown? I have demonstrated by my behavior that the last BTU of electricity I buy is worth twice as much as the last BTU of oil. (Some of the earlier units of either may be worth much more. Fortunately, the market rarely requires me to specify whether I would rather do without any oil or any electricity.) My behavior here also indicates that the last BTU of electricity is worth more than eight times as much as even one BTU of coal. Similarly, the last BTU of oil is worth more than four times as much as a BTU of coal. The household that heats mainly with oil will certainly buy some electricity, but it is unlikely to buy any coal. Of course, the demand for each energy input will depend on the prices of the others. With the increase in oil and gas prices in recent years, some households have begun using coal again.

It is easy to see why people will buy some electricity even when oil is half as expensive per BTU. Electricity is very convenient if you want light immediately or you want to power a small appliance, as well as for producing small, controlled, precisely located amounts of heat. If I want heat on my aching back, an electric heating pad is more convenient than an oil-fired one.

Figure 3.3 *Purely Competitive Firm*

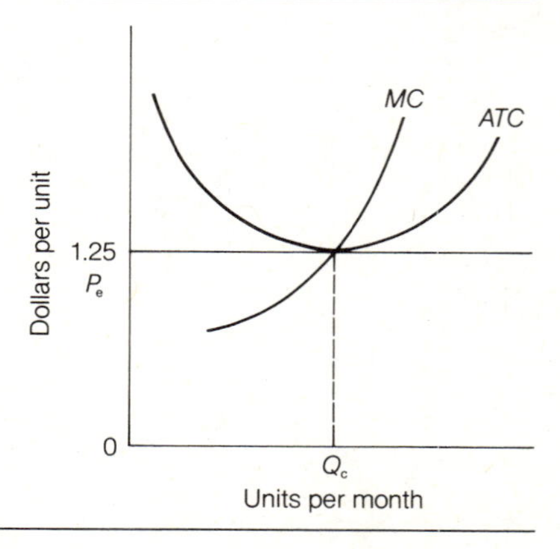

Markets, Prices, and Efficiency

The demonstration that unfettered markets maximize human welfare *under certain conditions* focuses attention on the degree to which actual markets approach those conditions. This is an exceedingly controversial area. Critics list violations of the assumptions of the competitive model and say it is inapplicable, whereas others assert that competitive forces eventually win out in the marketplace despite monopoly and government regulation.

The most obvious problem is that prices in energy markets are frequently not equal to marginal costs. For the regulated utilities, the regulatory goal is a price equal to average cost. Until recent years this diverged substantially from average cost, as we explain in Chapters 10 and 11. More generally, any firm that faces a downward-sloping demand curve and maximizes profits will set a price that exceeds marginal cost (Figure 3.4). Although there is a great deal of interfuel rivalry in particular markets, it would be difficult to describe some of the energy firms as price takers.

The bias of prices that exceed marginal private cost is opposite in direction to the bias from external costs. In the absence of government regulation, the costs as perceived by firms would be lower than the full costs to society because they would not include the cost of pollution. This topic is discussed in Chapter 14, but the implication for prices is that they will be too low. Briefly, if one gallon of oil sells for $1 because of the monetary costs incurred by the various firms in bringing it to market, but it also causes 10 cents' worth of damage to the environment from oil spills and so on, then the marginal social cost is $1.10, rather than $1. We can also treat the possible problems to national security that stem from dependence on the Middle East as an external cost of oil. Arguments of this sort suggest that the market prices are too low to lead to maximization of social welfare.

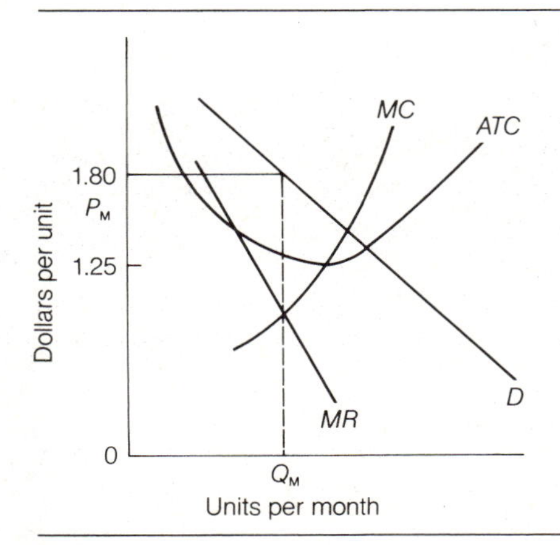

Figure 3.4
Monopolistic Firm

Only in a rare coincidence would the upward bias from market structure exactly offset the downward bias from neglect of externalities.

Welfare economics addresses a host of other market imperfections. Some, such as problems of distribution of income and the necessity for public goods, have no special relevance for energy and so will be ignored here. Others that are more directly relevant to energy include the long time required to adjust machinery and buildings to the new optimum after a major change in relative prices, the long lead time to develop many energy sources, and the uncertainties that are inherent in natural resource industries.

Despite the problems, this book generally treats prices as though they approximate marginal cost and marginal utility. The exceptions are only the most blatant ones, such as binding price controls, extreme environmental costs, or large subsidies. The two justifications for doing so are convenience and the substantial degree of rivalry apparent within most energy markets.

Rivalry among fuels is most intense in the electric utility and industrial sectors. Some industrial fuel requirements are highly specific, such as the use of coal with certain characteristics to make coke or the use of a particular petroleum product as a chemical input. Much coal, natural gas, and oil, however, is burned to produce large amounts of heat. In such uses firms can substitute one fuel for another easily, sometimes within a matter of days.[4] During the late 1970s, for example, coal regained some of the markets for boiler fuel that it had lost earlier to the cleaner and more convenient fuels. Such interfuel rivalry serves much the same function as would competition among firms producing identical products.

In such specialized applications as annealing ovens for the steel industry, the battle was fought between oil and natural gas during 1983. Electricity, with its versatility and ease of control, is also a technically feasible option in many applications. Although the high cost per BTU deters use, if the price differential were to narrow it could capture some applications. But why have oil and gas sold at such different prices as those shown in Table 2.1? This seems to be an indication that equilibrium has not yet been attained. Most homes do not use both gas and oil for heating, however, so few people are actually buying both and spending 20 percent more per BTU for oil than for gas. These are national averages that reflect the sorts of options open to different people in diverse locations. Before analyzing consumers' decisions, we must examine the choices open to them, as well as their uncertainty about future prices and availability.

Consider a homeowner in 1980 pondering conversion of his oil heating system to natural gas. His choice depended on his guess about future price levels, not only for those two fuels, but also for anything that might prove to be better than either within the next few years. In the extreme, the consumer may have believed the reports that gas supplies would be exhausted within a decade or so. In any event, he could guess that both oil and gas prices would rise as controls expired. Guesses about future prices and availability seemed to vary with the weather, the forecaster's disposition, and the latest speeches from OPEC ministers. Under such circumstances households and firms are reluctant to convert to the use of a different fuel as long as their existing equipment works. It is hard to believe that consumption patterns have adjusted fully to the particular set of prices observed in the market.

The government also has delayed adjustment by creating uncertainty. Firms that might otherwise have converted to cheap local coal have delayed because of the continuing vacillation about the maximum permitted sulfur emissions and the technological uncertainties about the efficacy and reliability of pollution control equipment. Although the requirements could sometimes be met with more expensive low-sulfur coal, using that reduced the advantage of converting from oil or gas. Furthermore, it is often not simple to change from one type of coal to another once the boilers are installed because the appropriate equipment is specialized for the type of coal. The general uncertainty about what will be permitted makes delay appear to be more profitable than rapid adjustment.

The government has also created uncertainty over natural gas by first forecasting that supplies were dwindling rapidly and next, in the Natural Gas Policy Act, moving to reserve the available supplies for residential use. As a result, energy users have been reluctant to convert from oil to gas.

For these and other reasons, the existing set of quantities and prices should not be assumed to be at equilibrium. In particular, when all abnormal uncertainties about supply and government regulation are eliminated, it seems likely that consumption of natural gas, and perhaps coal, will increase substantially. The key problem is to guess the impact that this will have on relative prices. It would appear that gasoline, diesel oil, and jet fuel will continue to command a large premium because of their unique advantages in transportation. Electricity will be expensive because the upgrading to an easily usable form has already been accomplished. Coal will always be cheap because it is widely available and use is restricted by inconvenience in handling. The one real question mark is natural gas. This is a topic to which we must return in Chapter 10, but the issue concerns the long-run supply of natural gas. Is it so scarce that it will be a premium fuel, or will it be cheap enough to compete with residual oil and coal in stationary applications for large users? The answer to this will have repercussions throughout the energy industries.

Although economists have generally continued to use conventional economic analysis of costs and benefits, the alternative approach of net energy analysis (NEA) has attracted support even in Congress. Under the Federal Nonnuclear Energy Research and Development Act, new technologies are to have an evaluation of their "potential for production of net energy."[5] *Net energy* is defined as the "energy remaining for consumer use after the energy costs of finding, producing, upgrading, and delivering have been paid." Superficially, this seems like a useful requirement. After all, shouldn't Congress know whether solar cells or gasohol increase the energy supply before being asked to subsidize them? It is conceivable that growing and preparing the crystals for a photovoltaic cell consumes more energy than it will ever repay.

The standard economic response is that the economy functions most efficiently to meet the preferences of people if economic efficiency is the criterion for investment decisions. The counterarguments in favor of NEA are of two sorts: One is that economic calculations are flawed by the departures of actual prices from those generated by the pure competition and perfect markets assumed by economic theory. The second argument for NEA is that energy is somehow more basic and fundamental to life than anything else. Gilliland, for example, suggests that the laws of thermodynamics lead to the following three implications: "(1) energy is the only commodity for which a substitute cannot be found, (2) potential energy is required to run every type of system or production process, and (3) energy cannot be recycled."[6]

From an economic point of view, the interesting aspect of these implications is that they are all wrong. First, it makes no sense to say there is no substitute for energy; nearly every production or consumption activity can be carried out with less energy if other inputs are substituted. Second, the energy to run economic processes can sometimes come from the sun or from waste and thus is not a cost to the relevant (economic) system. Finally, energy is often recycled, in the economically relevant context, without violating thermodynamics. This occurs whenever a way is found to extract more use from the same amount of fuel (for example, using the heat from a furnace flue pipe to warm the cellar).

Even if we accept the premise that energy is the most important input, actually performing the NEA runs into conceptual problems. The basic approach of NEA seems simple enough: We should try to calculate the amount of energy that will be available to consumers after deducting all the energy required to find, produce, upgrade, and deliver the gross energy of a project. But the labor and capital used in the project also require energy to produce, so that amount must also be subtracted. The difficulties accumulate rapidly: What is the energy cost of labor, or for that matter of a ton of steel, since steel can be produced by different processes and labor can live at different standards implying different amounts of energy?

The advocates of NEA also specify that the energy must be weighted according to "quality," that is, 1 kwh of electricity is worth more than the 3,412 BTU of heat it contains – something closer to the 10,435 BTU of coal it would take to produce it. But, of course, that relationship is an economic one, not a physical constant. The ambiguity of weighting by quality becomes apparent when Odum and Odum describe information as "energy of very high quality." It appears that either the term *energy* is used in different ways or that an analogy has been confused with a technique of analysis.[7] When it comes to determining the appropriate trade-off between coal and oil, the concept of quality becomes strictly economic. It loses any moorings in physics or engineering practice. NEA is not any more securely based on physical constants than is economic analysis.

More cogently, Huettner asks what kind of guidance we can expect from NEA in grappling with the complex questions that confront decision makers.[8] Huettner's novel analysis proceeds by asking when economic analysis and NEA will give decision makers the same answers. His result is that they will be the same if and only if the energy coefficients are exactly proportional to money prices. This implies either that all commodities are produced with identical proportions of energy and other inputs or that energy is the only scarce resource. If the first were true, there would not be much point in worrying about energy specifically, since our decisions would not have any influence on its use. It is the latter position, that only energy matters, into which NEA would nudge us. But this, of course, is totally unreasonable in view of the fact that capital and labor are substitutes for energy in many common activities.

The other question that arises when two forms of analysis are used is which one will prevail when they differ. A net energy analysis of any energy conversion process, such as liquid fuel from coal, is likely to suggest that energy is being lost. Yet someday a clever design may increase economic value while decreasing the available energy because people will pay more per BTU for liquid fuel than for coal. Faced with such differing results, the analyst could (1) reject the market, saying that people do not know what they should have, (2) adjust the NEA to correspond with the market by giving the liquid fuel a sufficiently high quality weight (which would turn out to depend on the price differential between coal and oil), or (3) reject NEA totally. Will NEA improve the process of making policy decisions?

For the present it seems reasonable to accept the existing prices as measures of both the consumers' evaluations of additional supplies of energy and the costs of additional energy to the economy. This implies that the economic measure of energy and other inputs to the economy is more fundamental than the engineering measure of heat content. The main drawback of measuring in dollars is that dollar values change over time with inflation. The advantage is that at any moment of time all inputs and outputs can be put into comparable, dollar, terms so that input, output, and overall efficiency can be measured.

THE MEANING, PURPOSE, AND PROBLEMS OF ENERGY EFFICIENCY

Efficiency is generally a slippery concept in economic activities. Often we can find well-established engineering measurements of efficiency; for example, we could determine what percentage of the heat content of the fuel consumed actually was transmitted to the occupied parts of a house. This could be one measure of the energy efficiency of the heating system. Suppose that we were analyzing a super-efficient airtight stove that transmits 90 percent of the heat content of the wood to the room in which it is placed, but (1) burns only carefully seasoned maple and hickory cut 24 to 26 inches long and split to 4 inches diameter, (2) requires that the windows be left open to produce a sufficient draft for combustion, (3) is too hot to sit near, and (4) costs $1,000 by the time it is installed. Is it more efficient than a simple old stove that extracts 50 percent of the BTU from household rubbish, wood scraps, odd logs, and old newspapers; starts easily; and costs one-fifth as much? From an engineering standpoint the first is more efficient. It may also be more efficient from an economic point of view, especially if it is the primary source of heat and the fuel is supplied according to some orderly system. But the economic advantage is not clear-cut, and we can think of many situations where the stove of 50 percent efficiency would be preferred to one of 90 percent efficiency.

The situation becomes more complex if we try to define the efficiency of an automobile. From an engineering viewpoint we could measure the fuel consumed and its BTU content, then measure the work done by the engine in overcoming various sorts of friction and wind resistance. That would say something about the efficiency of the engine, or of engine and drive train if we looked at the power delivered to the wheels. But an automobile of different shape would have less wind resistance and therefore would travel further on a gallon of gas. Surely that has something to do with efficiency, even if the ratio of power out/power in declines. So we could shift to a simple analysis of miles per gallon of fuel.

This measure leads to absurdities in two different directions, however. First, we can increase miles per gallon by shifting to more expensive fuel (for example, increase compression and therefore use premium higher octane gasoline). In the extreme we can imagine lightening and fine tuning an automobile so much to save gasoline that it requires a complete overhaul after each trip. Surely the criterion of efficiency for the typical automobile owner should be cost per mile, not fuel consumed per mile. Even cents per mile is a sophisticated calculation, however, since it involves the conceptually tricky problem of weighing operating costs (gasoline) against fixed costs (the price of the vehicle), as well as probability and costs of breakdowns. Cents per mile is not an adequate measure of cost in my view at least, because I would happily buy some extra gasoline to have a vehicle that starts reliably and does not give up on the highway. Unpredicted repairs often cost a lot of time and worry, as well as money. I might even be willing to pay extra for a safer car, one that is easy to repair, or perhaps one that damages other people's lungs less.

These latter characteristics, the nonmonetary costs, begin to blend into the output side of the formula. It is important to know not only the cost of the vehicle but also the output. Output can be measured in miles, but the miles per dollar measure can always be increased by making the vehicle smaller. Perhaps passenger-miles per dollar is a better criterion, but then a large bus is superior to a car. If we restrict the analysis to personal transportation vehicles, we see that some people are willing to pay for the space and other features of a luxury car, even though the tiny fuel miser may have seats for as many people as are usually carried. This suggests that output too is not a simple matter but must be determined by individuals according to their own preferences.

If we proceed to reduce cost per mile by reducing the size of the vehicle, we note that a motorcycle is more efficient than a car and a motorbike is more efficient than a motorcycle. But why stop there? With a bicycle, miles per gallon of gasoline becomes infinite. A bike must be the most efficient form of transportation! Some busy people contemplating a trip from Cleveland to Boston in January might disagree.

It is possible, of course, to calculate total BTUs expended per mile in various forms of transportation including planes, trains, oxcarts, walking, cycling, and so on. The comparison between bicycling and walking on short local errands in pleasant weather illustrates the perversity of calculations based on simple notions of efficiency. Walking uses about three times as many BTUs, or calories, per mile as pedaling a good modern bicycle. But in an era when the average American adult consumes too many calories, is the added energy expenditure of walking a cost or a benefit? For most of us the energy cost can be assigned a negative value. After all, if I burn up more energy I can eat another dessert, which is pure pleasure.

This example is frivolous, but the central points are that efficiency is a slippery concept and that for economic activities it is essential to rely on an economic definition of efficiency rather than to be misled by the apparent rigor of a thermodynamic or engineering definition that is peripheral to the economic issues. Economic efficiency can be attained by allowing firms to maximize profits and allowing consumers to make their own choices unless we can identify serious imperfections in the markets.

NOTES

1. The best place to start a more thorough investigation is any standard textbook in intermediate microeconomic theory, such as Edwin Mansfield, *Microeconomics* (New York: W. W. Norton, 1982).
2. The subjective nature of costs is explained in James Buchanan, *Cost and Choice* (Chicago: Markham, 1969).
3. For a guide to the literature on government failure, see William S. Peirce, *Bureaucratic Failure and Public Expenditure* (New York: Academic Press, 1981).
4. For a detailed discussion of competitiveness, see Thomas D. Duchesneau, *Competition in the U.S. Energy Industry* (Cambridge, Mass.: Ballinger, 1975).
5. For a detailed discussion with copious references, see Ernst R. Berndt, "From Technocracy to Net Energy Analysis: Engineers, Economists and Recurring Energy Theories of Value," *Studies in Energy and the American Economy*, Discussion paper no. 11, MIT-EL-81-065WP (Cambridge: Massachusetts Institute of Technology, November 1981).
6. Martha W. Gilliland, "Energy Analysis and Public Policy," *Science*, September 26, 1975, pp. 1051–1056.
7. Howard T. Odum and Elisabeth C. Odum, *Energy Basis for Man and Nature* (New York: McGraw-Hill, 1976).
8. David A. Huettner, "Net Energy Analysis: An Economic Assessment," *Science*, April 9, 1976, pp. 101–104.

CHAPTER FOUR

Energy and Economic Growth

The difficulties in weighting energy inputs and in defining efficiency at the micro level have their counterparts at the macro level, which become painfully obvious in a discussion of energy and economic growth. Any quantitative analysis requires both an aggregate measure of energy input over time and a measure of economic activity. The standard technique is to relate BTUs to GNP. This ratio can be called the *energy intensity* of the economy. The *Monthly Energy Review* publishes a regular series, which shows, for example, that the energy consumption (BTUs) per dollar of GNP (deflated to 1972 dollars) declined from 59,500 in 1973 to 47,800 in 1982. Before pondering the meaning of such data, it is useful to consider some of the problems of concept and measurement involved.

THE THEORY OF PRODUCTION

The standard microeconomic theory of production is depicted in Figure 4.1. In this simple illustration the isoquant $Q = 100$ is to be taken as the set of all of the technically efficient ways to produce 100 units of output; that is, we can produce the same quantity of output using a lot of energy and little labor, or we can cut back on energy consumption but continue to produce the same output by adding labor. The isoquant illustrates the assumption that substitution of one input for another proceeds smoothly over a wide range of possible input combinations. In the real world, of course, there may be only a few known methods for producing quantity Q. What we would see then is just a few discrete points, rather than the smooth curve shown here. Also, it is quite likely that only one or two techniques are really well mapped out—those that correspond with standard practice today. If we were really interested in trying something far removed from current practice, it might require a great deal of detailed design and development work, not just dusting off an unused blueprint from the shelf.

In Figure 4.1 it is assumed that the technical choices open to the firm (the isoquant) remained unchanged from 1970 to 1980 but that the prices of energy and labor both increased substantially between the two years. With labor costing \$3 per hour in 1970 and energy available at \$2 per 10 million BTU (which describes a large industrial user located in the natural gas fields of Texas in 1970), the different mixes of energy and labor that could be purchased for \$24 are shown by the isocost line for 1970. Similarly, the isocost line for 1980 represents the combinations of inputs that could be purchased for

Figure 4.1 *Substitution of Labor for Energy in Production*

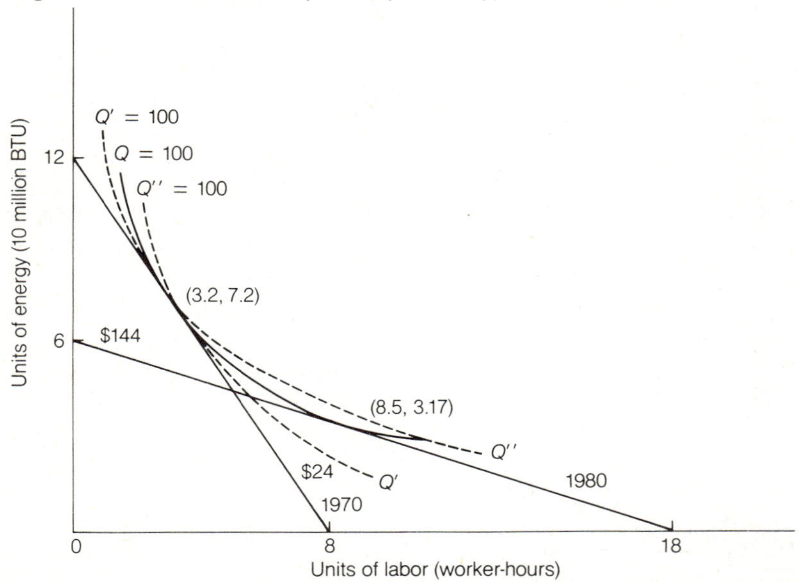

$144 at the 1980 prices of $8 per hour of labor and $24 per 10 million BTU. An isocost line, as well as an isoquant, can be drawn through every point on the diagram. This particular isoquant was selected arbitrarily, then the isocost lines for the two years were drawn just tangent to it.

The point at which the isocost line is just tangent to the isoquant shows the least expensive combination of energy and labor to produce the chosen output in the given year. In 1970 the cheapest approach was to use 3.2 hours of labor and 7.2 units of energy. With the change in relative prices, the cheapest technique by 1980 had shifted to 8.5 hours of labor and 3.17 units of energy. Naturally, this example was constructed to yield a very dramatic change in the optimum input combinations. Many industrial processes do not permit such easy substitution of one input for another—at least not without a substantial amount of engineering work and major capital outlays. Even in such cases, however, the managers would certainly feel the pressure for change, even if they could not respond to it immediately. Buying the 1970 combination of inputs at 1980 prices would cost $198.40, which would make the cost of the 1980 combination at $144 seem like a bargain that was worth some exertion and investment to attain.

It is also worth asking whether the twelvefold increase in energy prices assumed here is realistic. It reflects a slight telescoping of experience but is not far from the mark. In 1952 electric utilities burning natural gas paid an average of 14.5 cents per million BTU. By 1970 this had increased to 27 cents and by December 1980 to $2.33. Over the same time period, the price of coal went from 27.3 cents per million BTU in 1952 to 31.1 cents in 1970 and $1.38 in December 1980. Residual oil went from 33.1 cents per million BTU in 1952 to 36.6 cents in 1970 and $5.22 in December 1980—a more than fourteenfold increase during the decade.

Changes in price of the magnitude of those that occurred in the markets for industrial energy during the 1970s are rare. It is to be expected therefore that such extreme changes would exert pressure on management to adapt by trying to reduce the use of energy and to switch to cheaper forms of energy. It may not be quite as obvious that the extreme changes in prices also produce distortions in the economic statistics that presume fundamental stability in important economic relations.

By the usual economic definitions, the illustration constructed here displays no technological change and no change in productivity. The isoquant, which shows the minimum amount of labor needed to produce a given level of output for the different amounts of energy, has remained unchanged, which is the standard definition of static technology. Since the firm was on the isoquant to begin with (that is, it was technically efficient) and now has moved to a new point on the isoquant, there is no indication of new knowledge—just a substitution of one input for another in response to a change in relative prices.

The popular literature, however, often considers productivity to be output per man-hour (or worker-hour). That measure declined from 100 units of output divided by 3.2 worker-hours = 31.2 units per worker-hour to 100/8.5 = 11.8 units per worker-hour over the decade. Conversely, of course, energy efficiency, which is defined as output per unit of energy input, rose from 100/7.2 = 13.9 to 100/3.17 = 31.6 over the same period. The direction of change in both partial productivity measures is implicit in the statement that labor was substituted for energy. All of this suggests that the recent stagnation in output per worker-hour and the increase in output per energy unit are about what we would expect with the given changes in relative prices.

The real world is not as simple as the picture, however. According to production theory, a technological change has occurred only if the 1970s output could be produced in 1980 on a different isoquant from that which describes 1970. If the 1980 isoquant is closer to the origin (in some range) than the 1970 isoquant, then total factor productivity has improved. Unfortunately, the only part of the isoquant that we ordinarily observe is the point itself; and we can draw any number of isoquants through the 1970 point, some of which are above, some below, and some through the 1980 point. If the 1970 isoquant had been $Q' = 100$, then the 1980 point would indicate a decrease in productivity. If the 1970 isoquant had been $Q'' = 100$, then the 1980 point would indicate that productivity had improved. When one input increases while another one decreases (for the same output), the analyst needs additional information to decide whether productivity has improved.

INDEXES OF OUTPUT AND PRICES

A more important consideration is that most economic data are highly aggregated; that is, usually analysts discuss output per worker-hour or per BTU for an entire industry or the economy rather than for a single homogeneous product. In these cases, output cannot be measured in physical terms. Various products have to be combined, usually by using prices as weights. For example, the Index of Industrial Production, regularly reported in any good newspaper, combines shoes and ships and sealing wax and automobiles and almost anything else you can think of.

Table 4.1 sets out a simple example using an economy consisting only of Product X and Product Y. Only Product Y has benefited from technological change. The main assumption is that the products use only labor and energy, so that the total cost can be found by adding expenditures on labor and expenditures on energy. The data in the table are based on the assumption that consumers will react much as the firm did in Figure 4.1. As the relative prices shift, consumers will buy less of the commodity that is becoming relatively more expensive. In this example Product Y was initially eight and a half times as expensive as Product X, but by 1980 the relationship had switched, with Product X being 44 percent more expensive than Y. The response of consumers was to change from a consumption mix of 100 X and 60 Y in 1970 to 50 X and 110 Y in 1980. The responsiveness of consumers to changes in relative price can, of course, be shown with indifference curves, which look like isoquants. But it is not easy to imagine consumer tastes remaining unchanged for a decade, as is implied by an unchanged indifference curve.

The data shown in Table 4.1 suggest a more serious problem of interpretation. Consumers in the two years were shown as spending $144 in 1970 and $182 in 1980. Did this indicate that they were paying more for what they got or that they were consuming more? In fact, they consumed more Y and less X in 1980, but did the increase in Y fully compensate for the decrease in X? Obviously, it does no good simply to count the number of items because that would be equivalent to saying that three new cars is equivalent to three new potatoes. Nor can we weight by prices because the prices were changing.

The crux of the matter is that the 1980 output, if valued at 1970 prices, would be 50(24 cents) + 110($2) = $232, suggesting a substantial improvement over the $144 bought in 1970. If the 1970 output were valued at 1980 prices, it would total 100($1.44) + 60($1.00) = $204, which indicates a decline in output from 1970 to 1980. This is not surprising because consumers buy more of the goods that are relatively cheap. If the analyst uses the prices of a different time (or place) to weight current purchases, items that are given a low weight

Table 4.1 *Price and Quantity Changes in a Two-Commodity Economy*

	Product X	Product Y	Whole economy
1970			
Price	$0.24	$2.00	
Quantity	100	60	100 + 60 = 160
$P \times Q$	$24	$120	$144
1980			
Price	$1.44	$1.00	
Quantity	50	110	50 + 110 = 160
$P \times Q$	$72	$110	$182

because they were once cheap may be bought in small quantity. Consumers who adapt quickly to a new set of relative prices suffer much less from price changes than do consumers with highly rigid preferences (square corners on their indifference curves).[1]

That does not help the statistician, however, who is charged with presenting two numbers: the growth in real output and the increase in prices. The Consumer Price Index takes the "market basket" of goods and services bought in the base period, values them at the later prices, then calculates the index. Here the 1970 output cost $144 in 1970 and would have cost $204 in 1980. The price index for 1980 is thus $204÷$144=1.42. Prices have increased 42 percent. This is usually written as an index of 142. If the output of the economy in 1980 (valued in 1980 dollars) of $182 is deflated by the price index of 142, real output (valued in 1970 dollars) in 1980 is $182÷1.42=$128, which is a decrease from the $144 recorded for 1970.

Such simplified illustrations serve mainly as a reminder that during periods of rapid economic change the standard statistical measures of the economy should be interpreted with great caution. In particular, this example has GNP declining, although with large internal changes as some sectors expanded and others contracted. With the concurrent incentive to substitute labor for energy in individual products and with consumer substitution of other products for those that are particularly energy intensive, it is not surprising that output per unit of energy increased, while output per unit of labor input declined.

THE RATIO OF ENERGY TO REAL OUTPUT

Much more surprising than the decline during the recent period in the ratio of energy to GNP is the fact that it is apparently a trend of long duration. Figure 4.2 shows the series peaking before 1920 and then declining into the 1940s. Upturns in the mid-1940s and late 1960s were just fluctuations in a generally declining trend. The net result was a decline of about 50 percent between 1920 and 1980. This is particularly interesting because it occurred during a period when energy prices were declining in real terms and frequently even in dollars. (These data are reviewed in more detail in the chapters dealing with separate energy industries.)

The phenomenon does not have any single accepted explanation.[2] Instead, it reflects a combination of things including the growth of the more refined manufacturing and service industries following the great upsurge of heavy industry and basic transportation in the half century following the Civil War. As these metal fabricating, manufacturing, and service industries grew, their contribution to GNP was large relative to the energy consumed.

In addition to the sectoral shifts, the application of electric power to replace steam often resulted in a net reduction in energy consumption, even after allowing for the losses in generation. This reduction in energy use is particularly marked in applications requiring intermittent power or small amounts of power. A steam shovel, for example, might operate only during the daylight hours, but the boiler would burn coal twenty-four hours a day. The electrically powered replacement draws only as much power as it is actually using at each moment of the day.

Just because of the way terms are defined, moreover, a switch from direct burning of coal—even at 100 percent efficiency—to indirect consumption of coal in the form of electric power would increase GNP more than it would increase energy consumption. Assume that roughly two-thirds of the energy is lost in conversion. If the coal had been used at 100 percent efficiency, the use of electricity would triple coal consumption. GNP would increase even more, however, because electric power sells for far more than triple the price of coal per BTU. Since GNP is measured as gross national cost, the substitution of more expensive energy increases GNP more than BTUs. For large industrial users in 1982, electric power was about eight times as expensive per BTU as was coal. The combination of the efficiency and convenience of using electricity makes it worthwhile, but the impact on the ratio of BTUs per dollar of GNP is to reduce it.

In a less dramatic way, the same shift occurs for the same arithmetic reason every time anyone substitutes an expensive form of energy for a cheap one. Certainly, one of the distinctive trends in energy use during the twentieth century has been the substitution of fuels with relatively high cost per BTU for fuels with lower

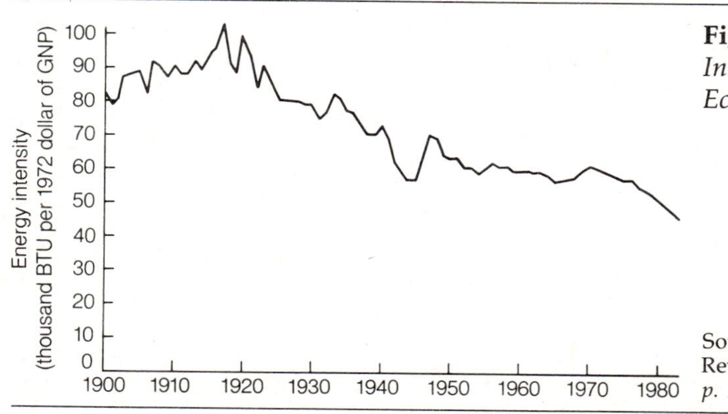

Figure 4.2 *Energy Intensity of the U.S. Economy, 1900–1983*

Source: Monthly Energy Review *(September 1984), p. 18.*

monetary cost. Thus wood and coal were displaced by electricity and oil. As energy prices fell relative to wage rates, consumers were willing to pay for the convenience of electricity and refined petroleum. But a shift to electric heat from coal, for example, creates a decrease in the BTU/GNP ratio, just because the electricity is so expensive per BTU.

ENERGY AND RISING INCOMES

As this discussion suggests, we can view energy as a consumer good. At some early stages in economic development, it is evident that people will increase energy consumption rapidly as they acquire the means to do so. A warm home is a great luxury after a drafty cave. Similarly, the mobility of automobile or jet travel is a highly prized addition to a frugal but adequate standard of life. Beyond some income level, however, it is certainly possible that the income elasticity of demand for energy is low.

Furthermore, energy is an intermediate good, not something desired in itself. It is possible that in trying to make homes draft free, consumers will actually reduce energy consumption; or by buying sportier automobiles, they may reduce gasoline consumption. When all the consumption and investment activities are aggregated, however, it is difficult to imagine that increased income will generate a decrease in demand for energy. If energy is viewed as a consumer good, we could expect the quantity demanded to increase with income, but the exact rate is an empirical question. Table 4.2 shows the modest difference in energy consumption by income levels in the United States.

Table 4.2 *U.S. Energy Consumption by Households, May 1980–April 1981 (by income ranges)*

Family income 1979 (dollars)	Consumption per household (millions of BTU)	Expenditures per household (dollars)
Under 5,000	98	754
5,000– 9,999	104	807
10,000–14,999	102	837
15,000–19,999	112	900
20,000–24,999	126	986
25,000–34,999	123	1005
35,000 and over	143	1206

Source: U.S. Statistical Abstract, *1984, p. 577.*

It is also possible to gain some insight into the question of income elasticity by looking at a sample of different countries. Figure 4.3 shows GNP per capita and energy intensity for several advanced countries. The relationship is not a very strong one for such industrialized economies. A chart that included extremely poor countries would show a stronger relationship, but the fact that much variability is observed among rich countries suggests that the relationship between energy and economic growth is not simple.

The traditional argument is not that energy is merely a consumption good but that it is an essential input into a technologically advanced society. Substitution of machines for human muscle is an integral part of the process of economic growth and necessarily requires energy inputs. Hence we can view the consumption of large amounts of energy as either a cause or a symptom of economic growth. Surely, though, the amount of energy consumed will depend in part on its price and availability. In an era of expensive energy it will become a priority for managers and researchers to devise ways to economize on energy. These efforts may not bear fruit immediately, but over time, on average, they will.

Fundamentally, this sort of argument suggests that capital is a substitute for energy. It is possible to insulate better or build smoother roads or more efficient machines to save energy. We can also use immense amounts of capital to decrease the loss of energy in conversion processes such as generation of electric power. Similarly, with enough capital we can capture more solar energy or produce liquid fuel from oil shale. The main problem is not energy

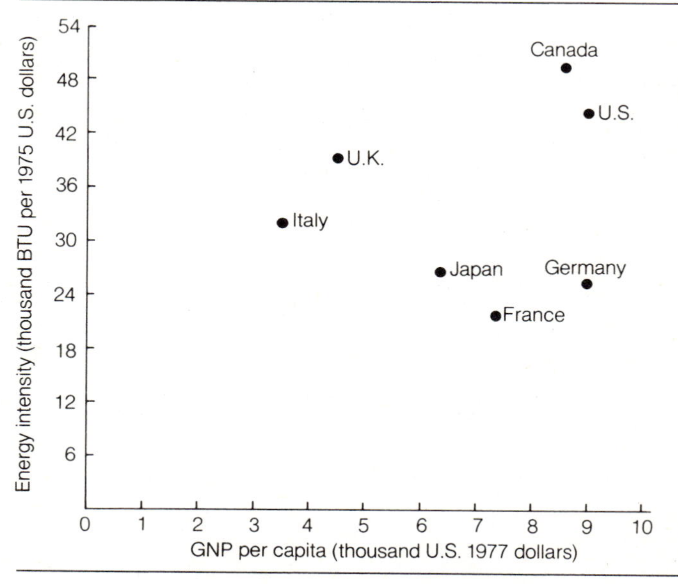

Figure 4.3 *Energy Intensity versus Income per Capita, Various Countries, 1978*

Source: U.S. Statistical Abstract *(1981), pp. 583, 878.*

per se but the growth in wealth and knowledge that will enable us to conserve existing sources of energy and to develop new ones. This is the reason that economic efficiency, which is directed toward using all resources to create the greatest income and wealth, merits greater attention than energy efficiency.

NOTES

1. "That man is the richest whose pleasures are the cheapest" (Henry David Thoreau, *Journal*, March 11, 1856). Susan Staples reminded me of this quotation. He who would be rich in pleasure should buy the pleasures that are the cheapest at the time and place. A modern resident of Greater Boston who adapts easily to relative prices would not choose the simple pleasure of living alone in a small cabin beside a pond in the woods.
2. Sam H. Schurr, "Energy Efficiency and Productive Efficiency: Some Thoughts Based on American Experience," *Energy Journal* 3, no. 3 (July 1982), pp. 3–14. For additional studies of the experience of the 1970s, see Sam H. Schurr, Sidney Sonenblum, and David O. Wood, eds., *Energy, Productivity, and Economic Growth* (Cambridge, Mass.: Oelgeschlager, Gunn & Hain, Publishers, 1983).

CHAPTER FIVE

Guessing about Future Consumption and Production

Consumers and producers alike are more concerned with future prices and availability of various forms of energy than with their history. The discussion of past trends in consumption, prices, and efficiency is useful mainly to suggest the range of conditions that the economy has endured and to give some hints about the adjustments that are plausible as the relative prices shift in the future. Many people crave more definite forecasts, however, and whenever a demand is so well articulated in the market an industry springs up to supply it.

MODELS, PROJECTIONS, PREDICTIONS, FORECASTS, AND GUESSES

Distinctions are sometimes drawn among models, projections, predictions, forecasts, and guesses, but in this book speculations about the future will be called guesses to encourage the reader to treat them with the appropriate respect. This is not intended as a judgment on the usefulness of guessing about the future. Indeed, we cannot make any rational plans without doing so. Nevertheless, even the most elaborately mathematized efforts to foretell the future are founded on guesses about what will remain constant and what will change. Before using a forecast for any serious purpose, the decision maker should examine the underlying guesses.

Models usually involve an explicit structure of relationships. The implication of the term in the literature of economics is that we have set down a number of interrelated equations, although a supply and a demand curve sketched on the back of an envelope also constitute a model. Models occasionally involve only one equation, but sometimes hundreds are included. Building a model can be very useful because it forces analysts to make their assumptions explicit. A formal model can be run through a computer to check numerous possibilities. It must be borne in mind that no matter how elaborately constructed, a model of the future is still a web of guesses. If the model is carefully estimated from past data, the guesses are of two types: (1) that the relationships in the future among variables (say, between price in the market and the quantity consumed) will be unchanged and (2) that the specific assumptions on which the forecast is based will hold true (for example, no war in the Middle East, population growing at the assumed rate, and no additional environmental restrictions on strip mining).

Projections are identical to models, but the emphasis is on the specific assumptions about underlying conditions in the world (we can, for example, project the implications of a growth rate of 2 percent in energy consumption). Models, by contrast, emphasize the structure of relationships that yields the specific guesses. Many projections are based on the thoroughly naive model that lines on a graph can be extrapolated a few years by eye, ruler, or equation.

Forecasts and *predictions* need not have an underlying formal model. The implication of the forecast is that the forecaster is expressing an opinion about the way things will really work out. It is not the "if this, then that" of the projection but "here is how it is going to be." Forecasts can be based on complex mathematical models, tea leaves, satellite pictures, or the width of the stripes on woolly bear caterpillars. They are more comforting to the uninitiated than are projections because they convey a greater sense of certainty and authority.

In the final analysis, however, all attempts to state what the future will bring are guesses. Some guesses are based on better information about the past and present than are others. Some make better use of available information by employing useful theory. Some use a large number of small specific guesses, whereas others use a few aggregate guesses. But guesses they all are.

THE UNCERTAIN FUTURE

Although any discussion of the future is a guess, subject to some uncertainty, there are differences in the various approaches. First, some guesses build up an aggregate forecast by adding up the guesses about individual sectors. Others focus only on aggregates. A detailed sectoral forecast has more uses, but it is an open question which is more accurate. If we try to make separate guesses about the output of coal, oil, natural gas, and nuclear power a decade hence, the whole may sum to a totally unreasonable number, just as the expansion plans of separate firms may add up to something that is impossible for the industry. The equivalent problem may characterize consumption forecasts by fuel, but there the reasons why the aggregate may be more reliable are more evident. Every adoption of coal as a principal fuel means one less user of natural gas or oil. It may be easier to estimate overall consumption than the relative prices and other considerations that induce switching.

Models differ from one another, not only with respect to the level and direction of aggregation, but also with respect to the underlying assumption about technology and human behavior. Economists typically assume that people can consume widely varying mixes of goods and services and, similarly, that the economy can produce a variety of goods and services. Economists also usually assume that decisions of both consumers and producers are strongly influenced by relative prices. A quite different view is that certain requirements have to be met.[1] The difference in views is illustrated in Figure 5.1. Someone who forecasts on the basis of fixed requirements for energy is essentially arguing that no substitution is possible. If 100 units of output today are being produced using 50 units of energy and 50 units of labor, then some forecasts assume that producing 200 units of output will require 100 units of energy and 100 units of labor. In effect, such a forecast assumes that all expansion must be along the ray *OR* and that there are no economies of scale.

The economic view is to see substitution possibilities everywhere. The particular combination of inputs in use today was chosen on the basis of relative prices. If energy becomes very scarce, its price will rise relative to that of labor. The new (dashed) price line is tangent to the isoquant at a different input combination. The economy adapts to the changed conditions without serious difficulty. Consumers can also substitute less energy-intensive activities, so the economy has even more flexibility than this simple example indicates. Thus there are no dire consequences of failure to meet the requirements.

The feeling from reading much of the requirements literature is that failure to meet the requirements means that industry grinds to a halt and the public languishes in misery. The market view, in contrast, suggests that modest reductions in energy use can be accomplished with relatively small disturbance to human welfare and industrial output as long as the market price is free to vary so that the least-valued uses are priced out of the market first. If prices are free to vary, then shortages and surpluses are both self-correcting, with quantity supplied, quantity demanded, or both adjusting enough to eliminate the problem. It is, of course, the equality between supply and demand that is crucial to market equilibrium—not the meeting of some particular requirement. In fact, once supply and demand have been equilibrated by price, it is not clear what could be meant by saying that requirements were not being met.

GUESSING IN PRACTICE

All decisions are based on guesses about the future. Unfortunately, however, forecasts always reflect the time in which they are made, rather than the time for which they are made. If it is a period of

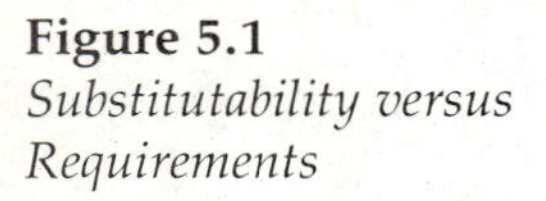
Figure 5.1
Substitutability versus Requirements

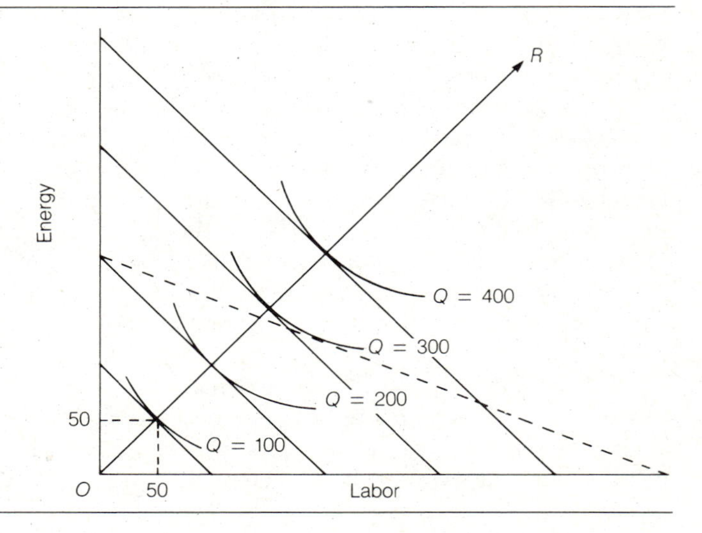

sharply rising prices, forecasts for the decade ahead will not really be statements about what to expect ten years from now, but rather will be exercises in compounded growth rates: What will prices be in ten years if they continue to grow at the rate they have grown for the past five? Similarly, a forecast made during a period of excess supply will concentrate on the favorable trends in discovery and conservation and suggest that prices are about to decrease further.

Sometimes trends do continue, and the starting point for serious attempts to visualize the future is to project a continuation of present trends. But it is only a starting point. The next step should be to look for the impending changes. For many years electric energy consumption in the United States grew at 7 percent per year. Any long-term projection will prove that such a rate of growth cannot be sustained for many decades without absorbing the entire GNP. Yet in planning for capacity, the industry assumed that the growth would continue at a rapid rate despite the price increases of the 1970s.[2] The failure to predict a turning point makes a tremendous difference to the industry because it takes about a decade to put a new plant in service. If output grows 7 percent per year, capacity must be doubled each decade; at the recent growth rate of 2 percent per year, an increase of 22 percent in capacity is enough for the decade. The stakes are high, but who will defy the trend to forecast the turning point?

In a comparison of seventy-eight energy projections for the year 1985, Brodman and Hamilton showed the influence of the conditions prevailing when the forecast was made.[3] To cite one of their many examples, the forecasts for imports of oil by the United States in 1985 averaged 764 million tons for forecasts made before 1974. After the embargo forecasts became lower, averaging 463 million tons for forecasts made between 1974 and 1978. In 1982 the United States actually imported 251 million tons of oil, so the average of forecasts made in that year would undoubtedly be even lower than those made in 1974–1978.

Not only did the average forecasts of imports decrease over the decade of the 1970s but also the standard deviation (a measure of how much the forecasts differed from one another) increased. This is not really surprising in view of the immense changes in energy markets during the decade. In energy, as in so many other areas, changes in consumer preferences and in technology, as well as in the politics of the Persian Gulf and the success of minerals exploration, can make a substantial difference to the tone of the market in a relatively short time. It is difficult to imagine a forecast for several years in the future being within 2 percent of the actual figure. Yet the difference between a surplus of 2 percent and a shortage of 2 percent is immense in terms of the way the markets behave.

Small differences in quantity have an especially strong impact in energy markets because of the maze of government regulations and the low short-run elasticities of supply and demand. When supply and demand are inelastic with respect to price, a small shift in either curve requires a large price change to achieve equilibrium. If the government does not permit the price to change, then the pressure will spill over into the other energy markets. This sort of problem occurred in the United States, where oil and natural gas prices were regulated in an uncoordinated way by the federal government while electricity and natural gas were regulated by separate state public utilities commissions.

Because economists are trained to ignore the transition problems to focus on the higher long-run elasticities, economists' forecasts are typically less pessimistic about the consequences of unforeseen changes than are those made by other specialists. In any formal model, the degree of optimism depends on the assumptions about the elasticities of technical substitution and consumer demand in response to price changes, as well as the lags in adjustment and the degree to which market prices reflect impending scarcities. Economists have been conditioned by their professional training to be relatively optimistic about all these matters. Technical and consumer substitution have already been discussed. Moreover, markets often do anticipate future shortages, as we discuss in connection with the theory of the mine in Chapter 7.

WHY FORECAST?

Since forecasting always raises great difficulties and forecasts are generally wrong, the question that always arises is, Who cares? Why do we forecast anyway? The answer is that the forecasts serve a variety of purposes. The forecasts that receive the most publicity are those released by the government. Ironically, the quality of such forecasts cannot be judged by the accuracy of the predictions because they are political documents.[4] As political documents they must be judged by their effectiveness in supporting the policies of the party or agency that issues them. If the agency were advocating a gigantic program of subsidies for shale oil, the agency would not issue a forecast of a glut in conventional oil supplies. A successful forecast would be one that helped to increase funding, not one that gave an accurate description of the world to come.

More disinterested scholarly studies of aggregate phenomena are prepared by organizations outside government, such as Resources for the Future and various research centers associated with universities. Forecasts of the total amount of energy expected to be used are of interest mainly in assessing whether modern industrial society will face major changes in amounts and patterns of energy consumption. Projecting current trends for a decade or two can be useful in showing whether such trends are sustainable over this time period.

Table 5.1 provides examples of aggregate forecasts. The guesses for 1985 and 2000 were made when 1974 data were the latest available. The data for 1983 have been added to suggest how the numbers might be changed if the forecasting were to be repeated a decade later. It is interesting that the familiar items did not change much between 1974 and 1983. The forecasts were not too far off for domestic coal and natural gas and ordinary domestic oil. The problem arose from the consumption forecasts, which were very high. The only way such demand could be met in the model was to expand the supply of nuclear power, imports of oil and gas, and increased production of domestic oil. In retrospect, the big jump in oil prices in 1979 induced consumers to cut back energy consumption about 25 percent compared with the forecasts.

Figure 5.2 offers another example of an aggregate forecast. This was part of a presentation made to a group of executives in May 1973, before the embargo. Although the specifics were inaccurate in retrospect, it is worthwhile to consider the consequences of accepting the central message. The presentation of the guesses was accompanied by a strong warning that such a degree of dependence on foreign supplies was dangerous. The executives were told to expect severe shortages of energy. If they had acted on that advice by stockpiling oil and moving toward greater reliance on domestic coal,

Table 5.1 *Comparison of Energy Forecasts (quadrillion BTUs)*

			Forecast for 1985		Forecast for 2000	
	1974 base	1983 actual	Bureau of Mines	EEI	Bureau of Mines	EEI
Domestic coal	13.1	15.8	21.3	16.3	34.8	19.5
Domestic oil	20.3	18.3	29.3	32.8	27.4	32.4
Imported oil	13.1	11.6	16.3	10.4	23.8	7.1
Domestic gas	21.0	16.3	18.8	21.0	17.0	14.7
Imported gas	1.0	1.1	1.3	6.6	2.6	3.2
Nuclear	1.2	3.2	11.8	14.2	46.1	32.5
Other	3.3	4.1	4.7	[a]	11.8	[a]
Total consumed	73.0	70.4	103.5	101.2	163.4	109.5

Source: *Compiled by the U.S. General Accounting Office in "U.S. Coal Development—Promises, Uncertainties" (EMD-77-43), September 22, 1977, p. 1.11. The forecasts are from the U.S. Bureau of Mines and the Edison Electric Institute.*

[a] *"Other" includes geothermal, oil shale, and hydropower. The Edison Electric Institute estimate includes other with nuclear.*

their companies would have benefited greatly during the 1970s. The projection was economically naive in its failure to incorporate price changes, but it did serve a useful purpose.

The most meaningful forecasts are highly specific. Particular firms and consumers want to know whether to invest in equipment that uses gas, oil, coal, or electricity, which means they want to know relative prices of various fuels, the price of the cheapest relative to conservation measures, and the reliability of supply. Similarly, producers want to know the price at which they can sell and the size of the market. For these sorts of questions the aggregate data do not supply many answers. If you are interested in the market for coal a decade hence, you do not need aggregate forecasts but rather such specifics as (1) the rate of growth of electric power production, (2) the success of nuclear power in overcoming its current difficulties, (3) the availability of natural gas, and (4) the adoption of refining processes that decrease residual oil output.

For public purposes as well, little use can be made of aggregate forecasts. It could be argued that public policy is concerned only with the rough equivalence between aggregate supply and demand, not with the details of who consumes what. But if reliance is to be placed on the capabilities of consumers and firms to substitute one fuel for another, we might as well recognize that substitution of other goods and services for energy can bring about the equivalence between supply and demand in response to changes in relative prices. Forecasts are most useful when they deal with specific commodities and locations where shortages could be damaging and

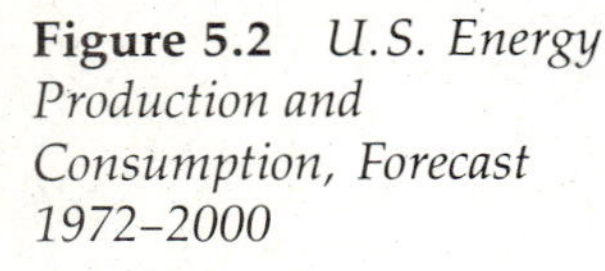

Figure 5.2 *U.S. Energy Production and Consumption, Forecast 1972–2000*

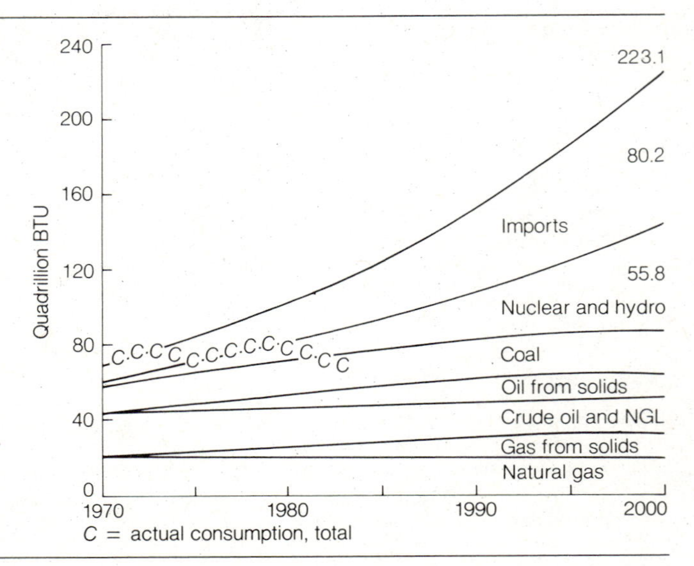

Source: *John J. McKetta, "The Broad Energy Outlook," the Charles M. Schwab Memorial Lecture, presented to the Eighty-First General Meeting of the American Iron and Steel Institute, May 23, 1973, p. 10. Reprinted by permission.*

industrial and consumer options for substitution are few. But these are the sorts of forecasts that alert users of specialized materials make for themselves.

FORECASTS OF ENERGY AND GNP

For large oil companies, of course, the specific forecasts necessary for their plans include the aggregate data. Figure 5.3 shows forecasts of U.S. oil imports as compiled by British Petroleum Company in

Figure 5.3 *Estimates of U.S. Imports*

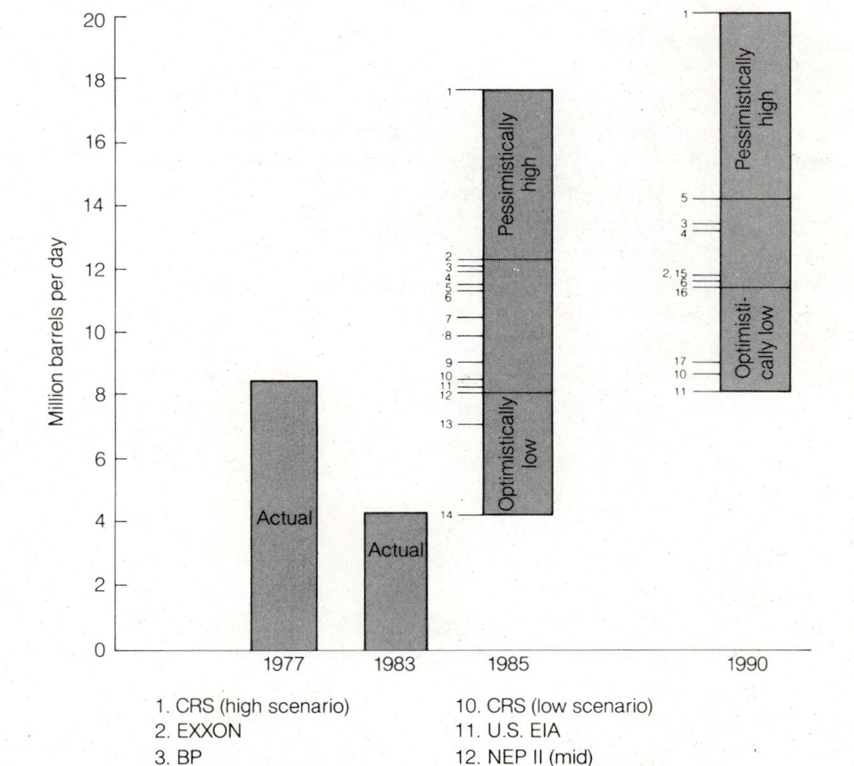

Source: *"Oil Crisis . . . Again," British Petroleum Company (2nd printing, October 1979), p. 20. Reprinted by permission.*

1979. The actual imports for 1983 were about half those for the 1977 base year and below all but one of the forecasts for 1985. The lesson is that even those who have a direct financial stake in such forecasts have had difficulty in keeping up with the changes that have occurred so rapidly in recent years. In this particular case, OPEC prices were sharply increased just after the forecasts were made.

It is a brave forecaster indeed who makes an explicit guess about prices. To the extent that demand is inelastic, consumption is not influenced by price. Yet the empirical studies suggest that inelasticity is a characteristic only in the short run. This can be taken into account in a long-run forecast by guessing that the ratio of BTU to GNP will decline in the future. Since that ratio was declining even when real prices were going down, however, we are faced with the necessity of forecasting the rate of the decline.

Even the GNP forecast itself is a risky enterprise, although one that numerous commentators undertake. The main conceptual problem is that labor and energy are both inputs that help to determine the level of GNP. From a theoretical view, the only conceptually correct approach is to construct a complex model relating GNP to all inputs including labor and energy and also embodying demand equations that make energy consumption a function of per capita income, labor force participation, family size, wage rates, and other variables. It is not difficult to sketch out such models, but we do not have the information to write down even the relationships that now prevail—to say nothing of the ones that will prevail in the future. The relationship between labor and energy input and GNP is particularly intractable. Gross National Product can be written as a function of energy, labor, and capital. Whether such aggregate production functions are meaningful at all is an open question.[5] In any event, we do not know the terms on which labor or capital can substitute for energy currently. The attempt to forecast them must be an even more questionable activity.

Two main findings come out of this analysis. First, forecasting is an extremely uncertain business. To be useful, energy forecasts must be highly specific about sources and uses of energy. Nothing in the apparent precision of having a physical measurement for energy makes that degree of specificity easier to obtain. Partially offsetting the problem posed by the difficulty of forecasting energy consumption is the second implication: The relationship between energy use and GNP is not rigid. Energy intensity has varied substantially over time and between countries. Since it will undoubtedly continue to vary, forecasting is both more difficult and less important.

NOTES

1. For an example of the requirements approach, see U.S. Federal Power Commission, Bureau of Natural Gas, "Natural Gas Supply and

Demand 1971–1990," Staff report no. 2 (Washington, D.C.: FPC S-218, February 1972).

2. For a discussion of the neglect of prices in this and other studies of energy, see Hendrik S. Houthakker, "Whatever Happened to the Energy Crisis?" *Energy Journal* 4, no. 2 (April 1983), pp. 1–8.
3. John R. Brodman and Richard E. Hamilton, "A Comparison of Energy Projections to 1985," International Energy Agency monograph (Paris: Organization for Economic Cooperation and Development, January 1979).
4. For a detailed case study of the way in which an ostensibly technical report becomes a political document, see Bela Gold, *Wartime Economic Planning in Agriculture* (New York: Columbia University Press, 1949), p. 542.
5. For a discussion of production functions and an analysis of the problems of using them at the aggregate level, see Robert N. Baird, "Production Functions, Productivity, and Technological Change," in *Research, Technological Change, and Economic Analysis*, ed. Bela Gold (Lexington, Mass.: D. C. Heath, 1977).

PART TWO

The Extractive Industries

CHAPTER SIX

Reserves and Resources

Although the different sources of energy are good substitutes for many uses, the similarity does not extend to the supply side. The various energy industries are different in many ways. These differences affect the speed of response to changing market conditions, as well as the long-term prospects for the separate industries. To deal with questions of energy supply, it is necessary to consider specifically the industries based on exhaustible minerals like oil and uranium, those based on renewable resources like rain and sunlight, and those that convert one form of energy to another such as electrical generation.

The energy industries include some activities that can be analyzed like ordinary manufacturing activities, but the mineral industries share typical characteristics that set them apart from manufacturing. Since the behavior of such extractive industries as oil and natural gas furnishes an essential background for understanding even something as different as solar power, we begin this study of the energy industries with a brief discussion of topics usually associated with the theory of the mine—that branch of economic theory that deals with using up a fixed stock of natural resources.

As the theory is developed, it will become clear that the geologic and technological aspects of the industries are not adequate to explain behavior. Efficient production, just like efficient consumption, is inevitably related to the preferences of consumers. On the production side, for example, the determination of what mineral deposits are reserves, and hence of the total quantity of reserves, is an economic question as much as a geologic one.

DEFINITIONS

To keep the discussion clear let us define some terms that are used consistently throughout this book. The definitions used here are common in recent literature on natural resources but are by no means universal, especially in popular discussions and earlier writings.

The two general themes in an analysis of mineral resources are uncertainty about what might be underground and the varying qualities of the minerals that are there. It is reasonable therefore to order whatever information might be available according to those characteristics, as is shown in Figure 6.1. The general notion is useful and simple. The resources are arrayed from left to right according to decreasing certainty and from top to bottom according to increasing cost of the finished product produced from the mineral. Reserves include only those minerals in the top left corner: those that can be extracted and sold at a profit under current market conditions and that have already been identified, which means that

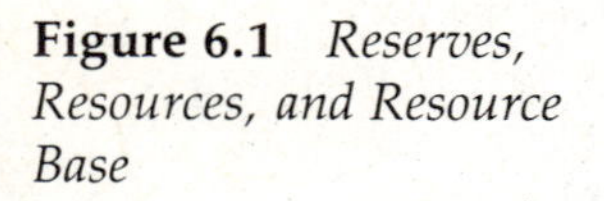

Figure 6.1 *Reserves, Resources, and Resource Base*

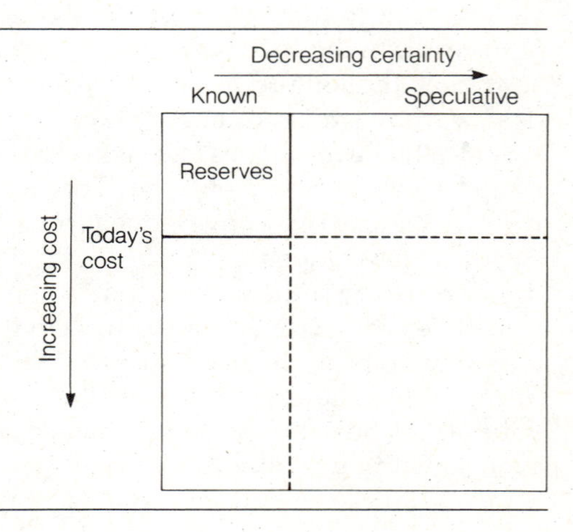

their existence and location are known.[1] Sometimes these are called proved reserves (meaning that someone has drilled into the mineral deposit to prove its existence), but the *proved* is redundant here because it is part of the definition of *reserves*. The entire big box consists of resources. The lower left part of the box may be known with a considerable degree of certainty; yet it is not a reserve because it is too expensive at present. The top right of the box is expected to be inexpensive, but the deposits have not yet been found.

Sometimes the cost dimension is subdivided into recoverable (at today's price), paramarginal, and submarginal, where paramarginal could be profitable to recover if prices rose 50 percent and submarginal is lower quality than that. The key idea, however, is the continuum of quality. Quality of a mineral has numerous aspects, which can be summarized as cost of extraction, cost of transportation, and cost of use. The vertical axis of the box in Figure 6.1 is the market price at which it is just worth mining a particular deposit. This depends on the grade of the mineral (such as viscosity of oil or heat value of coal) and the extraction cost (how expensive it is to remove from the ground), as well as various other characteristics (transportation cost, impurities that hinder use, amenability to mass processing techniques) that can also be significant determinants of cost or market value.

The horizontal dimension of the box can also be marked off according to the degree of uncertainty: Are the guesses being made about geologic formations that are known to be similar to those already being exploited? Have drills actually penetrated the strata that are supposed to contain the resource? Here again it seems more useful to think of a continuum of uncertainty, rather than emphasizing discrete categories.

At what points along these two axes should we locate the box labeled reserves? From both a practical view of availability of data and the conceptual standpoint, it is most useful to take the boundaries used by the industry itself; that is, the analyst might as well consider reserves to be whatever the firms themselves are willing to count as reserves. This creates certain difficulties noted later, but the advantages are overwhelming. In the first place, no one is better prepared to estimate the costs of producing particular deposits and their present market value than the people who succeed in making a living by doing so. The guesses by the firms on the cost dimension are most likely to be realistic and based on the most recent generally accepted techniques. Similarly, in assessing the degree of certainty necessary to count something as a reserve, the outside analyst could engage in a learned discussion of the statistical questions involved in choosing the best pattern and spacing of exploratory drilling. The fact remains, however, that the firms in the industry have bet good money on the existence of reserves when they develop mines or wells, transportation links, and processing facilities. Hence the information on which firms act must be as reliable a guide to what is in the ground as we are likely to find.

The disadvantage of attempting to use data generated by firms is that firms may prefer to distort or conceal them. In particular, states have often taxed reserves, with the result that firms understate them.[2] Similarly, when governments regulate maximum prices that the industry can charge, it is advantageous for firms to minimize stated reserves and to use their declining magnitude as an argument for higher prices. In the absence of taxation and regulation, reserves would often be overstated for the ordinary commercial purpose of making the firm appear strong, reliable, creditworthy, and not hungry to acquire more reserves. Because of these biases, it is the actual operating information of the firm that we would like to have, rather than the numbers that it wants to give us.

Assume for the moment that firms do provide the outside analyst with their best estimates of reserves. What do the numbers mean? If reserves are 100 million tons and the firm produces 5 million tons per year, should we expect reserves to be zero after twenty years? The answer, of course, is no. The firm continues to add new reserves as it exhausts the old. Reserves can be considered a kind of inventory. Firms do not try to maximize or minimize inventory but rather try to find the appropriate amount to facilitate maximization of profits (or the pursuit of other goals). Reserves also are expensive to buy, prove, and hold, but permit the firm to make long-range plans for processing and marketing. Thus we would expect firms to aim for reserves that match their long-range planning, which means something in the vicinity of fifteen years. The notion of what is a reserve and how much is adequate, however, turns out to be much more complex, as we discuss in connection with particular industries in Chapters 8, 9, and 10.

When economic theory treats reserves in the same way that it treats manufacturers' inventory, the assumption is that reserves can always be augmented if a firm is willing to pay the price. In terms of Figure 6.1, reserves are supposed to be created from the resources in the larger box by sufficient investment. The firm can acquire or prove reserves by exploring new areas for indications of the mineral, looking more intensively at areas thought to contain the mineral, developing techniques for using submarginal resources, or extending transportation facilities into inaccessible areas.[3] Any particular firm can also buy reserves from other firms, but ordinarily that will not increase the total reserves of the economy. Sometimes, however, the buying firm will know the techniques for exploiting a deposit profitably, whereas to the selling firm that same deposit is a submarginal resource.[4]

Although it is certainly reasonable to suppose that investments will increase reserves, it still leaves the question of how we know the dimensions of the big box—total resources—if no one has an incentive to stake out the boundaries. This is the topic of the next section. The key point to bear in mind is that the true dimensions of total resources are unknown. This applies with as much force to the cost dimension as to the certainty one. Until people have an incentive to explore the boundaries of mineral deposits, they can only guess at what is there, but there is little incentive to measure submarginal resources.

HOW DO WE KNOW WHAT IS THERE?

This book deals mainly with reserves and the amount of investment required to increase reserves; these are the important economic variables. Still, there are some reasons for being curious about the boundaries of the big box of Figure 6.1—total resources. This section asks why people are interested and examines the techniques used to estimate total availability. Although the estimate could be based on tea leaves or animal entrails, in the minerals area it has become more common to start either from the scraps of geological evidence or from the results of past searches.[5]

The main reason that people think about total resources is fear that the economic calculations are too shortsighted. If firms are proving out enough resources for their own consumption over the next couple of decades, does that mean total exhaustion of the resource might overtake us suddenly in thirty years? To alleviate this fear, however, it is not necessary to estimate the exact dimensions of the resource box. It is enough to know that beyond the boundaries of reserves there are geological formations that look promising or deposits that can be exploited at somewhat higher cost. The techniques for estimating total resources therefore should be judged according to their usefulness in satisfying this somewhat modest objective.

The three principal techniques are volumetric, crustal abundance, and behavioristic. The *volumetric* approach is the simplest to describe. In its crude form it requires only some estimate of the amount of mineral per unit of some geological structure and an estimate of the extent of that structure. If we assume, for example, that a cubic mile of sediment will contain 50,000 barrels of oil, it is only necessary to obtain a geological estimate of the amount of sediment to guess resources of oil. The technique can be refined by recognizing differences in the richness and accessibility of different sediments. This would also require more detailed geological work to identify and locate various sediments.

Nonfossil minerals, including uranium, can be estimated as a function of their *abundance in the earth's crust*. The general idea is that an area as large as the United States can be considered a random sample of the entire earth. If, for example, a mineral is believed to constitute 0.01 percent of the earth's crust, then the reserve that is estimated to be found in the United States is 0.0001 times 1 billion tons, or 100,000 tons. Total domestic resources might be ten times as large. This gives only a rough idea of what to expect. It does not tell where to look for undiscovered resources. Since it is not applicable to coal and oil, which were laid down in large beds as a result of the growth of plants and animals, we do not consider it further here.

The *mathematical* or *behavioristic* approach is to look at the history of human exploitation of the mineral, rather than the geology. The most famous example is the forecast by M. King Hubbert.[6] The technique consists essentially of plotting cumulative production against time (Figure 6.2). This yields a set of points to which some arbitrary growth curve can be fitted. The curve is then extrapolated and its asymptote indicates the total amount of the resource that will be exploited. The change in cumulative production during any year is the annual output. It is generally expected to increase up to some peak level and then decrease as exhaustion approaches.

This technique has two great advantages and one disadvantage. The first advantage is that King Hubbert used it to forecast the decline in oil production in the United States during the 1970s. The second advantage is that it makes use of accurate data that are readily available. The main disadvantage is that the data are irrelevant. It is much like using the accurately known scores of Cleveland Browns games last year to forecast the scores of Cleveland Indians

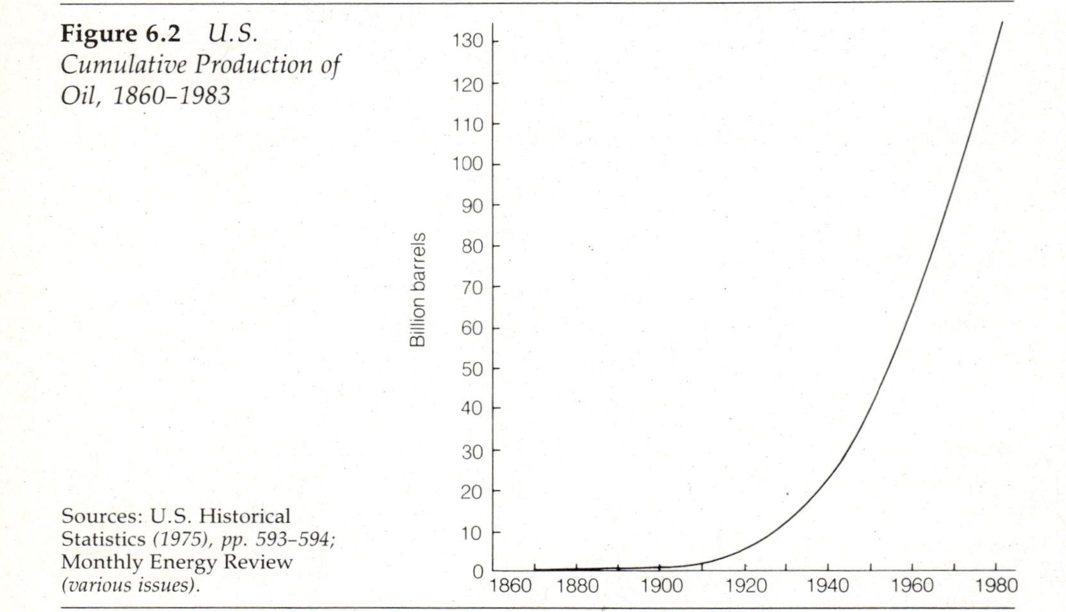

Figure 6.2 *U.S. Cumulative Production of Oil, 1860–1983*

Sources: U.S. Historical Statistics *(1975), pp. 593–594;* Monthly Energy Review *(various issues).*

games next year. Even an occasional success should not inspire much confidence in the technique. The task at hand is to estimate what remains in the ground. This is a geological question that can be answered only by using geological data. No amount of information about what people have found in the past is directly relevant to answering the question. Annual production depends on many variables such as cost of production and demand; it is not just a proportion of the remaining resources as the behavioristic approach implies.

We could construct theoretical underpinnings for the behavioristic approach by assuming either that exploration is totally random or that deposits are accurately known from the beginning. If holes were drilled at random locations, the larger deposits would be disproportionately represented among the deposits intersected by the random grid of exploratory holes. Since developmental drilling clusters around successful exploratory holes, we could presume that the largest deposits are, on average, encountered and exhausted before the smaller ones. Subsequent exploratory drilling on a finer random grid would encounter smaller deposits that are more costly per ton to discover. If, in contrast, information were perfect, exploitation would begin with the best deposits and, under some geological conditions, move to ever smaller and more isolated deposits. Under other conditions, however, the deposits exploited later might be massive and uniform but just more costly. Oil shale and formations in which gas is tightly bound are examples.

The real world enjoys neither perfect ignorance nor perfect knowledge. Exploration is guided by the imperfect knowledge embodied in geological theories and wildcatters' guesses. Although theory and the intuition of experienced people provide a higher probability of success than random drilling, they do not guarantee that the best class of deposits is discovered first. Thus the behavioristic approach, although it can be useful in analyzing past experience, does not foretell the future.

This suggests that there are no real substitutes for detailed geological exploration to estimate what is under the ground. The data on proved reserves therefore are the most meaningful figures. We should, however, interpret the reserve figures as the inventories of raw materials carried by producers of minerals. It is an error to interpret reserves as an indication of total availability.

RESERVES AND RESOURCES OF MAJOR FUELS

The data on petroleum reserves (Figure 6.3) offer a good illustration of these points. Reserves grew almost steadily with production until 1967. Although there were occasional years in which reserves declined, the general trend was upward even during the war years.

The reserve/output ratio rose during the earliest period. The earliest discoveries happened to be relatively high-cost fields. This meant that new reserves could still be worth the investment necessary to find them if they were inexpensive to produce.[7] Once the industry had adjusted to the accidents of early discovery, the ratio of reserves to output settled to a more reasonable level.

After 1950, when the magnitude and quality of Middle Eastern reserves were becoming known, the situation changed again. It then became difficult for a company to justify hunting for the high-cost domestic oil when foreign reserves were much cheaper. Production from domestic wells continued, but the inventory of oil reserves in the United States was not fully replaced during the 1960s.

Table 6.1 gives an overview of reserves, resources, and consumption for coal, oil, gas, and uranium. Since these data compare domestic consumption with domestic reserves, the oil situation appears quite bleak, with reserves equivalent to only about six years' consumption. With imports accounting for about one-third of consumption, domestic reserves will be conserved. Moreover, they will continue to be augmented as resources are identified more precisely.

The resource data give an excessively optimistic impression in one respect, however. The implication of a resource/output ratio is that, for example, uranium resources would last 175 years at the

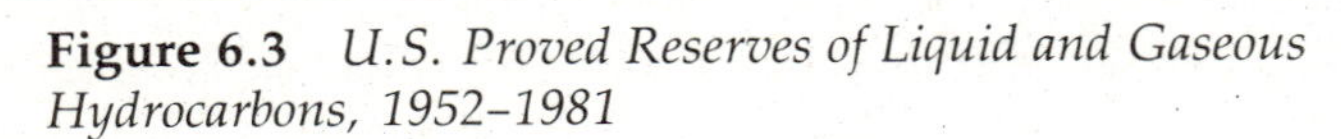

Figure 6.3 *U.S. Proved Reserves of Liquid and Gaseous Hydrocarbons, 1952–1981*

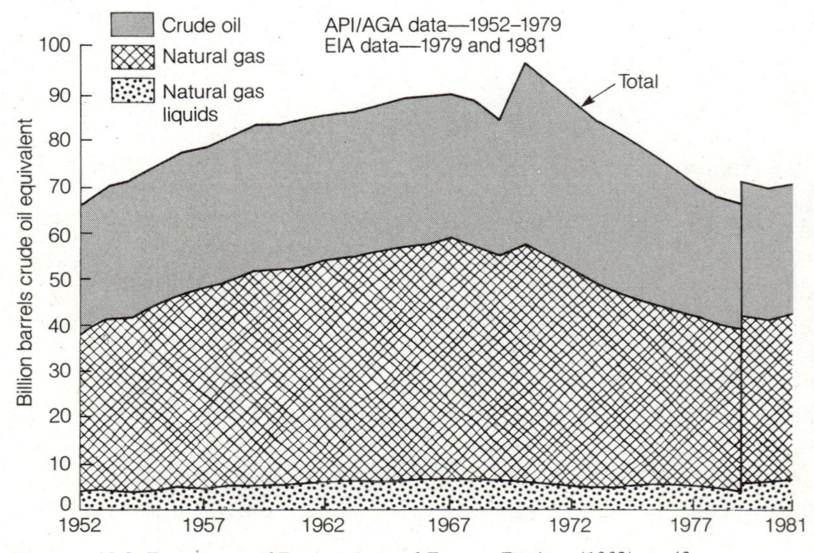

Source: *U.S. Department of Energy,* Annual Energy Review *(1982), p. 42.*

current rate of consumption. In fact, we would have to consider recovery rates (the proportion of the mineral that is extracted from the deposit in the ground) before making that type of statement. The recovery rate for coal has traditionally been estimated to be 50 percent; but it will vary depending on the physical conditions in particular deposits, the mining technique, the price of the coal, and other factors.

Although we must recognize that reserves will be augmented from resources, we must also avoid the error of assuming that more of a mineral can always be found by spending more on exploration. Finding and exploiting a mineral in more remote and difficult locations will require spending more, but spending more will not necessarily produce the mineral—that depends on the geological fact of whether it is there.

NOTES

1. For discussions of these concepts, see Donald A. Brobst and Walden P. Pratt, "Introduction," U.S. Geological Survey, *Professional Paper* 820 (1973), pp. 3–5, and V. E. McKelvey, "Mineral Resource Estimates and Public Policy," in the same volume, pp. 13–14.
2. For a discussion of efforts to tax reserves and the results, see Warren Aldrich Roberts, *State Taxation of Metallic Deposits* (Cambridge, Mass.: Harvard University Press, 1944). For an account of the sources of oil and gas resource estimates, see Aaron Wildavsky and Ellen Tenenbaum, *The Politics of Mistrust* (Los Angeles: Sage, 1981).

Table 6.1 *U.S. Reserves, Resources, and Consumption of Fuels, 1981*

	Fuels			
	Oil[1]	Gas[2]	Coal[3]	Uranium[4]
Consumption[5] (quantity/year)	5.84	19.4	728	19
Proved reserves (quantity)	34.7	199		470
Reserves/consumption (years)	6	10		25
Resources (quantity)			3,224,372	3,336
Resources/consumption (years)			4,400	175

Sources: *U.S. Department of Energy,* Annual Report, *vol. 2 (1981), pp. 7, 43, 173; U.S. Geological Survey,* Professional Paper *820 (1973), p. 137.*

Notes: *(1) Oil data are in billions of barrels and include natural gas liquids at crude oil equivalent. (2) Natural gas data are in trillion cubic feet. (3) Coal data are in millions of short tons. Resources are given for 1972. (4) Uranium data are in thousand short tons of* U_3O_8. *(5) "Consumption" is actually domestic production plus imports minus exports.*

3. Accounts of noteworthy technological and geographical explorations for nonfuel minerals are provided by E. W. Davis, *Pioneering with Taconite* (St. Paul: Minnesota Historical Society, 1964) and Forbes Wilson, *The Conquest of Copper Mountain* (New York: Atheneum, 1981).
4. This was cited as justification for the acquisition of Belridge Oil Company by Shell Oil in 1979. See George Getschow and Roger Thurow, "Shell-Belridge Merger Thrives on Technology, Avoids Most Pitfalls," *Wall Street Journal*, November 5, 1981, p. 1.
5. McKelvey, "Mineral Resource Estimates," pp. 14–17.
6. M. King Hubbert, "Energy Resources," in Committee on Resources and Man of the Division of Earth Sciences, National Academy of Science–National Research Council, *Resources and Man* (San Francisco: W. H. Freeman, 1969), pp. 157–242.
7. Indeed, a sufficient quantity of new cheap reserves could even lower the price enough so that what was once a reserve becomes a resource. This occurred in the iron ore industry when the new taconite process displaced some of the traditional iron ore reserves. See Davis, *Pioneering with Taconite*.

CHAPTER SEVEN

The Economic Theory of the Mine

The economic theory of the mine is one of the most intriguing areas of economics because it combines the theoretical problems of uncertainty about the resources remaining underground, the progressive exhaustion of particular mineral deposits, and the durability of mining investments. When the differences in quality of different deposits, the impact of technological changes, the complex welfare issues posed by exhaustion, and the fugitive nature of some resources are considered, the theory invites an unending display of theoretical techniques. By the same token, however, it is difficult for the uninitiated to grasp the essential features of the mineral industries by the dazzling glare of the theoretical fireworks.[1]

Most of the complications of the theory of the mine therefore are introduced along with the discussion of particular energy sources for which they are relevant. The focus is on understanding the particular industries rather than on the economic theory itself. Nevertheless, it is helpful to begin with a sketch of the rudiments of the theory to provide a framework for considering the separate industries.

The basic problem addressed by the theory of the mine is the timing of sales to maximize profit. To abstract from all complications, think of a dealer who owns 10,000 barrels of crude oil that he could sell for $35 per barrel or $350,000 today. Is it more profitable to do that or to wait? If he is trying to compare the profitability of selling at two different times (now and one year from now), he obviously needs to know the price today and to guess about the price next year. Suppose that the dealer expects the price to be $45 per barrel next year, that the cost of moving the oil from storage to market is $1 per barrel, and that the dealer owns the storage tanks and has no other use for them. These data can be summarized as follows:

	Sell now	Sell next year
Gross receipts	$350,000	$450,000
Operating cost	10,000	10,000
Net receipts	$340,000	$440,000

By waiting one year for his profit, the dealer can expect to receive $440,000 instead of $340,000. This is a return of 100,000/340,000 or 29.4 percent. Unless he can think of an alternative use for the money on which he expects a higher rate of return, he should withhold the oil from the market. Of course, he is not bound to a one-year decision. At each moment he has only to ask himself whether he expects the net returns to rise faster than the rate of interest he is paying on bank loans or the rate of return he expects in the best alternative investment. This is the same behavior that maximizes profits for a speculator or investor in any commodity or financial instrument, such as stocks or bonds.

If the dealer expects the net receipts to decline, stay constant, or increase only slightly, he will prefer to sell the oil now. If, for example, he has an opportunity to earn 15 percent by putting the money in the bank, then he must expect the market price to rise to more than $40.10 per barrel in a year's time to induce him to hold the oil. Otherwise, he could sell now, invest the receipts of $340,000 at 15 percent and have $391,000 total after one year's time.

TIMING OF EXPLOITATION

Deciding when to exploit a mineral deposit is much more complicated. The most tractable problem is that developing a mine or well requires a large initial investment to produce a stream of returns lasting many years. The optimum decision requires both geological knowledge and economic information about prices and costs far into the future. These facts can only be guessed when the decisions are made. If the numbers were known, it would be a standard economic exercise to analyze them within the framework of investment analysis.

Table 7.1 gives a simple illustration. Mineral ventures can have a great variety of different patterns of investment and payout, but the example in the table illustrates a common one. The first point to notice is that receipts from sales of the mineral do not begin until several years after the start of large expenditures on development. In this case, it is the fourth year before any sales are made, but many large projects require investments for even longer periods before any output can be sold. The project is dated from the beginning of substantial development expenditures, although significant amounts may have been required earlier for detailed exploratory drilling and planning before the decision to proceed with development.

As each year of development work passes, the total investment in the project increases, both because more is being spent and because the mining company must allow for interest on the capital that has already been invested. In the table the cash flows are shown in row 3. This is the amount that the company can take out of the project in that year, which is receipts minus all expenditures. During the investment phase it is negative, of course, because expenditures are large and receipts are negligible. The fourth row shows the cumulative investment year by year. At the end of the first year it is

Table 7.1 *Cumulative Investment in a Mine (millions of dollars)*

	Year 1	Year 2	Year 3	Year 4	Year 5	Year 6	Year 7
Gross receipts from sales of product	0	0	0	1	4	5	5
Expenses for mining and development	1	3	6	10	2	0	0
Cash flow	−1	−3	−6	−9	2	5	5
Cumulative investment at 20 percent interest	1	4.20	11.04	22.25	24.70	24.64	24.57

$1 million. At the end of the second year that $1 million must be considered to have grown to $1 million times $(1+i)$, where i is the interest rate. The relevant interest rate is the greater of that paid by the firm to borrow additional funds or the return that the firm could earn on the best alternative investment. In this case it is assumed to be 20 percent.

During the second year the initial $1 million grows to $1.2 million and an additional $3 million is added, for a total investment of $4.2 million. During the third year the $4.2 million grows to $5.04 million and an additional $6 million is spent, so the total is $11.04 million. In the fourth year some output is sold, but expenditures still exceed receipts by $9 million, so total investment grows to $22.25 million. In the fifth year net receipts are finally positive, but interest charges increase the total investment to $24.7 million—the maximum that it attains in this example. Subsequent net receipts of $5 million per year exceed the interest of $4.94 million and hence permit the investment to pay off if output can be continued long enough. (It would take twenty-four years to pay off a debt of $24.7 million at 20 percent interest by making payments of $5 million per year.)

This simple illustration demonstrates several important features of the mineral industries. In the first place, the investment decision is just like that in any other industry where durable assets must produce a relatively small stream of returns for a long time. Yet although investment decisions in every industry must be taken in the face of uncertainties about the market prices of inputs and outputs several years in the future, the mineral industries also face the particular uncertainties of exploiting raw materials that cannot be perfectly measured in quantity or quality until the materials are extracted from the ground. It would not take a very large miscalculation about investment cost or the quantity or price of output to turn the project of Table 7.1 into a financial mistake. However, a doubling of output prices after the investment is in place could produce a very handsome return.

Of greater long-run significance is the question of how the mining process comes to an end. A manufacturing firm can look forward to rebuilding or replacing its plant in twenty years or so while continuing in business at the same location, unless better choices are available. The firm that exploits a mineral deposit can look forward to the exhaustion of that particular deposit. Just as the manufacturer can build a replacement for a worn-out plant, so a mining company can search for a new deposit to replace the one that is exhausted. Nevertheless, the usual assumption is that the cost of mining rises as the particular deposit approaches exhaustion. Furthermore, it is reasonable to expect the cost of replacing reserves to rise as exploitation progresses. This means that *economic rent* is an important concept in the mineral industries.

The Economic Theory of the Mine

An ordinary manufacturing enterprise must earn a normal rate of return on its capital; otherwise in the long run the capital will be transferred to more profitable activities. In addition, if management is especially clever or if the firm has valuable patents, other strong monopoly privileges, or a string of good luck, it may earn something more, which economists call economic profit. The mining company must also earn interest on its total capital and may, with luck or skill, earn a profit. Besides interest and profit, a third type of return is particularly important in the mineral industries, so it is singled out for attention here, although it is often neglected entirely in analyses of manufacturing operations. This return is named economic rent. (The usage is unfortunate because of the familiar use of *rent* to describe the payment people make to a landlord for an apartment.)

Economic rent, in general, is defined as "any return to a factor of production in excess of the minimum required to attract it to a particular activity." For a mineral deposit it is any surplus after all the costs have been met, including the minimum compensation that the landowner requires to permit mining, as well as wages of labor and interest on capital. The boundary line between profits and rents is often fuzzy, and for most purposes the common practice of describing as profit all returns above the minimum for operation does not lead to confusion. Nevertheless, the concept of economic rent is useful because of its focus on the key features of mineral industries. Rent derives from the characteristics of the land. Some mineral deposits are richer and better located than others. Profits arise from the superior insights, timing, or luck of the entrepreneur.

The four principal economic characteristics of a particular mineral deposit are the cost of extraction, cost of processing, cost of transportation to market, and the price of that mineral in the market. If, for example, the market price of a particular coal is $25 per ton, the extraction cost is $15 per ton, cost of processing is $1 per ton, and the transportation cost is $4 per ton, then this leaves an economic rent of $5 per ton. Obviously, each deposit will have its own set of costs and a selling price, so different deposits will yield different amounts of rent. Nevertheless, some systematic relationships can be observed, as in Table 7.2.

The data in the table show each mine earning economic rent, that is, each mine is apparently profitable to operate now. Mine C, however, is a high-cost operation that survives only because its product commands a premium in the marketplace. It might, for example, be located in a remote area and serve the local market that is cut off from cheaper choices. It might also produce a premium product such as coal of low sulfur content or good coking qualities (the premium for such qualities is often much higher than that shown). The first point to note is that a premium price will attract research, development, and investment effort as managers of firms buying coal strive to substitute cheap inputs for expensive ones. Whether they will prove successful is, of course, an empirical question. Some quality differentials have persisted for long periods. Development of an inexpensive sulfur removal process, however, would certainly cut into the premium for low-sulfur coal.

If we look only at Table 7.2, it seems that all three deposits should be mined now. If we consider the owners' expectations for the future, however, the example assumes the same form as that of the oil dealer discussed at the beginning of the chapter. Although the mine owners can earn rent by selling coal today, they may be able to make more by deferring the sale. Suppose that prices are expected to increase $1.50 per ton in a year and costs to stay unchanged. With the data given in Table 7.2, Mine A has a return of $1.50 for deferring $15 for a year, or 10 percent. Mine B receives the $1.50 for deferring $5, or a 30 percent return. Mine C, similarly, receives almost 19 percent for waiting. If the prevailing market rate of interest is 20 percent, then the owner of A should sell now and invest the proceeds at 20 percent. By next year the owner would have $18, rather than the $16.50 that would be netted from the coal next year. For B waiting is much more profitable than selling today. For C it is a tossup, although selling today offers a few more cents. The general rule is that the deposits for which rents are the greatest are the ones that are most profitable to exploit now. Low-rent mines (and especially those that would have to sell at a loss today) find that the rate of return to waiting is higher than the rate that can be obtained by selling the mineral and investing the proceeds.

If the deposits are all identical in quality and cost of transportation—differing only in extraction cost—then the situation is particularly clear: The cheapest deposits to exploit should be mined first. When different qualities and transportation costs must be considered, then it is necessary to construct a table like Table 7.2 for the next year as well and calculate the rates of return. All of this is straightforward.

The implication of this analysis is that economic rents will rise over time at the rate of interest. The reason is that if rent promised to increase faster, the resource owner would defer exploitation. This

Table 7.2 *Price, Cost, and Rent for Three Mines (dollars)*

	Mine A	Mine B	Mine C
Sales price/ton	25	25	35
Total costs/ton	10	20	27
Rent/ton	15	5	8

would decrease current supply (raising its price) and increase future supply (lowering its price). This process would continue until the present and future price had adjusted enough to bring about the appropriate rate of increase in rents. Conversely, if resource owners do not expect rents to increase as rapidly as the interest rate, they will dump more minerals on the market today with the plan to invest the proceeds elsewhere. But as they sell more today, the price today will decrease and the expected future price will increase until rents again are expected to follow the path predicted by theory.

Figure 7.1 shows this in a simple two-period model with a fixed total quantity. In the figure everything that is not sold this year will be sold next year, so prices in the two periods are directly linked once a guess is made about the demand curves. Indeed, the same vertical line can serve as the supply curve for both years because supply and demand in year 1 are drawn in the usual way, whereas supply and demand for year 2 have their origin at the lower right corner of the diagram, Q_{total}. (This is the feature that restricts this diagram to the simple case where the mineral is exhausted in two years.) The vertical line, showing how the fixed supply is used in the two years, can be moved back and forth until the price differential is just adequate to give the proper pattern of rents, that is, rent in year 2 equals $(1+i)$ times rent in year 1.

An alternative way to express the rule by which owners of mineral resources maximize their wealth is to say that they must reallocate production until the present value of rent per ton is the same, regardless of when it is received. *Present value* is the amount that a sum of money scheduled to be received in the future is worth at the present. The dollar amount to be received in the future must

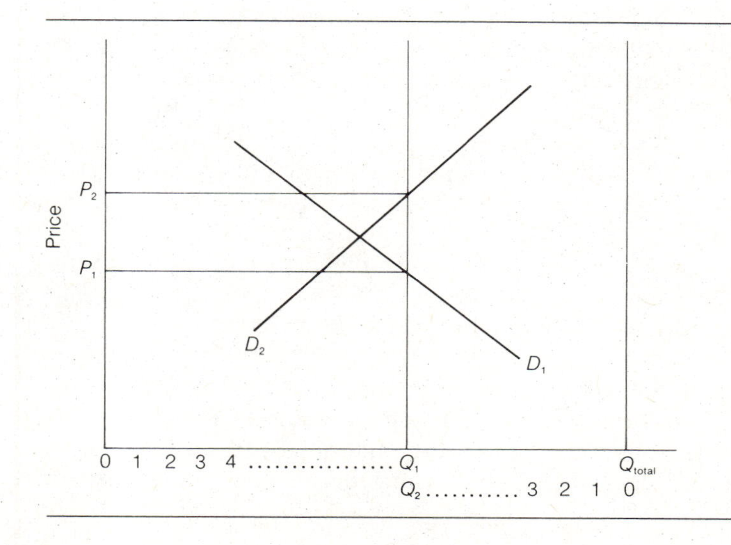

Figure 7.1 *Two-Period Model of Mineral Exploitation*

be discounted to arrive at the present value, because if you had a smaller amount now (the present value) you could invest it at the prevailing rate of interest to accumulate to the future value. If you are scheduled to receive $110 one year from today, you should be willing to settle for $100 today if you have the chance to put it in a secure investment yielding 10 percent (or perhaps to pay off a loan on which you must pay 10 percent interest), because the $100 today will accumulate to $110 in one year. More generally,

$$\text{Present value} = \frac{\text{Future value}}{(1+i)^n}$$

where i is the rate of interest and n is the number of years until the money is received.

When the mineral markets are in equilibrium such that the rent per ton rises at the rate of interest, the following pattern prevails:

$$R_o = R_o\frac{(1+i)}{(1+i)} = R_o\frac{(1+i)^2}{(1+i)^2} = R_o\frac{(1+i)^3}{(1+i)^3} = \ldots.$$

R_o is the rent per ton today. If resource owners delay exploitation for a year, they will receive $R_1 = R_o(1+i)$. The present value of $R_1 = R_1/(1+i)$. Since $R_1 = R_o\ (1+i)$, the present value of $R_1 = R_o\ (1+i)/(1+i) = R_o$. An equivalent relationship applies to every time period, with the same factor being used to increase the rent received in the future year and to discount its value to the present. The result is that the present value of the rent, regardless of when it is received, is R_o.

Because rent appears to be so important in the mineral industries, it seems worthwhile to examine it more closely at the level of the firm. Figure 7.2 shows the standard cost curves for two different mines. Under some conditions, the supply curve for the industry can be obtained by adding the quantities (given by the marginal cost curve) that each firm will find it profitable to produce at any given price. If the demand curve is such that the price is P_1, then the higher-cost firm (of the two shown here) earns no rent. In a routine and stable industry where profit is negligible, the lower-cost firm does earn rent (the shaded area), but earns no rent on the marginal unit sold. The firm, of course, is following the standard profit-maximizing rule of choosing the quantity at which marginal cost equals marginal revenue. In this example it is not possible to distinguish between profit and rent, but the term *rent* is appropriate when the superior returns of the firm can be attributed to a superior mineral deposit.

The diagram indicates one of the oversimplifications of the tables used as illustrations. In the tables output does not vary with price. In effect, this assumes that the marginal cost curve becomes vertical at full capacity, as shown in Figure 7.3. Although this is unrealistic, it can be a useful approach in analyzing the initial decision to invest in opening a mine. In that case, the appropriate calculation compares total revenues with total costs to obtain total profit (including economic rents). These numbers can all be divided by output to obtain average revenue, cost, and rent figures. Once a mine is in production, however, the owner maximizes profits by setting annual output where marginal cost equals marginal revenue.

There are many different margins in mining. In discussing marginal minerals it is necessary to distinguish between the activities that affect the total amount of mineral that will ultimately be extracted from the deposit, on the one hand, and activities that affect only the timing of extraction, on the other. The former include the decision to mine or abandon small seams of coal lying above or below the main seam and to use or discard low-grade mineral (mixed with waste) from the edges of the seam. A higher price for the mineral pushes out all such margins of exploitation with the result that ultimate recovery is increased. By the same token, the opportunity cost is zero for the unmined mineral that would otherwise be lost forever.

The situation is quite different when the choice is made to extract today a mineral that might be extracted later. In such cases, the opportunity cost of the mineral deposit is the present value of

Figure 7.2 *Cost Curves for Two Mines*

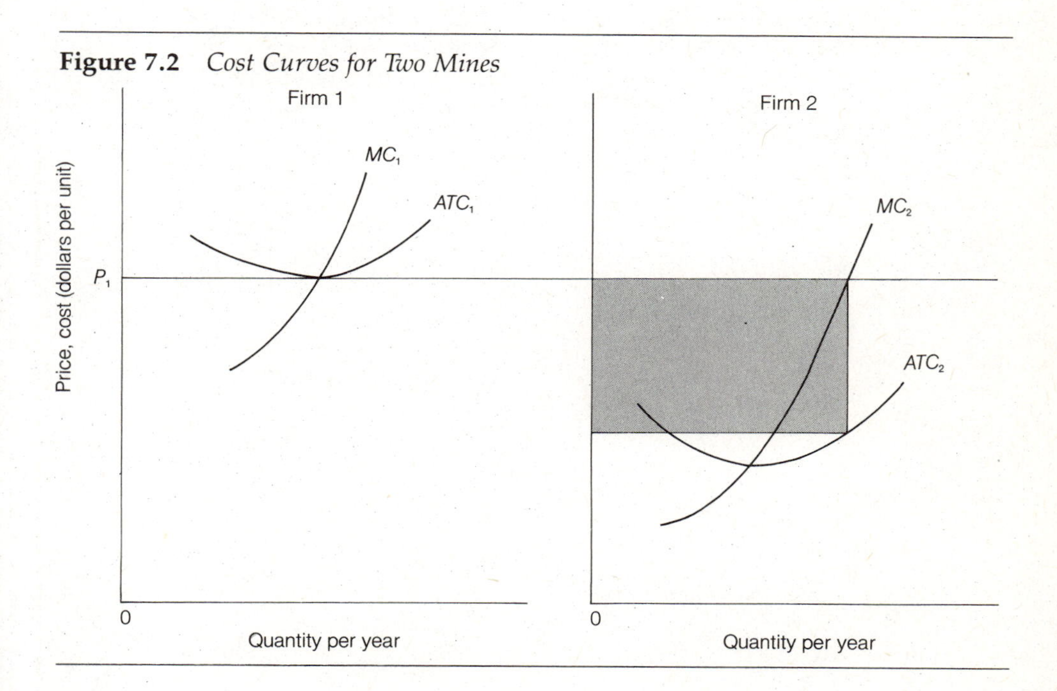

the rent that would otherwise have been received at the time the mine was exhausted. In Figure 7.4, for example, the planned timing of exploitation is indicated by the solid curve. In the sixth year, however, the mine owner reacts to unexpectedly high prices by trying to increase production. In effect, the mineral that would have been produced in the twentieth year is produced in the sixth. The cost of the increased production is not just the labor, materials, and so on, but also the user cost, that is, the present value of the rent that will not be received when the mine is exhausted. Thus the standard $MC = MR$ criterion is correct only when marginal cost includes user cost. If rent really is rising at the rate of interest, then user cost will be equal to the current economic rent.

This sketch barely suggests the complexity of investment decisions for minerals in the real world. In particular, decision makers are uncertain about geology, costs, and future market prices. They must nevertheless commit resources for a long investment period and still longer payback period. Furthermore, many mining projects are characterized by such high fixed costs relative to the operating costs that mines are kept in operation long after the original investment is seen to be a mistake.

Despite the complications, we would expect market pressures to lead to an equilibrium such that economic rents increase at the rate of interest. As Table 7.2 indicates, however, rents increase faster for the low-rent deposits. Equilibrium is therefore possible for only one class of deposits—the one yielding the highest rent—whereas others will be withheld from production.

The simple theory, although elegant and suggestive, is inadequate to describe the real world. In fact, mineral deposits yielding

Figure 7.3 *Cost Curves for the Firm with Fixed Capacity*

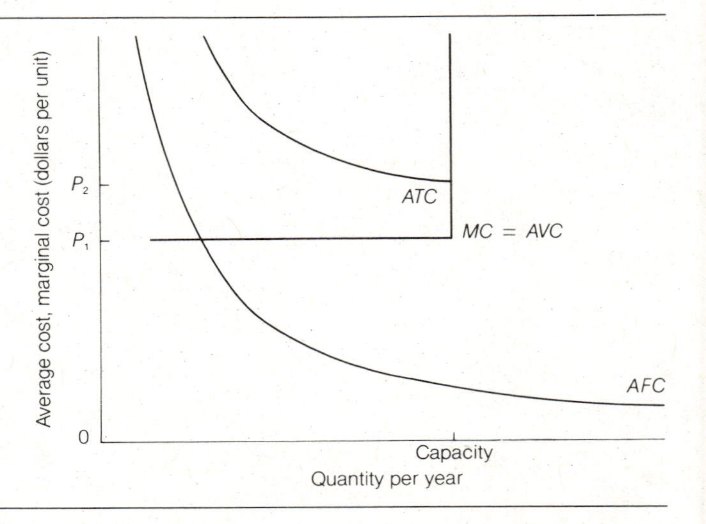

different rents are exploited simultaneously, and there has been no discernible tendency for rents to increase at the rate of interest. Why does such a reasonable theory not describe economic behavior? The explanations are developed in more detail as we proceed—indeed, they provide much of the essential background for understanding the mineral industries. Some of the main themes are (1) the inherent uncertainty in the rate and pace of discovery, which leads to the development of some inferior sources before superior ones are discovered, (2) the long-term nature of much mineral development, in which an investment that takes several years to complete is expected to yield output for decades, (3) imperfections in property rights, about which more is said later, (4) imperfections in the markets for labor and capital, (5) fragmentation of control over reserves, (6) the different comparative advantages of different scales of output, and (7) the impact of technological change.

EFFICIENCY AND EQUITY

Before beginning the analysis of such complications, let us consider the broader implications of departure from the optimum pattern of exploitation of mineral deposits. The advantage to investors of exploiting the highest-rent deposits first is that they will be wealthier at the end of their lives (or be able to consume more during them) than they would if they followed any other procedure. Under the assumptions spelled out in standard microeconomic theory, the economy's total wealth increases fastest if individuals make such wealth-maximizing decisions on their own. As long as other forms of capital can easily be substituted for natural resources, the decision

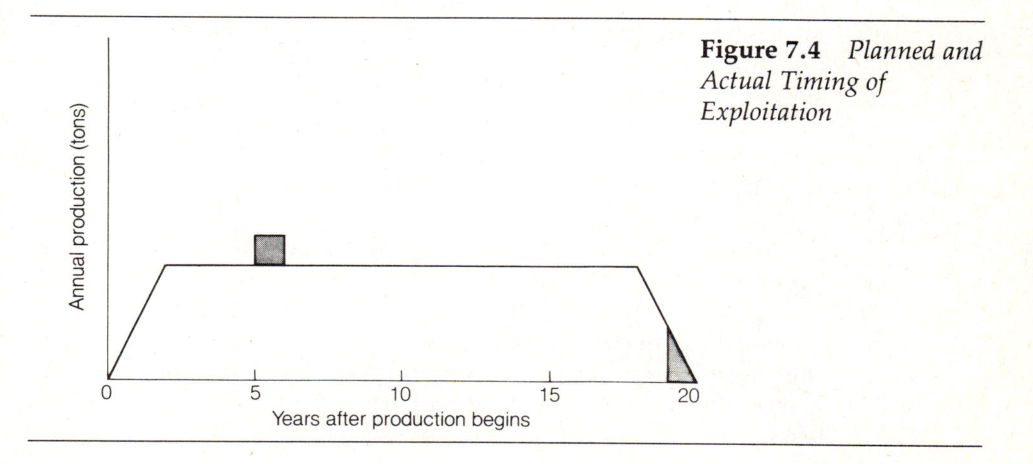

Figure 7.4 *Planned and Actual Timing of Exploitation*

process that seems rational and efficient for the individual maximizes the economy's wealth as well. If the substitution of capital for energy is expected to become increasingly difficult in the future (after the easy insulating jobs have been done), then investors will buy and hold reserves of fossil fuels in the expectation of higher prices. This would bid up current prices, inducing individuals to save more energy.

The efficiency question is worth a closer look when markets for capital and mineral resources are imperfect. In particular, owners of a mineral deposit may not be content to hold it unused while its value increases if they need cash now and find it difficult to sell a share of property or to borrow against the wealth. Such forces tend to distort income to match desired consumption and consumption to match actual income. It is possible to find examples of the excessively rapid extraction of fossil fuels that the conservationists deplore, but this is not the general case because of opposing forces, which are discussed later.

So much for efficiency. What about equity? Is it fair to future generations to use up the richest sources of fossil fuels now? The easy answer is yes, because they will have other forms of capital to compensate for the scarcity of natural resources. In fact, we have no rigorous theoretical grounds for dealing with questions of fairness among different individuals and especially with the question of intergenerational equity.[2] Economic theory teaches that we have no basis for comparing the welfare gains and losses of different people or even of the same person at different times. Furthermore, the connection between income and welfare is nebulous, especially when the items being included in income change, as they do over time.[3] My own value judgment, which others may not share, is that we do not have to worry about how rich future generations will be as long as per capita income is rising. Maximizing wealth today is the best that we can do for future generations. The questions of intergenerational equity become acute only if our increased consumption means that our grandchildren will be poorer than we are. Nevertheless, this point of view presumes that other forms of capital are good substitutes for natural resources.

RENTS, ROYALTIES, AND TAXES

The discussion of the optimum rate of extraction assumes that the decision maker actually receives the economic rent, but under current institutional arrangements that may not be so. This certainly makes a difference because people are more likely to try to maximize the present value of their own income than of some economist's abstraction such as economic rent. Why might the two differ? At least five decision makers may be significant to the control of a mining operation and may share the economic rents. These include the owner of the mineral rights (who may, but need not, also own the surface rights), the mining company, the mine manager, the government (or several different governments), and labor. Any of these may be a single individual or a group with different internal interests as well. Often the same individual will play more than one role, but the key point is that the roles often are played by different people with diverse interests.

The mining company contracts to pay a royalty to the owner of the mineral rights to compensate for extracting the mineral.[4] The maximum amount of royalty that the mining company would be willing to pay (assuming perfect information and analysis) is equal to the economic rent. That still leaves the mining company with a normal rate of return after paying all costs, so it should not be induced to transfer its capital elsewhere. Why should the landowner settle for anything less than the entire rent?

The main difficulty in negotiating the theoretically perfect royalty agreement is the ignorance of both parties about the true geological conditions and about the technology and the market conditions when extraction takes place. Many different royalty arrangements have been devised for dealing with the uncertainties; some of these are discussed in connection with particular minerals and also in connection with various taxing arrangements, which involve the same sorts of problems as royalties. The traditional way to deal with the problem, however, has been to assign to the landowner a set fraction (traditionally, one-eighth) of gross output. This is only a starting point in actual negotiations, but at least it provides that. Government oil land is typically leased for a royalty of one-sixth of the oil produced. Since some terrain is more promising than others, the government auctions off the right to drill, produce, and pay the one-sixth royalty. The company making the highest bonus bid—a fixed dollar amount payable at once—obtains the right to extract the oil. (The government has experimented with other leasing agreements, but this is the most common arrangement.)

Private landowners would not ordinarily trouble with a formal auction, but they could bargain for a good bonus to complement the standard royalty rate, or they might insist on a higher royalty rate. Mineral leases also frequently include a fixed annual payment or a minimum dollar amount of annual royalties or some other device to encourage immediate development. Another possibility is to work some sort of inflation adjustment into payments fixed for long time periods. This is often done when the royalty is specified in dollars per ton rather than as a percentage of gross. Finally, it is possible to devise royalties based on net revenue rather than gross. Net revenue comes close (if defined properly to allow for a normal rate of return on the mining company's own capital) to the economic concept of economic rent. Hence, landowners should theoretically be able to extract up to 100 percent of the net without eliminating the incentive for the mining company. Using net income as the basis for either royalties or taxes, however, requires that landowners or the government keep close control over the business operations of the mining company. Anyone who shares net revenues becomes, in effect, a partner in the firm and so must monitor closely the behavior of the other partners.

Governments find economic rents to be a convenient and popular source of tax revenue.[5] As we noted, governments can share economic rents using arrangements similar to the lease agreements of landowners. The particular legal traditions of various countries differ, however, so the exact forms of taxes in use show great variety. In come countries the distinction between government and landowner (or mining company) is meaningless. Even in the United States the government frequently functions as landowner. In some countries the government has retained the mineral rights even when the surface rights are held by a private landowner. In such instances the government can capture economic rent by drawing up a clever royalty agreement.

When the government is neither the mining company nor the landowner, it must obtain its portion of the economic rent by taxation or by expropriation. At the extremes the two are quite different: If a government announces a tax structure before the mining companies sign royalty agreements and then retains the same structure and rates, the companies can mine or go away without complaint. If a government waits until a mining project is operating satisfactorily and then nationalizes it without compensation, the company has clearly been wronged and the country will find new investments hard to attract. In the middle ground is a wide range of possibilities including subtle rewriting of tax laws to extract a larger share of the economic rents, major increases in taxes that squeeze from the mining company not only the rents but also part of the normal return on invested capital, partial expropriation, or total expropriation with partial compensation. The attempts to extract economic rents via taxation differ only in degree from taking the entire property. But it is important to remember that (1) mining ventures may generate economic rent, (2) the amount of rent is difficult to predict in advance, (3) attempts to renegotiate the sharing of rents after durable investments are in place put the company at a great disadvantage, and yet (4) when projects are successful, the public pressure to preserve some rent for the community where it originates is overwhelming.

If the landowner and the state do not extract all of the economic rent, labor may get some. The most striking instance of this is the coal-mining industry in the United States, where union wages were about twice the opportunity cost of labor during the three decades after 1933. This was a period of decline (except for World War II) in industry fortunes but a time of great union strength. Nevertheless, the union and nonunion firms often sold coal in the same markets. Although some of the nonunion firms were large and paid high wages to head off unionization, many were tiny mines working thin seams of coal and paying low wages. The workers in the best seams were receiving a significant share of the total rent.[6]

The implication of this sharing of economic rent is that the pressure to maximize the present value of rent is attenuated. To the extent that the mining company receives the rents, it may try to time the exploitation of various deposits in accordance with the theory; that is, it will attempt the best prospects first (but errors of judgment will occur and a deposit that has been developed will be exploited unless the firm misjudged to an extraordinary degree). The firm is usually more than an entrepreneur with some mining equipment. Generally, it will include office personnel, some processing functions, and a sales department. All such downstream investment and personnel, as well as the customers, are dependent on an adequate supply of the mineral. Hence although the firm will use the best available resources first, it may have to exploit some deposits that are inferior by industry standards.

Landowners might be supposed to have the traditional motive of maximizing the present value of economic rent. Even if they have signed away some of their rents in a lease with the mining company, it could be argued that they still would like to receive their royalties at the time when the present value is highest. The two problems with such an argument are as follows: First, landowners usually give up control of the mining at the time they sign a lease, so their preferences do not matter. Second, landowners will find it difficult to borrow against future royalties until production is well underway because of the uncertainty about what is there. So if they want cash now, they will be in favor of immediate exploitation and may write the lease to encourage that. Alternatively, they could sell the mineral rights outright, but many people would rather participate in the great mineral lottery, even if they are in a hurry to see it begin.

Naturally, to the extent that government and labor succeed in extracting the economic rent, rent becomes irrelevant to the mining firm or the owner of the mineral rights in guiding the timing of exploitation. In fact, the decision maker who expects heavier taxes or outright expropriation will be eager to extract as quickly as possible. Similarly, the prince or president who doubts the stability of his or her regime will want to convert immobile reserves into Swiss bank accounts as quickly as is seemly.

We must bear in mind also that reference to a mining company is misleading. The interest that we have identified as the company is actually that of the stockholders, who presumably benefit from the maximization of the present value of profits and rents. Employees and managers are interested first of all in the continuity of production. In addition, employees of the firm may benefit more from a high-cost operation, which implies more employees and more demanding managerial functions, than from a low-cost, high-rent operation. Tapping the "impossible" mineral deposit (at a profit) will certainly furnish the mining engineer and manager with more professional esteem than will routine production of high-rent deposits.

For whatever combination of reasons, we do observe deposits being developed simultaneously that yield different amounts of economic rent. Once the costs of development have been borne, the deposit will most likely continue in production for many years, even if development proves to have been a mistake. The criterion for retaining a facility in production once it is built is that average revenue must exceed average variable costs; that is, are you selling enough to pay whatever extra costs you incur by operating the facility? The facility will show a loss unless average revenue exceeds average total cost (average variable cost plus average fixed cost). The fixed costs are there even if the facility is not operated, however. In effect, returns exceeding variable costs are quasi-rents, because the firm would continue to produce in the short run even if it did not receive them just as it can produce in the long run without receiving any economic rents. Of course, losses as measured by the accountant are a sign that the firm made a mistake by putting its capital into the project and should warn that firm and others to avoid similar mistakes in the future. Although the large investment and long life of a mine explain why particular mistakes are not abandoned earlier, they are not the fundamental explanation for the simultaneous development of mineral deposits of a wide range of qualities. After all, mining companies do, on average, make profits; so their mistakes should not be used to explain the general characteristics of the industry.

Furthermore, nothing that has been said so far explains why competition among efficient producers of high-rent deposits does not drive down the price enough to eliminate the producers relying on low-rent deposits. At the very least we would expect all new mines to exploit the highest-rent deposits, but this does not seem to have occurred.

THE COST CURVE FOR EXTRACTING MINERALS

The problem with crude illustrations like that in Table 7.2 is that they are either ambiguous or misleading about the shapes of the cost curves assumed. Taking the numbers at face value, it would seem that they imply a curve of the sort shown in Figure 7.3. The central feature of this is a single capacity level of output, at which point further increases in production are impossible (marginal cost becomes infinite). Since fixed costs are heavy in the mineral industries, the average fixed cost decreases noticeably even in the vicinity of capacity. That means that average total cost (*ATC*) is declining to full capacity, at which point it too becomes infinite. Under these assumptions only two levels of output make any economic sense for a competitive firm: Produce at full capacity and sell at the market price or shut down. In this example, the firm will produce at capacity as long as price exceeds OP_1 (minimum average variable cost) and will make a profit if price exceeds OP_2 (*ATC* at capacity).

These are the assumptions that permit us to specify levels of cost and to calculate rent per ton once the market price is known. The real world is generally more complex, however. In particular, it probably makes sense to think of a range in which marginal cost is increasing as in Figure 7.5, rather than a fixed capacity at which formerly constant costs become infinite. Usually, management has a preferred level of output at which various activities function smoothly and attention can be given to routine maintenance to avoid future costs. Often, however, output can be expanded beyond that level by adding additional crews, working overtime, adding shifts, pushing machinery harder, skimping on maintenance, and so on. The equivalent in oil and gas extraction is to exceed the rate that has been established to maximize long-run returns. (This is discussed in Chapter 9.)

If the true cost curves reflect the increasingly costly possibility of expanding output as in Figure 7.5, then it is much harder to define *capacity*. For every market price above OP_1, there will be a level of

Figure 7.5 *Output of the Mine with Increasing Marginal Cost*

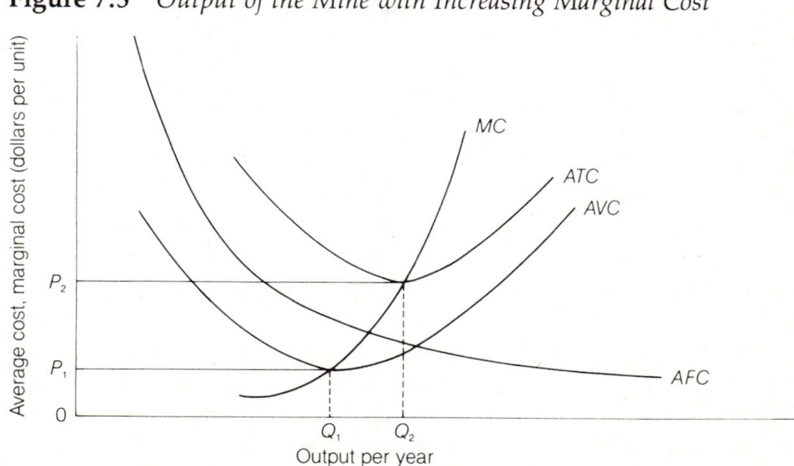

output that will maximize profit (or minimize loss). Since the preferred output is determined by market conditions as well as cost conditions, however, it is misleading to call it capacity.

The more significant point is that once the optimum level of output is determined by finding the output where $MR = MC$, then the economic rent on the marginal unit is equal to zero. In the short run this explains why the low-cost producers do not expand enough to squeeze out the high-cost producers: At the margin each firm has the same cost and it is equal to price (in a competitive industry). There are no high-cost or low-cost producers, only larger and smaller producers, if we are considering marginal cost (see Figure 7.2). It is still possible to find differences in average cost of the sort shown in Table 7.2, but they do not suggest that output should be expanded with existing facilities. Rather, the situation is a conventional one analyzed in the theory of the firm. Although marginal cost equals marginal revenue, which equals price for each competitive firm, some have lower average cost curves than others and therefore earn profits or receive rents. The short-run problem is solved. The rent-earning firms cannot profit by producing more currently.

What prevents the low-cost firms from expanding in the long run? This is always one of the sensitive points in the theory of the firm, but it appears at first glance that the theory of the mine provides the answer: After all, the firms that developed the best deposits are earning rent, just as the theory specified. We would expect any new establishments to be limited to inferior resources and hence to be higher-cost operations. Such an explanation would indeed be plausible and consistent with the theory, but it would lead us to expect prices to rise consistently over time as consumption increased and the best deposits were exhausted.

Bringing the concept of exhaustion into the analysis is crucial, for this is the characteristic that distinguishes the minerals industries from ordinary manufacturing industries. The implication is that we must pay more attention to the dynamics, that is, the changes that occur over time. Figure 7.6 sketches a stylized cost history for a particular mine. After an initial decline in average total cost, as measured at the rate of output where average cost is a minimum, the *ATC* begins a long and monotonic increase. The high initial value reflects the costs of starting up the mine and overcoming any special problems. In effect, it is part of the investment in development. Once efficient operations have been attained, costs start to rise again. The most accessible minerals are mined out, haulage within the mine becomes more expensive as distances increase, and pumping water out becomes more costly as the mine deepens.

It is possible, of course, that some favorable geological surprise will offset the tendency sketched here, but such random variations can work in either direction. It is also possible that a technological innovation can offset the increase in costs, but typically an innovation is much easier to adopt in a new mine than in an existing one.

This explanation focuses on the process of decline in a particular deposit, not on the quality of the natural resource available to the industry. The former is the more important phenomenon, and whereas other particular investments may also decline in economic value, the mineral deposit remains distinctive. Although a truck or a ship or even a building exhibits the same pattern of increasing operating cost relative to the value of the service produced, an entire plant site is never in quite the same situation, because a plant can be demolished and production continued at the same location with new capital facilities.

This distinction between the temporary nature of minerals extraction (which may, however, last several decades) and the permanence of other uses of land is of interest mainly to landowners and local communities. Both of these can face the certainty that someday the particular deposits will be economically exhausted, and long before that costs may rise enough so that economic rents are sparse and production perhaps confined to boom years. Yet aside from the local aspect, the similarities between the mineral and other investments seem more pronounced than the differences.

Although particular deposits are exhausted, the reserves of the firm as a whole or of the entire industry may increase. If reserves for the industry increase, it means that particular firms are locating and proving the existence of minerals or learning how to exploit inferior grades. Firms do not ordinarily take on such expenses until they have acquired the right to extract the minerals. Thus firms have

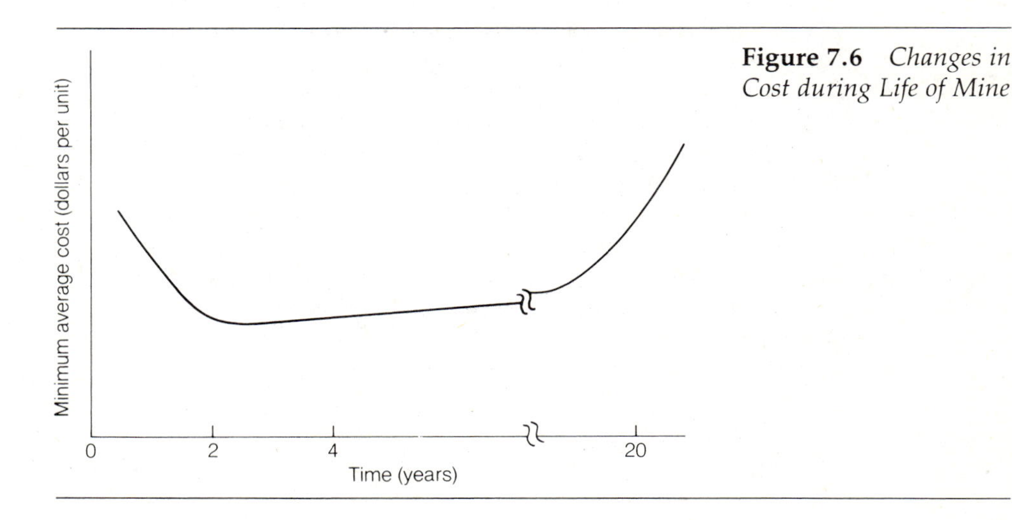

Figure 7.6 *Changes in Cost during Life of Mine*

control of resources that can be exploited at today's price and with today's technology and must decide how quickly they will develop them.

In a world of pure competition and perfect markets, the firms would base the decision on the present value of economic rents. Markets are not perfect, however. As outlined earlier, decision makers consider the rent they actually receive, as distinct from the amounts that the landowner, government, labor, and others take. Moreover, firms are constrained in their expansion plans by scarcity of capital and of managerial talent. These are generally constraints on the rate of growth, rather than on size itself, except possibly for the largest firms. But they are constraints that are exceedingly important in the dynamic world. In other words, the firm can manage some rate of growth. If it consistently grows faster than other firms, it eventually becomes one of the giants of the industry. If the average growth rate for firms in the industry is greater than the growth in demand, the industry will be faced with a decreasing price for its output. Growth is eventually limited by the difficult access to capital when profit margins are squeezed by the declining prices.

If the firm is not purely competitive, then marketing the product becomes a dominant consideration in the decision to expand. The world oil industry, during all but a few years of shortage, has had excess production capacity. The major companies have devoted a great deal of attention to marketing the ample supplies and trying to increase market share without bringing about significant price decreases. The coordination between development of the mineral deposit and final consumption is even more evident in the modern coal industry. The dominant and growing use of coal is to generate electric power. Typically, today, the mine and transportation link are planned and built concurrently with the power plant itself.

At some point as new capacity comes on stream and the old deposits become more expensive to exploit, the firm will cease production at the old wells or mines. This behavior too is like that of any other firm as it phases out old equipment in favor of new.[7] The decision to abandon production, of course, requires that the operating costs of the old facility exceed the revenue attributable to it. The new facility will be developed only if it is expected that the total costs of the new will be less than the revenues. It is the cost of the capital that must be committed to new facilities that protects the old from displacement by the new.

It is important therefore to consider the trend in the cost of new facilities. This turns out to be a complex topic; in part because of questions of scale and in part because of changes in the quality of deposits that can feasibly be exploited. The scale question is particularly intriguing because of the finiteness of each deposit. If the market appears to be permanent, a manufacturing plant can be planned for the size that seems best with respect to the proportion of the total market it will supply and the costs per unit of output expected from plants of various sizes. These factors are important for mineral developments too, but in addition the scale must be influenced by the speed with which the deposit will be exhausted. Is it better to invest a small amount to extract at a slow rate for a very long time or to invest a huge amount to extract at a very rapid rate and exhaust the deposit quickly?

These possibilities are sketched in Figure 7.7, which plots the annual extraction rate against the years it will take to exhaust a deposit of 100 million tons. If the deposit of fixed size is exhausted, the curve shown in Figure 7.7 will be a rectangular hyperbola, since tons per year times years equals tons in the deposit. Thus 100 million tons can be extracted in 10 years at 10 million tons per year, in five years at a rate of 20 million tons per year, or in 100 years at a rate of 1 million tons per year. What rate and scale should the firm choose?

One recurrent theme in the mining literature is that larger scale is associated with lower average cost. For particular types of equipment over a wide range of common capacities, it is easy to document the decline. Examples include trucks, excavating equipment, pumps, and pipelines. Opposing this, of course, is the fact that rapid extraction requires equipment with much higher capacity. Developing a mine with capacity of 10 million tons per year requires more capital than developing a mine with capacity of 1 million tons per year—even though it does not require ten times as much. Of course, some equipment can be moved to other sites and even something as immovable as an underground pipeline may have

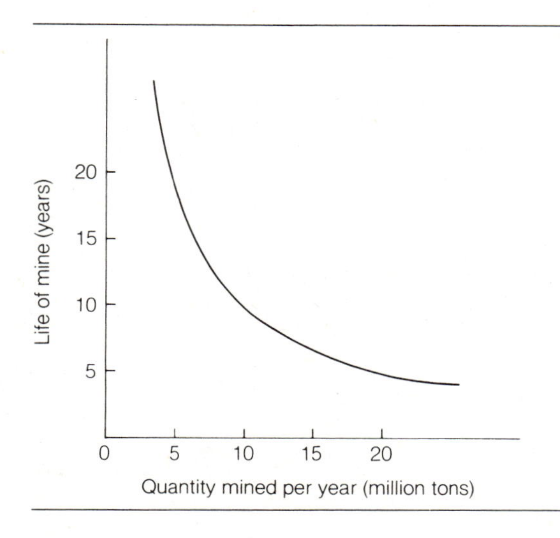

Figure 7.7 *Life of Mine versus Rate of Production*

some salvage value, especially if more mineral development is expected in the vicinity of the original deposit.

These general considerations are illustrated in Figure 7.8, which shows the present value of the expected net returns of the project (total receipts minus total expenditures each year, discounted to the present) plotted against the capacity for which the mine is developed. The curve as drawn shows a sharp peak at 5 million tons per year (twenty-year life) and negative values for rates departing much from that. Although the basic idea is simple (some mine size is most profitable), actually to find that most profitable size is very difficult. Obviously, it depends on the capital requirements and operating costs with different extraction rates, which people in the industry can estimate tolerably well. It also depends, however, on the rate used for discounting future costs and receipts, as well as the expected path of output prices. It is not clear that an attempt to estimate each of these elements separately will produce an overall guess that is superior to some rough rule of thumb. This skepticism is strengthened by the acknowledgment that the guess about the total quantity of mineral in the deposit is also highly uncertain.

Under such conditions the crude approach of designing the mine to extract the measured reserves in ten or twenty years has much to recommend it. If the investment seems likely to be paid back well within that time, then some profit will be gained. In the meantime, some additional reserves will probably be found, so the later phases of exploiting the deposit may yield substantial returns. (Oil and gas, as fugitive resources, pose special problems that are considered in Chapter 9.) Such crude approaches may seem offensive to the economist's sense of theoretical exactitude. The relevant

Figure 7.8 *Profit versus Rate of Production*

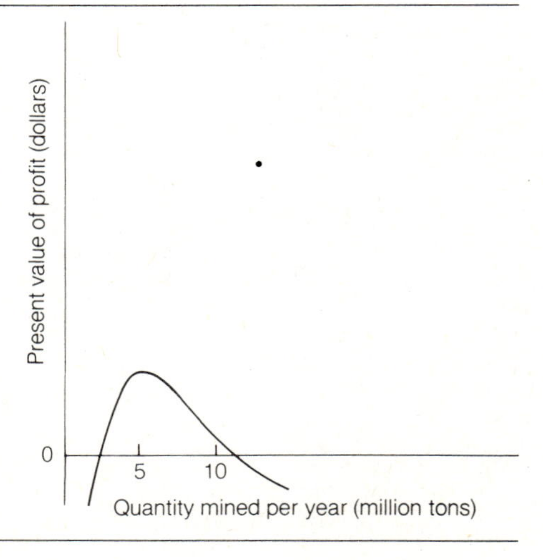

questions, however, concern the sensitivity of the decision to differences in the assumptions and the practical matter of whether refined methods are in fact more accurate. The most important question marks in most such guesses are the price at which the mineral can be sold and the quantity that is really there. No amount of refinement of the techniques of calculation will answer these.

This discussion suggests, however, that large, regular deposits will be more valuable per ton than small deposits because they permit the use of high-capacity techniques, which are cheaper to use if the deposit lasts long enough. If large deposits yield higher rent, theory suggests they will be exploited first. The smaller deposits would rise in value too rapidly to be exploited currently. The main problem with this straightforward approach is that it ignores the differences among firms in capabilities and in opportunities. This is a point to which we must return in Chapter 8. If low-rent reserves are the only ones that can be afforded, they will be used to keep the firm operating, even if postponing their exploitation promises to increase profits.

Huge uniform deposits, even though of low grade, may be cheaper to exploit than small isolated pockets of a rich material. In coal mining, for example, seams of different depth or thickness being exploited concurrently may actually have similar total costs because the economies of scale in exploiting the uniform low-grade deposit offset the apparently richer quality of the other. The interesting aspect of this is that economies of scale have been extended by successive technological innovations in equipment and organization. This means that new mines under apparently poorer conditions have often had costs that were comparable to those in the locations that were richer by traditional standards. Not only does such technological change forestall the cost increases that are expected to increase economic rents in the long run but it may also move large amounts of resources into the reserve category by making them cheap enough to compete at the current price. This is more likely for the nonfuel minerals where the curve of geological availability generally shows a sharp increase as we move toward lower grades. It may also apply, however, to such low-grade plentiful sources of hydrocarbon as oil shale and tar sands.

VALUATION OF RESOURCES

As is true of other goods, mineral resources could be expected to end up in the hands of those who value them most highly. But why should valuations be expected to differ? The most obvious reasons are divergent opinions on the future costs and prices and divergent views on the amount of mineral in the deposit. Those who expect economic rents to increase rapidly will try to buy reserves with whatever capital they have or can borrow. Those who think that prices have stabilized will sell off their reserves.[8] The beliefs about the actual qualities of particular deposits can also differ. It is worth hypothesizing that those who have worked a particular deposit or one nearby will generally be most optimistic about the costs of exploitation and the total quantity of reserves. There can be exceptions: Firms may specialize in particular types of resources and decide that a particular property falls outside that category; but in general, the specialized knowledge of particular deposits will cause the existing operator to place the highest value on them.

The other factor that influences a firm's valuation of mineral or other property is the cost and availability of capital. This differs for different firms. It enters very directly into the valuation formula:

$$\text{Present value} = \frac{G_o - E_o}{(1+i)^o} + \frac{G_1 - E_1}{(1+i)^1} + \frac{G_2 - E_2}{(1+i)^2} + \cdots + \frac{G_t - E_t}{(1+i)^t}$$

where G_1 is the gross income, and E_1 is the expense in year 1. The higher the interest rate used to discount future returns, the lower will be the present value of the property to that firm. The present value of the net returns is the maximum that the firm can afford to pay for it.

What tactic is available to the firm that has poor access to capital but wants to stay in the business? It might allow financially stronger firms and individuals to carry the reserves, just acquiring mineral rights as it uses them. This is not a comfortable situation conducive to long-run planning, especially if a competitor holds the reserves.[9] Another technique is to acquire low-rent reserves. These may be acquired just for security in case the richer firms block the reserve-poor firm from access to good reserves at a later time. Yet once the firm owns resources of any quality, it will acquire more information about them and may even acquire such an advantage in using them that the distinction between high-rent and low-rent disappears or the direction is reversed.[10] This is yet another explanation for the fact that deposits of greatly different apparent quality are exploited simultaneously.

IMPERFECT MARKETS AND INTERGENERATIONAL EQUITY

The economic theory of the mine, with its rigorously logical derivation, provides a framework for organizing our observations of the mineral industries. The explanations for the failure of specific predictions as sketched here are so numerous as to offer only a checklist of items to consider when examining each industry. This is the task of the next few chapters.

From the viewpoint of the entire economy, however, it is necessary to ask how equity and efficiency are affected by the departures of the real world from the theoretical ideal. The answer is complex, but at the least it involves these elements: First, some of the best resources, which theory predicts will be used up first, are reserved for future generations. This may not be as beneficial to the future as it is costly to the present because as the technology of production and consumption changes, the best resources may lose some of their relative value. Second, research into the use of inferior resources begins earlier than the optimum time as firms restricted to such resources by the high cost of carrying better ones struggle to make them useful. In view of the well-known difficulty of forecasting the success of research and development activities, the early research efforts are useful insurance to society against the very real possibility that the superior resources are scarcer than expected. Third, the nicely predictable increase in prices that the theory tells us to expect does not occur. Whether this is an advantage or a disadvantage is not so easy to see. At the very least, however, the record of prices and availability in the minerals industries suggests that we are not cheating our grandchildren, whatever we may be doing to each other.

NOTES

1. For a comprehensive introduction to the theory of the mine that is accessible to those who know differential equations and intermediate economic theory, see Allen Kneese and Orris C. Hirfindahl, *Economic Theory of Natural Resources* (Columbus, Ohio: Charles E. Merrill, 1974). The easiest place to start on the theory is Robert Solow, "The Economics of Resources or the Resources of Economics," *American Economic Review* 64, no. 2 (May 1974), pp. 1–14; the classic article is Harold Hotelling, "The Economics of Exhaustible Resources," *Journal of Political Economy* (April 1931), pp. 137–175; Mason Gaffney, ed., *Extractive Resources and Taxation* (Madison: University of Wisconsin Press, 1967) offers both a review of the theory and interesting empirical information; some of the voluminous recent theoretical literature is discussed in Michael G. Webb and Martin J. Ricketts, *The Economics of Energy* (London: Macmillan, 1980), chap. 3.
2. The most publicized recent attempt to solve the problem is John Rawls, *A Theory of Justice* (Cambridge, Mass.: Harvard University Press, 1971).
3. For a strong statement of the view that growth does not improve welfare, see E. J. Mishan, *Technology and Growth: The Price We Pay* (New York: Praeger, 1970).
4. Owners of the surface rights, if they are not owners of the mineral rights, may not feel compensated when their farms are torn up or

their homes subside because of the mining. Nor will they be mollified if we tell them that they should have read the deeds more carefully before buying to see why the surface rights were such a bargain.

5. Government efforts to extract economic rent are discussed in Gaffney, *Extractive Resources and Taxation*; Raymond F. Mikesell, ed., *Foreign Investment in the Petroleum and Mineral Industries* (Baltimore, Md.: Johns Hopkins University Press/Resources for the Future, 1971); Warren Aldrich Roberts, *State Taxation of Metallic Deposits*, Harvard Economic Studies, Vol. LXXVII (Cambridge, Mass.: Harvard University Press, 1944); and Raymond F. Mikesell, *Petroleum Company Operations and Agreements in the Developing Countries* (Washington, D.C.: Resources for the Future, 1984).
6. For a detailed discussion, see C. L. Christenson, *Economic Redevelopment in Bituminous Coal* (Cambridge, Mass.: Harvard University Press, 1962).
7. W. E. G. Salter, *Productivity and Technical Change* (Cambridge, England: Cambridge University Press, 1960) provides a careful analysis of the manufacturing firm.
8. Ashland Oil sold off most of its reserves in 1979. See, for example, *Wall Street Journal*, January 25, 1979, p. 7.
9. See David D. Martin, "Resource Control and Market Power," in Gaffney, *Extractive Resources and Taxation*, pp. 119–138.
10. The classic case is the replacement of natural iron ore by taconite. See William S. Peirce, "The Ripple Effects of Technological Innovation: The Case of Iron Ore Pelletizing," *Omega* 2, no. 1 (February 1974), pp. 43–51.

CHAPTER EIGHT

The Coal Industry

INTRODUCTION

As the energy source for both transportation and industrialization, the coal industry was growing rapidly around the beginning of the twentieth century. By the mid-1920s, however, conditions had changed. The transportation sector began to rely more heavily on petroleum, while many industrial users converted to electricity, oil, or natural gas. The efficiency with which coal was used in generating electrical power also increased dramatically. Following World War II, homeowners shifted from coal to cleaner and more convenient fuels for home heating. The net effect of all these trends was a long period of stagnation in the coal industry. Production around 1960 was about equal to levels attained a half century earlier.

Parts of this chapter are reprinted by permission of the publisher, from *Technological Progress and Industrial Leadership: The Growth of the U.S. Steel Industry*, by Bela Gold, William S. Peirce, Gerhard Rosegger, and Mark Perlman (Lexington, Mass.: Lexington Books, D. C. Heath and Company, copyright 1984, D. C. Heath and Company).

Figure 8.1 shows the long decline in the relative importance of coal as a source of energy.[1] At the beginning of this century, coal accounted for about 90 percent of the mineral and electrical energy consumed in the United States. This proportion decreased almost without interruption to a low of 17 percent in 1972. Aside from the upturn of the past decade, the most prominent interruptions in the decline were associated with the First and Second World Wars, which placed extraordinary pressure on the railroads and the heavy manufacturing industries at a time when those sectors relied heavily on coal. The basis for the recent upturn in the relative importance of coal is the great upsurge in coal consumption by electric power plants.

This can be seen more clearly from Figure 8.2, which shows the total tonnage of coal produced and consumption by various uses. Electric utilities have been increasing their use of coal ever since 1933. Before that year the increasing generation of electricity from coal had been offset by the decreasing consumption of coal per kilowatt-hour of electricity produced. By 1981 the electric utilities were using more coal than the entire economy had consumed for all purposes in any year before 1976!

Until 1961 the growth in coal consumption by the electric utilities was offset by declines by other users. The railroad market ended precipitously in the decade following World War II as diesels replaced steam engines. The retail market (sales to residential and commercial customers) declined less abruptly after World War II, but by 1981 accounted for only 1 percent of the 733 million tons consumed. General industrial use, primarily for steam, had declined from the World War II peak of 165 million tons, but still amounted to 67 million tons in 1981.

The two other markets that remain important are coking coal and exports. To make coke, small particles of coal are heated in a

Figure 8.1 *Coal as a Percentage of Mineral and Electrical Energy Consumed in the United States, 1900–1981*

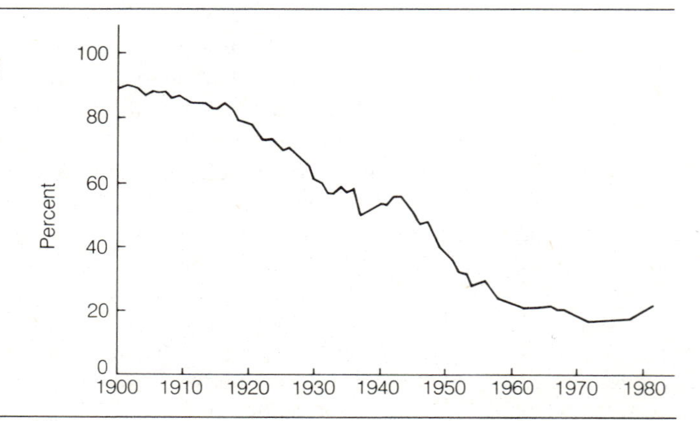

Sources: U.S. Historical Statistics *(1975), pp. 587–588, M 83–90;* U.S. Statistical Abstract *(1981), pp. 577–578;* Monthly Energy Review *(various issues).*

sealed oven. The process drives off volatile components and agglomerates the small particles into large chunks. Most coke is used as an input to the blast furnace, so the coal must meet stringent specifications regarding coking quality and impurities. Although it seems likely that most iron will continue to be produced in the blast furnace, this market is not likely to grow. Instead, we might expect the continued decline in the ratio of coke input to iron output of the blast furnace to offset any growth in iron output. The export market grew from 38 million tons in 1960 to 112 million tons in 1981, at which time the largest customers were Japan, Canada, Italy, and France. Further growth in this market remains a possibility, but the United States must meet the competition of low-cost suppliers such as Australia, as well as the subsidized mines of Europe.

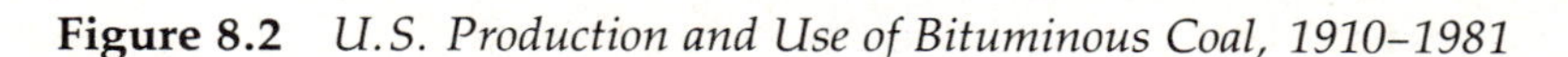

Figure 8.2 *U.S. Production and Use of Bituminous Coal, 1910–1981*

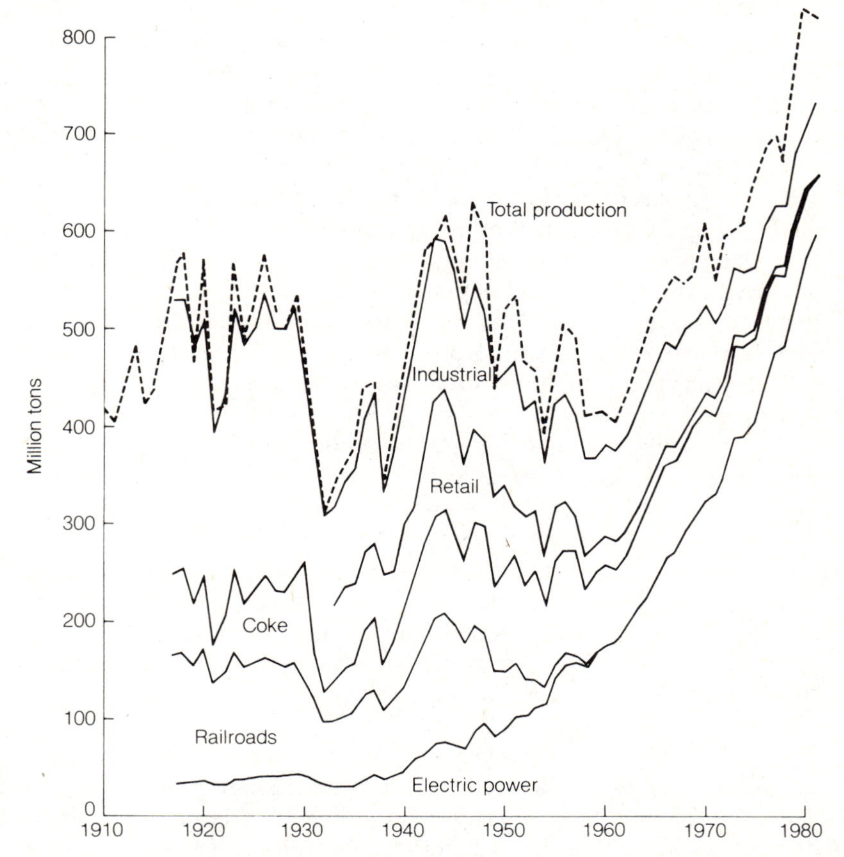

Sources: U.S. Historical Statistics *(1975), pp. 590–591, M 113–120;* Monthly Energy Review *(various issues).*

These changes in coal markets have had pervasive impacts on the industry, which are explored in the remainder of this chapter. One that is obvious from Figure 8.2 is the increasing cyclical stability of coal production. All of the major traditional coal markets except electric power have been highly sensitive to the business cycle. This has always created difficult problems for mine management and investment. Although instability remains a problem for coking coal and industrial use, the data are now dominated by the comparatively stable market for electric power.

A less obvious result of the change in markets is a change in the specifications of the product. The markets that have become very small—railroad and retail—were the ones that put a premium on careful handling of the coal because large lumps sold for a higher price. In steam and coking uses, in contrast, coal is usually crushed before it is fired so careful handling is not necessary. This means that highly mechanized methods of mining, which break up the coal more than careful hand mining, can be used without reducing the quality of the output.

In another respect, however, the quality requirements for steam coal stand at the opposite extreme from those for coking coal. (Much of the coal that is exported must meet the specifications for metallurgical coking coal.) Boilers can be designed to burn almost anything, although it may be difficult to switch types of coal once the equipment is in place. In the major markets therefore the emphasis is placed on securing an adequate supply of a known quality of coal for the life of the consuming plant. The plant can be designed to take advantage of whatever is cheaply available at the particular time and place. Reserves of coal that meet the stringent specifications for metallurgical coking are much more limited. The location of the steel industry has been strongly influenced by the location of coking coal reserves.

When the coal industry began to grow again after the trough of 1961, it was a different industry with a different output than the one that had entered its decline a generation earlier. It was highly mechanized and oriented toward supplying large amounts of cheap energy to a few large customers. The one sector that was growing throughout the period of decline—coal supplied to electric power plants—had come to dominate the entire industry. The demand for coal seems assured in the immediate future because people will continue to use electricity and no strong competitor seems likely to displace coal right away.

The coal industry, however, continues to be as troubled as it was during the long years of decline. The problems are reflected in the decline in output per worker-hour that began in 1969. The decline does *not* indicate that the industry has been forced to exploit more and more difficult seams as the best deposits have been exhausted (Table 8.1). Instead, the industry has worked with roughly constant natural conditions in a more and more restrictive social environment. The regulations promulgated in the name of mine safety and environmental protection have been particularly burdensome. The industry also had unusually great labor relations problems during the 1970s when the long-term decline in coal-mining employment was finally reversed.

Views of the industry's future have ranged from wildly optimistic to extremely pessimistic, depending on which of the characteristics of the industry seemed most significant at the moment. The vast reserves of coal in the United States bolster optimistic forecasts of its role in making the United States more nearly independent of foreign sources of energy. Not only does coal power electric-generating plants, but it can also be used to produce substitutes for natural gas and liquid fuels. The costs of gasification and liquifaction always seem to be on the verge of becoming competitive with imported fuels. On the negative side, however, are the heavy environmental costs of coal mining, processing, and consumption, as well as the high accident and fatality rates that seem inherent in the industry. Nevertheless, as long as coal remains significantly cheaper than oil and gas, it will be purchased by electric utilities and other heavy industrial users.

Coal is found in seams ranging in thickness to more than 50 feet, but three-quarters of all production in the United States comes from seams between 2 and 5 feet thick. When the vegetation from which coal is formed was laid down, the deposits must have been approximately horizontal. In the intervening years, the coal seams and

Table 8.1 *U.S. Coal Production Classified by Thickness of Seam Mined, 1920–1965*

Thickness of seam mined (feet)	Percentage of total production					
	1920	1945	1950	1955	1960	1965
Less than 2	0.4	1.2	0.8	1.0	1.4	1.4
2–3	5.6	5.7	6.9	8.0	9.9	11.1
3–4	18.5	24.0	27.2	24.8	24.3	21.1
4–5	22.4	23.6	21.0	21.8	20.0	22.7
5–6	22.3	13.0	14.5	18.1	17.5	17.7
6–7	13.5	14.5	14.8	12.2	11.4	9.9
7–8	11.2	8.9	6.8	7.8	8.0	9.5
8 and over	5.0	9.1	8.0	6.3	7.5	6.6
Total	100	100	100	100	100	100

Source: *U.S. Bureau of Mines,* Information Circular *8345, 1967, p. 9.*

associated rock strata have been subject to the usual geological processes.

The oldest coal—*anthracite,* or hard coal—often is found in steeply pitched and broken seams that result from the mountain-forming processes occurring after the coal bed was laid down. The next oldest—*bituminous,* or soft coal—is usually found in more regular seams, which nevertheless may still be pitched and are often cut through by rivers. *Lignite,* or brown coal (and the *subbituminous* that is intermediate between bituminous and lignite), is often found in thick, regular beds. The contemporary form of vegetation that became coal in the past is *peat*. In many countries chunks of peat are cut from the bog and dried to be used as fuel. Although the use of peat as fuel in the United States was never very important, as energy prices rose in the 1970s the resource was reexamined for its energy potential.

The data referring to coal include the first three categories—anthracite, bituminous, and lignite—unless otherwise noted. As a practical matter, bituminous coal has dominated the industry, with anthracite accounting for only about one-half of 1 percent of output in 1982. Lignite, however, has become increasingly important as a boiler fuel because many deposits are cheap to mine. Most government data now group bituminous coal and lignite.

THE EVOLUTION OF MACHINERY AND PROCESSES

The traditional coal-mining industry consisted of tiny underground mines, many of which sold to local markets. The mining functions consisted of undercutting the coal, breaking it off the face, loading it into mine cars, and transporting it out of the mine. In addition, auxiliary functions such as ventilation and roof control had to be performed, and the mine might need to be pumped out or the coal cleaned before marketing.[2] Coal was broken from the seam by drilling holes in the working face and inserting explosives, which were then detonated. To prevent the force of the explosions from shattering the coal into small fragments and weakening the roof of the mine, a groove was cut along the floor under the coal to be blasted. At one time all of the functions were performed by hand, but the trend toward mechanization was well under way by the beginning of the twentieth century. Nevertheless, significant parts of the traditional style of production persisted until very recently, playing some role in shaping the modern industry and raising some question about the margin of superiority of the newer techniques.

When examining changes in the technology of an industry, it is easy to be overwhelmed by a mass of technical detail. To avoid this we must focus on a few key trends. These are the advances in earth-moving equipment, the growth in demand for Western coal (Figure 8.3), the increases in capacity inherent in modern underground equipment, the consolidation of consumption in a small number of large electric power plants, and the development of specialized equipment to exploit inferior seams.

Earth-moving equipment has improved so much during the twentieth century that the dollar cost of moving earth has actually trended downward despite the general increases in wages and prices. Some coal has always been mined by stripping, but it was not until the development of caterpillar tracks in the experimental tanks of World War I that steam shovels became convenient enough to be very useful for coal mining. Conversion to electric power as well marked a big step forward in the efficiency of excavating equipment, but a constant succession of less visible changes also contributed to greater efficiency. These changes included better materials, controls, and designs that increased the speed and capacity of the machines. One obvious result of the improvements was the growth in the maximum capacity (as measured by the amount that can be dug in one bite) to 220 cubic yards in 1969. Size reached a plateau then, in part because of the problems encountered by the largest machine (named "Big Muskie") but also because of the diminishing returns to further increases in scale.

Shovels and draglines are not the entire source of stripping efficiency.[3] Equipment used today for excavating includes at one extreme huge bucket wheel excavators for moving immense masses

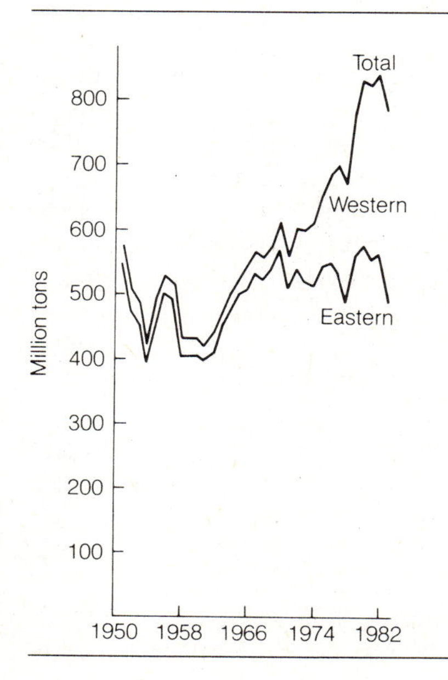

Figure 8.3 *Coal Production in the Eastern and Western United States, 1951–1983*

Source: *U.S. Department of Energy,* Annual Energy Review *(1983), p. 167.*

of relatively soft material. At the other extreme, standard construction equipment, including diesel shovels, bulldozers, and front-end loaders, have the flexibility and adaptability necessary to permit exploiting the small deposits in hilly terrain. Trucks too have increased in capacity and power to keep pace with the excavators. Indeed, at the beginning of the century a surface mine generally used railroad haulage, which imposes severe restrictions on layout and requires that time be spent moving tracks. Not until the 1930s were trucks durable and powerful enough to use for mining. The development of trucks has proceeded in two directions: first, the development of larger and larger specialized haulers for use with the very largest shovels and draglines and, second, the improvements in ordinary highway trucks that are adaptable to small stripping operations.

Many of the improvements in the efficiency of moving earth seem to be derived from pure increases in the scale of individual machines such as excavators and trucks. It is tempting to suppose that a 100-ton truck could have been made as early as a 10-ton truck if someone had just used larger components. In fact, such increases in scale are often not simple matters. The limitations may be technical; for example, the capacity cannot be increased because the materials are not strong enough. Limitations may also be economic; for example, you cannot buy large enough tires for the huge truck you want to build because tires are prohibitively expensive until many other people want to buy tires of the same size. These limitations suggest that capacity will usually increase step-by-step without any apparent difficulties but without great leaps beyond the known technology and assured markets.

Although the development of large excavating and hauling equipment has permitted the economical development of large surface mines, the trend toward surface mining has also been promoted by the growth in demand for Western coal. The coal available in many parts of the West is best mined by surface techniques. Indeed, in some places the seams exceed 50 feet in thickness and lie close to the surface, so open pit mining is the only feasible technique. We should note, however, that the growth in demand for Western coal is only in part due to the shift in population to the West and the use of coal for generating electric power. It is also promoted by the limitations on sulfur emissions from stack gases because much of the low-sulfur coal is concentrated in the West.

Another major trend in technology is the increasing capacity inherent in the mechanization of underground mines. The smallest scale on which it is feasible to develop a highly mechanized underground mine is enough to put it in a large size category by traditional standards. The association of scale with mechanization suggests that small mines face a continuing struggle to compete with large mechanized rivals. Nevertheless, the data on size of mine indicate that small mines did manage to survive despite apparently obsolete techniques.

The introduction to this chapter stressed that the coal industry has come to concentrate on serving a relatively small number of customers. At the other end of the process, however, it has increased its capability for exploiting a great diversity of resources. The earliest mining machinery worked to greatest advantage on the seams that were easiest to mine by traditional methods. But as the equipment has been developed, suppliers have produced specialized versions for difficult conditions such as thin seams or weak roofs. The result of this process of innovation has been a trend toward reducing the margin of superiority enjoyed by the best seams.

UNDERGROUND MINING

The preceding survey indicates some of the major themes in the technological history of coal mining, but here it is important to give a more organized description of the changes in underground mining. Surface mining is treated in the next section, after which some of the economic results and prospects are summarized.

Early in the twentieth century, mechanization was well under way in the task of undercutting the coal seam before firing explosives to break loose the lumps of coal. When the cutting operation was performed by hand, it was an exceedingly time-consuming and irksome task, since it involved pick and shovel work at ground level. It is not surprising therefore that many of the earliest efforts at mechanization were directed at this operation. Figure 8.4, which

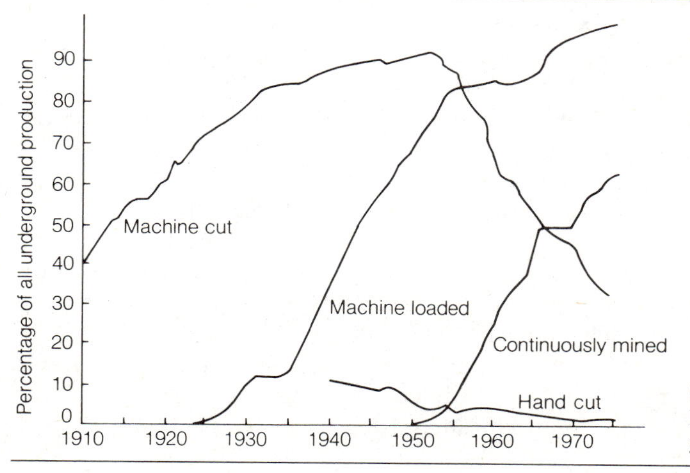

Figure 8.4
Mechanization of Underground Coal Mining, 1910–1975

Source: *U.S. Bureau of Mines,* Minerals Yearbook *(various years).*

shows the proportion of all coal mined using various mechanical methods, indicates the early diffusion of mechanical cutting.

The breaking operations were never such a great problem. Explosives have been used for centuries. Mechanical drills for the holes into which the explosives are inserted were available relatively early and gradually diffused through the industry. In recent years different types of explosives have coexisted with such alternatives as carbon dioxide–powered breakers.

Loading was a big bottleneck once the cutting operation was mechanized. Experiments had long been under way with various forms of mechanical and mechanically assisted loaders, but it was not until the mid-1920s that adoption was significant. The circumstance that triggered the use of loaders was the strength of the miners' union in Indiana after its collapse in the rest of the country. This left wage rates relatively high in the union fields, which in itself is an inducement to try to substitute machinery for labor. The low wages in the rest of the industry provided an additional impetus, because the union mines knew they would be driven out of business in the absence of drastic action. Fortunately, the unionized seams were thick enough to accommodate the awkward early loading machines, so the innovation was successful. As the machines were refined and adapted to more difficult geological conditions, they were adopted more widely. It is interesting, however, that even in the 1970s enough coal was loaded by hand to show up in the statistics.[4]

Haulage systems for removing coal from the mine have displayed great variety. Many types have been used simultaneously to meet the diversity of conditions encountered underground. The first efforts to replace animals for hauling involved rail systems, but the power source posed the big problem. Steam and internal combustion engines were obviously unsafe underground, although sometimes they could be stationed outside to operate some sort of cable haulage. Compressed air was tried but was bulky and of limited range. Standard practice quickly centered on the use of electric motors, but this still left the problem of conveying electricity to the motor. The three obvious approaches, which are all in use, are overhead trolley wires, trailing cables, and batteries.

Since trolley wires and rails are better suited to permanent installations, whereas battery-powered shuttle cars are highly flexible, a mine often has two haulage systems: the permanent mainline haulage and a separate system for moving coal from the constantly changing face to the mainline. As the quality of batteries improved, battery-powered shuttle cars became more important.

Mines established in recent years often use conveyor belt haulage on the mainline but still need some flexible link to the face. Experiments have been conducted for years to use some flexible bridge conveyor system directly from the coal face to the main haulage belt.

By the end of World War II, a conventional mechanized mine had machines to perform each of the separate functions of cutting, breaking, loading, and transporting the coal. The older hand operations persisted, especially in small old mines. The separability of the various operations also meant that a mine could be partly mechanized but continue with hand drilling or hand loading, for example.

During the entire period while conventional mechanical equipment was being developed, numerous attempts were made to bypass the traditional discrete tasks with one machine that could chew the coal off the unbroken face and deposit it in some haulage unit. The early attempts failed for a variety of reasons related mainly to the difficult working conditions of an underground mine and the limitations of the available materials. The development of mining machinery in general has owed much to the improvements in the strength of metals and the compactness of electrical motors (due to improved designs and improved insulating materials).

By 1949 technology had changed enough so that a successful "continuous miner" could be built. Because of the long-standing belief in the industry that this was the best way to mine coal, the innovation met with little resistance. It did not displace the conventional machinery, however, because the same improvements in materials and components that made the continuous miner feasible also made it possible to improve the conventional equipment. The two approaches coexist, each having the advantage under certain conditions. Figure 8.4 shows the diffusion of the continuous miner.

Other techniques are possible underground as well. In particular, long-wall mining has appeared to be ready for a major breakthrough for decades. This system requires that coal be mined along one long face—several hundred feet long—rather than in separate entries as is common in the United States. In modern long-wall systems the coal is sheared off the entire length of the face by a cutting device. It then drops onto a conveyor. When the coal face has been cut back, the steel props that hold up the roof above the face advance themselves, the roof caves in behind the props, and the cycle begins again. The system sounds extremely efficient for the particular resources to which it is suited. It has been used successfully in Great Britain and Europe, but somehow it has failed to make much impact in the United States.

Various other approaches to coal extraction are at different stages, ranging from speculation through experiment to occasional commercial use. Some success has been reported with hydraulic mining using high-pressure water jets to cut coal (or even harder minerals). Hydraulic mining overcomes the serious dust problems that result from the high extraction rates of modern equipment. It is also consistent with hydraulic transportation of coal from the mine; that is, grinding the coal, mixing it with water to form a slurry, and pumping it out.

Experiments have also been conducted with fully automatic machinery that can be sent far into the seam without people. This saves many of the health- and safety-related costs of current mining practice, such as the amounts spent on preventing roof falls and controlling dust. Similarly, most human exposure to the hazards of mining can be avoided by gasifying (or liquifying) the coal in place. It is not so clear that the environmental hazards can be avoided. Contamination of groundwater is a particular concern.

The auxiliary functions of roof control, dust control, and ventilation have taken on increasing importance in recent years because of the increases in the extraction rates and the greater stringency of recent mine safety regulations. These regulations are examined more closely later. First, however, it is useful to examine surface mining, which avoids most of the health and safety problems of the underground mines but has environmental problems of its own.

SURFACE MINING

The most primitive form of surface mining is simply to scrape the overburden from the exposed edge where a coal seam penetrates the surface of the ground and then to break and load the coal. This type of stripping has been practiced in a casual way for centuries. It was only after the development of heavy earth-moving equipment, however, that strip mining could be much more than a cottage industry. Strip mining became important after the 1930s (Figure 8.5). By the 1970s surface methods accounted for most of the coal mined.

Although the general idea remains to scrape off the overburden and scoop up the coal, differences in geology lead to three variations in the techniques and layouts used. In the level or rolling country of the Midwest where the seams are of ordinary thickness, area stripping is practiced. This involves removing the overburden from a swath of terrain, loosening and loading the coal, and then filling the ditch with the overburden from the next swath.

In the hilly terrain of the Appalachians, the coal seams are often discontinuous, being located under hills and broken off where the intervening valleys have washed away. The contour-mining operation consists of following the coal seam around the mountain where it outcrops. The equipment can continue to follow the seam into the hill up to the stripping limit, the point where the overburden becomes too deep. The stripping limit is not a fixed number, but rather depends on the type of equipment that can be used. Typically in the steep hills of Appalachia, the deposits are so small and access so difficult that only ordinary construction equipment can be used, and hence the stripping limit may be reached in "one trip around the mountain." Such contour stripping presents serious questions of reclamation, which must be considered later.

The third principal form of surface mining is open pit mining as practiced in the very thick coal beds sometimes found in the West. The techniques resemble those used for minerals such as iron and copper ore. In area stripping the seam might be 6 feet thick and would be removed in one pass. In open pit mining the seam might be 50 feet thick, so the mining proceeds along successive benches cut into the coal itself.

In addition to the three principal methods of surface mining, auger mining is a by-product that can be grouped with them. Once the mining equipment has made a cut, it leaves a vertical high wall of overburden with the seam of coal at the base. When the strip-mining operation ends (often because the machinery reaches its stripping limit), it is possible to extract more coal by boring a row of horizontal holes into the face of the seam with a large auger. Auger mining has very high output per work hour and uses little capital or other inputs, but it is obviously an adjunct to strip mining rather than a separate technique in its own right.

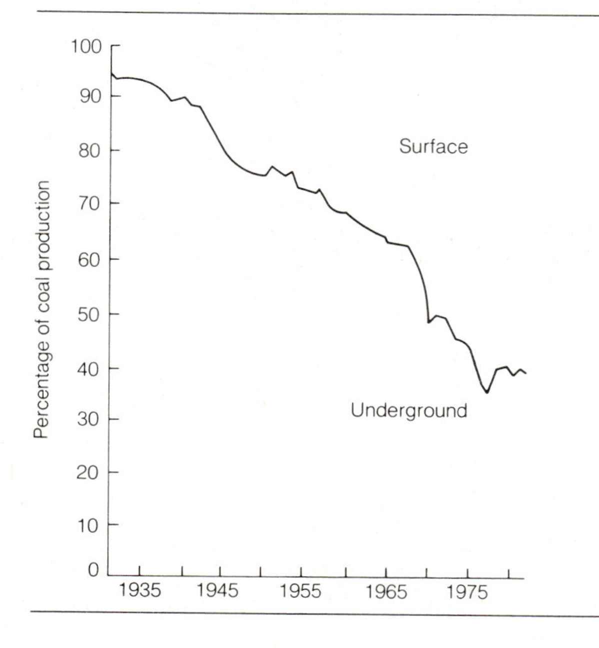

Figure 8.5 *The Declining Share of Underground Coal Production, 1932–1983*

Sources: Minerals Yearbook *(various years);* Annual Energy Review *(1983), p. 167.*

One result of the technological changes has been an increase in the annual tonnage extracted from a typical mine. The most dramatic growth in the share of output produced by large (more than 200,000 tons per year) mines had occurred by the end of World War II, at which time they accounted for 70 percent of output. The very large mines (greater than 500,000 tons) grew rapidly in importance from around 1940 to 1970. Despite the growth of the largest mines, however, those producing less than 100,000 tons per year have held their own since the late 1920s. The mines producing 100,000 to 200,000 tons have correspondingly been squeezed from 30 percent to 10 percent of output. This reflects two divergent technological trends. On the one hand, individual pieces of equipment have greater capacity, pushing mines into the higher output categories. On the other hand, changes in equipment have permitted the exploitation of small deposits where the individual mine will have low output.

During the decade and a half following World War II, the coal industry was often dismissed as a dying relic of a less sophisticated era. Output and employment were declining, leaving many coal miners and mining towns in an extremely depressed condition. Yet a closer look at that era indicates a very rapid growth of productivity, whether measured in the crude sense of output per person per day (Figure 8.6) or using more comprehensive measures of inputs and outputs, such as costs and prices.

Ironically, in the 1970s when the industry began to attract attention for its growth and prospective contribution to the nation's energy independence, its productivity plunged. Between 1969 and

Figure 8.6 *Output per Worker-Day in Coal Mining, 1949–1983*

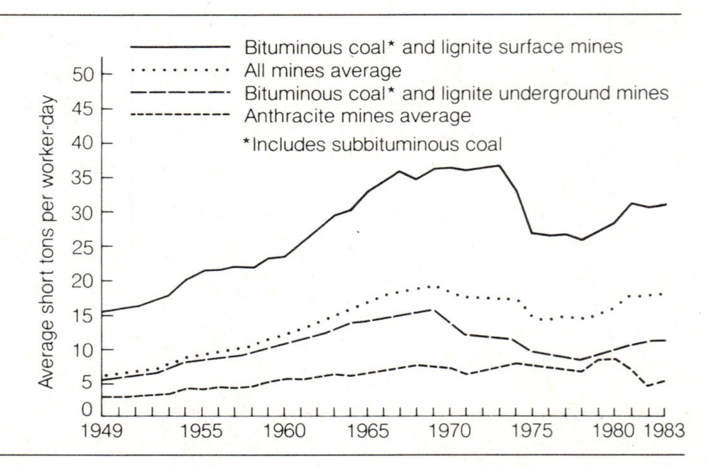

Source: Annual Energy Review *(1983), p. 176.*

1978, for example, output per worker-day in underground mining fell from 15.6 tons to 8.4 tons.

When productivity declines in the mineral industries, it is reasonable to check first to see whether the natural resource has been depleted sufficiently to make a difference. The theory of the mine predicts that the best resources will be used first, so it would be expected that eventually output per worker would decline as the worsening of seam conditions proceeded more rapidly than technological change could improve productivity.

No evidence of worsening natural conditions can be found in the coal-mining industry. The best single summary statistic for underground mining conditions is the thickness of the coal seam. It is an imperfect measure because it does not capture such important determinants of output as roof conditions, regularity and extent of the seam, floor conditions, pitch, amount of water and gas, and access to the mine. Nevertheless, the data as presented in Table 8.1 show essentially unchanged geological conditions.

When the technological history is considered, the situation is even better. One of the continuing trends has been the increasing compactness of mining machinery, which eliminates much of the penalty of exploiting thinner seams. Similarly, specialized equipment for dealing with other geological difficulties has extended the range of resources that can be exploited at the prevailing market price. In fact, the improvements in equipment have been so pronounced that thickness of seam has ceased to be a predictor of mining cost, although it was once considered to be the most important measure.

For surface production, summarizing geological conditions is even more difficult. We can look at the depth of the overburden that must be removed or at the relationship between the amount of overburden and the amount of coal that is uncovered (the stripping ratio). The three problems with such numbers are that they do not consider the extent of the area to be mined, the difficulty of mining, or the value of the material recovered. In particular, surface mining becomes substantially cheaper as larger stripping machines are used (up to the largest size range), but it is not worth assembling a huge machine unless reserves are adequate to keep it working for a couple of decades. As machines have been made bigger and stronger, the depth of overburden has become a less important determinant of cost than is the extent and regularity of the seam. A statistic showing that surface miners have moved more overburden per ton of coal does not necessarily mean that they had to—only that it was profitable.

If geological conditions are not deteriorating, why did output per worker-hour decrease? The best explanation is that the decrease has resulted from three main changes after 1970: (1) the increased stringency of environmental regulations, which have had a particularly expensive impact on surface mining; (2) the stricter laws (and enforcement) governing health and safety in the workplace, which have had a very strong impact on underground mines; and (3) the changes in labor-management relations. The first two reflected a change in social values, as did the third in part. The new laws can be considered as attempts to have consumers of coal bear the full cost of producing the coal, which includes the costs to the environment and the health and safety of miners. Whether the new laws impose the correct costs is a controversial question.

COMPETITION IN THE COAL INDUSTRY

At first glance the coal-mining industry appears highly competitive since 3,500 companies operate more than 5,000 mines.[5] Moreover, significant quantities of coal are traded internationally and coal must compete with other energy sources in virtually every one of its markets. Nevertheless, some have worried about the dominance of a few large firms, despite the fringe of small producers. Another concern is that the entire energy market will be controlled by a few oil companies that also control coal and uranium. Some have granted that the industry is competitive today but argue that it will become less so because ownership of reserves is more concentrated than is current output.

Despite the large number of mines, it is correct to say that output is heavily concentrated in a relatively small number of very large mines. Indeed, 15 percent of the mines furnish about 80 percent of the output. By the standards of most industries, however, 800 different establishments is not a high degree of concentration. More conventional definitions of concentration refer to firms rather than mines. In 1974 the thirty-one companies that produced at least 3 million tons apiece accounted for 58 percent of total coal output.

During the 1970s oil companies were actively diversifying into coal, uranium, oil shale, and even solar energy. Some political consideration was given to forbidding such expansion to preserve interfuel competition, but other analysts suggested that the transition from oil to other fuels would occur more smoothly if the big oil companies managed it. In an effort to acquire more information about the activities of the largest firms in the energy industries, the government requires specially detailed reports from the twenty-six largest crude oil producers—those that have at least 1 percent of the production or reserves of oil, gas, coal, or uranium or 1 percent of oil production, refining capacity, or product sales. Between 1977 and 1981 those companies increased their share of domestic coal production from 13 percent to 19 percent. In 1982 thirty-six U.S. petroleum companies mined 30.5 percent of all coal and controlled about 15 percent of coal reserves.[6]

Although the industry hardly seems to be in danger of monopolization, it is not the textbook case of pure competition either. Since transportation costs are high relative to the value of the product, the number of mines competing for a particular customer will be much smaller than the national data suggest. Moreover, differences in the quality of coal further restrict the number of potential suppliers. The most important departure from the textbook model of competition, however, comes from the importance of long-term contracts in the market for coal. An electric utility will generally arrange for the development of a mine when it builds a power plant. The price of the coal is negotiated, as will be the adjustments over the life of the mine. Coal is also sold in a spot market where prices fluctuate, but the long-term arrangements and outright ownership of mines by utilities and steel companies are important features of the market.

THE ENVIRONMENTAL COSTS OF COAL MINING

The coal industry departs from the competitive ideal in another respect as well: In the absence of government regulation, both the mining and the use of coal inflict heavy external costs on the environment. Environmental problems can be divided into three groups: temporary externalities, permanent externalities, and other permanent changes. Any cost (or benefit) inflicted on someone who is not a party to the transaction is an externality. Some externalities (for example, noise) are ephemeral. In the case of strip mining, the best example of a temporary externality is the inconvenience of having a road closed while heavy equipment crosses. It is a nuisance, but the road can be repaired and reopened. Obviously, such short-term external costs should not be a major concern of policy, except to see that the law creates mechanisms for fair assessment of costs against those who cause damages and fair compensation of those who suffer. At the same time we should recognize that some costs cannot feasibly be compensated. For example, the mining company (hence, indirectly, the consumers of coal) can be forced to pay for repairing a foundation damaged by blasting, but the nature lover will still lose some pleasure if the blasting drives away some rare bird. Similarly, the sheer ugliness of the mining process must be counted as one price of progress.[7]

Another class of externalities has more serious implications. These are the very long-term (or permanent) externalities, such as the acid mine drainage that can poison streams for decades after the mine closes. The acid drainage results from the contact of water with acidic rocks that are exposed and broken in the mining process. Similarly, poorly reclaimed strip-mined land can silt up streams for many years, as well as diminishing the quantity and quality of groundwater available to the neighbors of some mining projects.

The most interesting question relates to the permanent effects that are limited to the particular parcel of land. Assuming that the landowner also owns the mineral rights and makes the decision to exploit (when this is not the case the situation can be very different), then the condition in which the land is left should not ordinarily be treated as an externality. Obviously, if streams are poisoned by acid or clogged with silt or if the high wall left by stripping poses a hazard to life or property, these are externalities that should be a community concern if they are serious. But why should the way in which a landowner leaves the surface of the land be considered a matter for legislation? Much of the environmental legislation is justified in the name of serious externalities, of course. In particular, it can be very important to the surroundings to establish some vegetation quickly to reduce erosion, siltation, and rapid runoff. The law goes beyond this, however, to specify that the original contour of the land must be restored. This has created considerable controversy and opposition because it is often possible to increase the value of the land for building or agriculture by changing the contour. In particular, in the mountain areas of the East it has seemed more profitable to remove an entire mountaintop to get at the underlying coal, use the spoil to fill some of the surrounding lowlands, and end up with a substantial area of nearly flat land, which is scarce and valuable in Appalachia. The conditions under which such a procedure should be permitted have become a controversial matter.

What is the public interest in the contour of the land? We can argue that the public has some interest in every irreversible decision, but the extraction decision itself is an irreversible one and the public generally does not interfere greatly in that. In effect, the Surface Mining Control and Reclamation Act of 1977 (P.L. 95-87) is a statement that not only should externalities be controlled but also that agricultural uses of the earth and maintenance of the original contour deserve priority. Why should the strip miner pay $5,000 per acre to restore land to make it suitable for agriculture in an area where undisturbed farmland can be bought for one-tenth of that? Society has overruled the market, but the impact is not entirely to ensure that the land becomes as productive as possible since the bias is toward retaining existing contour, rather than creating the more productive level ground.

Although controls over environmental damage can clearly be justified as society's response to external costs, the implementation of the controls is a complex matter necessitating a great deal of administrative discretion for enforcement officials. Conditions differ so much that the regulations—even though they are already exceedingly numerous and unwieldy—cannot cover every contingency. Hence the discretion is inevitable. Administrative discretion, however, brings all of the problems associated with attempts to influence bureaucracy through political pressure or bribery. It is certain that this will be a chronic problem.

The attention given to the problems of surface mining should not obscure the environmental problems posed by underground coal mining. Serious problems of acid mine drainage and subsidence of the surface, as well as of waste disposal, characterize the underground sector. Moreover, when any coal is burned it can give off large quantities of air pollutants, but this topic is reserved for the chapter on electric power.

HEALTH AND SAFETY

Mining is a dangerous activity, and underground coal mining is especially hazardous. Figure 8.7 shows the fatal accident rates for underground and surface coal mining. The bad reputation of underground mining is well deserved because the typical miner is five times as likely to be killed at work as a typical manufacturing employee, and the underground coal miner faces far greater risks. However, the new mine safety legislation and stepped-up enforcement made a big difference in accident rates per person. When the Federal Coal Mine Safety and Health Act (MSHA) was passed in 1969, underground fatalities stood at 1.0 per million worker-hours, down from 1.6 in 1961. By 1978 underground fatalities had declined to 0.4 per million worker-hours, which seems substantial, even in view of the downward trend that was apparent before the new legislation.

When we examine the human cost of coal rather than the risk taken by particular miners, the recent history of the industry is less

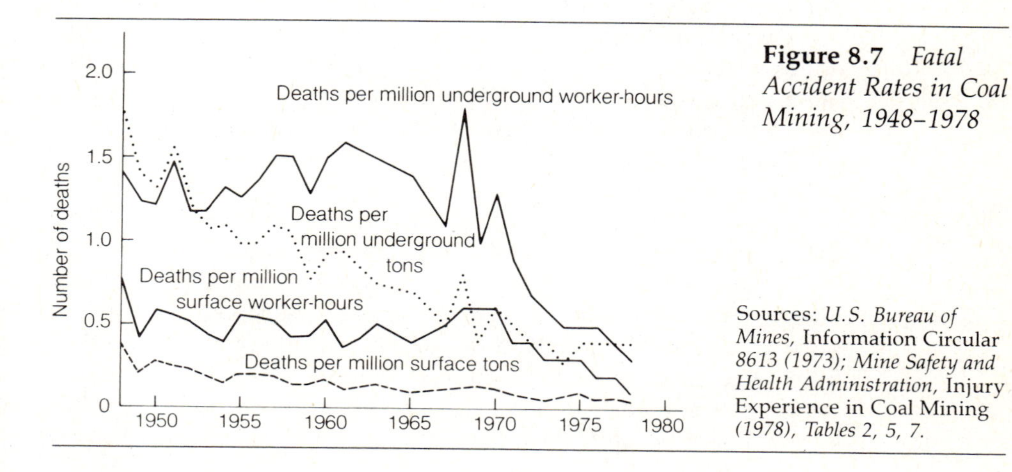

Figure 8.7 *Fatal Accident Rates in Coal Mining, 1948–1978*

Sources: *U.S. Bureau of Mines,* Information Circular *8613 (1973); Mine Safety and Health Administration,* Injury Experience in Coal Mining *(1978), Tables 2, 5, 7.*

impressive. The fatality rate per million tons was 0.4 in 1969, having declined from 0.9 per million tons in 1961. After 1969 the figure fluctuated near the 1969 level, ending up at 0.3 in 1978.

A critic of MSHA might allege that the expensive legislation had accomplished nothing for society! The additional people in the mines did just enough for safety to offset the greater exposure of additional people in the mines. This view captures some flavor of the truth but is somewhat exaggerated. Other factors affecting output per person were also operative during this era, so some of the additional human exposure resulted from factors other than the health and safety legislation. These include the stricter environmental standards and the changes in labor-management relations that accompanied the resumption in growth of the labor force. The influx of young, untrained (and perhaps obstreperous) new employees would have been expected to worsen accident rates in the absence of some strong offsetting force.

In any event, it is clear that the strongest force toward the reduction of fatality rates in coal mining has been the shift from underground to surface mining. The fatality rates in the latter were one-fourth that of the former in terms of worker-hours and were one-seventh as high per million tons in 1961. Despite all the changes, the advantage of surface methods remained about the same in 1978. The result of a shift toward surface mining therefore is to reduce fatalities. If, for example, the amount of coal mined in 1961 had been mined with the mix of surface and underground methods observed fifteen years later but with no change in the 1961 fatality rates for each mining method, the overall fatality rate would have been .46 per million tons, rather than the .68 actually observed. In other words, mining 57 percent of the coal by surface methods in 1961 instead of the actual proportion of 32 percent would have saved 75 lives in that year.

The efforts to reduce black lung disease and to compensate those who have been disabled, or their survivors, raise some additional questions. It has long been known by miners that many of them will sooner or later suffer from decreased lung capacity as a result of breathing coal dust in the mines. The severity of the condition ranges from a minor inconvenience to total disability depending on exposure and individual susceptibility. Since this is a well-known disadvantage of coal mining, economic theory predicts that prospective coal miners would require a wage differential high enough to compensate for that as well as other disadvantages of coal mining. Such well-known disadvantages as the high accident rate, the dark, wet working conditions, and black lung disease should be fully reflected in the pay of miners. (The issues are quite different in the case of someone exposed to a new substance that poses unknown hazards.) In fact, miners' weekly wages have generally exceeded manufacturing wages, despite the shorter work week, by about 20 percent. The differential vanished during the late 1920s and early 1930s but then increased steadily to exceed 20 percent in the early 1960s and grew to 50 percent by 1980. It would seem that the prediction of economic theory has been realized: Wages reflect the greater risks, discomfort, and effort in coal mining.

According to the theory there is no need for government action. The expenditure on safety by the mining companies would in theory be determined by their desire to avoid paying high wage rates. They could hire people more cheaply if they ran a safer mine. This, of course, is a point where the theory breaks down, for the wage rates in underground mining are largely determined by industrywide negotiations with the United Mine Workers union. Each firm (except for small inconspicuous ones) will pay the standard wage rate. (Even large nonunion firms generally pay at least the union rate to forestall unionization.) Since it cannot lower wage rates by improving safety, the firm will not have an incentive to improve safety conditions unless the incentive is supplied by an outside agency. (The one minor exception to this is that the firm with an excellent reputation will find it easier to recruit the best labor, whereas a firm with a very bad reputation will not have as many good applicants. But coal mines are usually the highest-paying employers in an area and hence usually do not have trouble in recruiting employees.) The outside agency today is the government, which sets certain standards and then must enforce them since each firm can save money by cutting corners. An alternative is to require each firm to carry private insurance to cover very large death and disability payments. In the normal course of events, insurance companies charge premiums based on experience and also apply pressure to reduce hazards.

As things have worked out in recent years, however, the cost has been paid three times. Wage rates have been more than high enough to attract people into the mines despite the risk of the work. The companies have been taxed to pay disability and survivors' benefits. Finally, the government has required the firms to reduce the hazards that necessitated the high wages and disability payments. The ease with which the large expansion in employment was accomplished during the 1970s suggests that coal miners are overcompensated. That does not imply, however, that the government should relax its efforts to improve safety, because individual miners cannot easily express their trade-off between income and safety.

SOME GEOGRAPHIC DIFFERENCES

When coal was the general industrial, commercial, and residential fuel, it was consumed wherever people lived and worked and was widely mined. For a commodity as heavy relative to its value as coal, transportation costs are a significant part of total cost to the consumer. Hence in the days of widespread consumption, there was always a trade-off between mining expensive or inferior coal nearby or paying large amounts to transport better or cheaper coal. As oil and natural gas became more generally available and the cost of electricity declined, the expensive mines serving local markets lost their customers. The years of contraction for the coal-mining industry (the 1920s into the 1960s) were also years of consolidation into bigger mines, cheaper and better seams, and locations for mines and customers that were served by specialized transportation links. After World War II it became increasingly rare to see coal delivered by men carrying sacks from a truck. It became increasingly common to have unit trains dedicated entirely to moving coal from a particular mine to a certain power plant in large volumes on a tight schedule. Even where small mines persisted, they fed into this mass network by selling to coal brokers near the mine site.

Thus the mines serving routine local use disappeared, to be replaced by other fuels. The uses of energy for which coal was particularly well suited—such as coking ovens—had often been located with special consideration given to raw material deposits anyway; and otherwise, a specialized transportation link was developed, such as the coal-carrying railroads or barges. Electric power presents a special case since, unlike the iron and steel industry that could be concentrated in a few locations, the electric power plants are scattered broadly throughout the country to be near the final consumers. Areas that enjoyed special advantages with other fuels (for example, the natural gas in the Southwest that would have been flared off in the absence of markets at the electric power plants) switched to them. Other areas such as New England were faced with unfortunate choices among coal moved long distances, residual oil from abroad, or experimentation with nuclear power. The coal industry increasingly concentrated on delivering huge quantities of coal cheaply to a few consuming points.

THE ECONOMIC OUTCOME

The combination of eliminating isolated, high-cost mines and expanding the lowest-cost segments of the industry gave the industry the appearance of great success in keeping costs down for a period of several decades. The impression is not totally unwarranted, but it is a little misleading, as is indicated by the data in Figure 8.8. The average coal price increased much less rapidly than any of the components. In effect, the industry was changing its product mix as the traditional markets disappeared and the cheap market expanded.

Table 8.2 indicates some of the changes in payments for various inputs. The dominant trend since 1909 has been the reduction of the importance of direct wage payments and their replacement by "other costs and profits" (OCP), the residual category after subtracting wages, salaries, supplies, and purchased materials from the value of net shipments. In the period from 1902 to 1939, wages accounted for roughly 60 percent of total costs and profits. By 1977 this had diminished to 23 percent. Other costs include some indirect labor costs, such as Social Security payments and the United Mine Workers (UMW) royalty (a fee assessed by the union on each ton of coal shipped). If these were transferred from OCP to wages, the effect would be a shift of about 10 percentage points from OCP to wages in 1977.

Even after making that adjustment, it is apparent that OCP has grown. Nor is that especially surprising, in view of the shift toward machinery from hand labor that characterized the entire period. OCP includes interest, profits, and depreciation, which should be related to capital inputs.

It is more surprising that the cost of materials has not increased in relative importance since 1954. Since wages are high and working conditions are poor in the mines, substitution of purchased inputs for on-site labor has been a continuing theme. Thus purchased roof bolts have replaced timber props cut and fitted in the mining area. Similarly, purchased electricity has replaced steam produced at the mine by burning coal dug there. Indeed, the trends of mechanization and of integration of mining communities with the industrialized economy by improved transportation should be expected to increase the importance of purchased inputs at the expense of on-site labor.

Figure 8.8 *Coal Prices, 1951–1981*

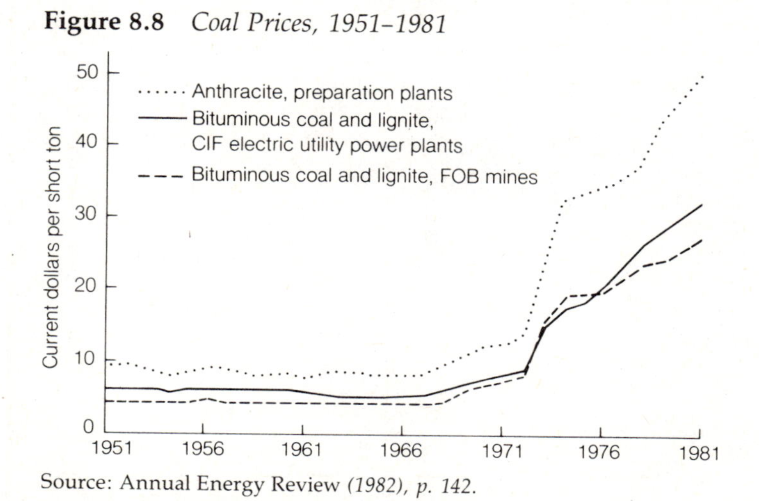

Source: Annual Energy Review *(1982), p. 142.*

When the data are restated in the form of average cost per ton of output (Table 8.3), the discontinuity of the most recent period becomes evident. The numbers suggest three separate eras in coal mining during this century. The traditional era ended with World War II. The era when technological progress offset major changes in input prices extended through the 1950s and 1960s. The 1970s were characterized not only by general inflation but also by the special problem of declining output per person and increasing wage rates.

These factors are examined in Table 8.4. The extraordinary increases in wage earners per ton and wages per worker-year combined to raise the index of wages per ton from 198 in 1967 to 703 in 1977. That this is more than the normal result of inflation in the decade of the 1970s is indicated by the increase in the ratio of wages in coal to wages in all manufacturing from 1.2 in 1963 to 1.5 in 1977. The decline in output per person during the recent period is also extraordinary. The most recent data indicate that output per worker-hour is once again increasing, suggesting that the industry has adjusted to the stricter safety and environmental regulations.

Table 8.2 *Various Costs as a Proportion of Sales, Bituminous Coal, 1902–1977*

	Value of net shipments and receipts		Supplies, fuels, and purchased electricity		Wages		Salaries		Other costs and profits	
	(1) ($1,000)	(2) (%)	(3) ($1,000)	(4) (% of Col. 1)	(5) ($1,000)	(6) (% of Col. 1)	(7) ($1,000)	(8) (% of Col. 1)	(9) ($1,000)	(10) (% of Col. 1)
1902	290,858	100	24,799	9	181,482	62	14,512	5	70,065	24
1909	401,816	100	57,672	14	270,666	67	20,501	5	52,977	13
1919	1,146,312	100	180,359	16	683,442	60	68,763	6	213,748	19
1929	966,694	100	144,828	15	574,919	59	58,647	6	188,300	19
1935	658,475	100	103,581	16	402,677	61	32,531	5	119,686	18
1939	727,358	100	118,141	16	430,564	59	44,183	6	134,470	18
1954	1,784,798	100	383,424	21	759,659	43	109,100	6	532,615	30
1958	2,065,892	100	447,363	22	753,320	36	151,721	7	713,488	35
1963	2,063,165	100	449,973	22	642,338	31	110,153	5	860,701	42
1967	2,598,500	100	611,000	24	767,300	30	148,500	6	1,071,700	41
1972	4,648,100	100	1,022,400	22	1,412,200	30	301,500	6	1,912,000	41
1977	13,816,400	100	2,979,300	22	3,203,200	23	801,300	6	6,832,600	49

Source: *Calculated from data reported in U.S. Bureau of the Census,* Census of Mineral Industries.

With all its problems, coal remains the cheapest form of energy in most parts of the United States, at least for generating large amounts of steam. The inconvenience of using it will probably continue to limit its use by small consumers but not by major industrial consumers. A major concern, however, is the air pollution that results from burning coal. This lies at the heart of the choice between coal and nuclear power.

In addition to cheapness, which commends coal to the large user, the main advantage of coal is the immense reserves that are widely distributed in the United States. During the paroxysm of policy making following the Arab oil embargo in 1973, attention naturally turned to the one fossil fuel in ample supply. It would seem relatively obvious to predict that coal consumption would surge during an era when other fuels were far more expensive and subject to disruption resulting from political and military events in the Middle East. But coal consumption did not leap ahead to the 1 billion tons per year that was forecast.

Table 8.3 *Cost of Materials, Labor, and Other Inputs per Ton of Bituminous Coal, 1902–1977*

	Value of output per ton		Cost of materials per ton		Wages per ton		Salaries per ton		Other costs and profits per ton	
	(1) ($)	(2) Index	(3) ($)	(4) Index	(5) ($)	(6) Index	(7) ($)	(8) Index	(9) ($)	(10) Index
1902	1.12	100	0.10	100	0.70	100	0.06	100	0.27	100
1909	1.07	96	0.15	150	0.72	103	0.05	83	0.14	52
1919	2.49	222	0.39	390	1.48	211	0.15	250	0.46	170
1929	1.80	161	0.27	270	1.07	153	0.11	183	0.35	130
1935	1.77	158	0.28	280	1.08	154	0.09	150	0.32	119
1939	1.86	166	0.30	300	1.10	157	0.11	183	0.34	126
1954	4.61	412	0.99	990	1.96	280	0.28	467	1.37	507
1958	5.00	446	1.08	1080	1.82	260	0.37	617	1.73	641
1963	4.49	401	0.98	980	1.40	200	0.24	400	1.88	696
1967	4.71	420	1.11	1110	1.39	198	0.27	450	1.94	718
1972	8.00	714	1.76	1760	2.43	347	0.51	850	3.29	1218
1977	21.22	1895	4.58	4580	4.92	703	1.23	2050	10.50	3889

Source: *Calculated from data reported in U.S. Bureau of the Census,* Census of Mineral Industries.

Table 8.4 *Wage Payments and Labor Input Requirements for Bituminous Coal, 1902–1977*

	Production, development and exploration workers		Wage payments		Indices			Wages per worker-year		Average weekly earnings		
	(1)	(2)	(3)	(4)	(5)	(6)	(7)	(8)	(9)	(10)	(11)	(12)
	Number	Index	($1,000)	Index	Physical output	Wage earners per ton	Wages per ton	($)	Index	Bituminous Coal ($)	Manufac-turing ($)	(10)/(11)
1902	280,638	100	181,482	100	100	100	100	647	100	n.a.	n.a.	n.a.
1909	488,307	174	270,666	149	145	120	103	554	86	11.70	9.74	1.2
1919	546,412	195	683,442	377	177	110	211	1,251	193	25.84	21.84	1.3
1929	458,835	163	574,919	317	207	79	153	1,253	194	25.11	24.76	1.0
1935	435,426	155	402,677	222	143	108	154	925	143	18.86	19.91	.95
1939	369,265	132	430,564	237	151	87	157	1,166	180	22.99	23.64	.97
1954	198,134	71	759,659	419	149	49	280	3,834	593	77.52	70.49	1.1
1958	161,908	58	753,320	415	159	36	260	4,653	719	97.57	82.71	1.2
1963	116,975	42	642,338	354	176	24	200	5,491	849	121.43	99.63	1.2
1967	107,600	38	767,300	423	212	18	198	7,131	1102	153.28	114.90	1.3
1972	129,600	46	1,412,200	778	223	21	347	10,897	1684	216.	154.28	1.4
1977	198,000	70	3,203,200	1765	250	28	703	16,178	2500	353.	228.9	1.5

Sources: *Columns 1 and 3 are from U.S. Bureau of the Census,* Census of Mineral Industries, *1963 and 1977, Table 1, p. 12A-8. Columns 2, 4, and 9 are calculated from Columns 1, 3, and 8, respectively. Column 5 is from Table 8.2, Column 2. Column 6 equals (Column 2 ÷ Column 5)100. Column 7 is Column 6 of Table 8.3. Column 8 = Column 3 ÷ Column 1. Columns 10 and 11 are from U.S. Bureau of Labor Statistics,* Employment and Earnings Statistics for the United States, 1909–1965, *Bulletin 1312-3, 1965, and* U.S. Statistical Abstract *(various).*

The failure to meet targets set by politicians did not indicate a shortage of productive capacity; the industry filled all its orders with capacity to spare, but demand did not grow as rapidly as forecast. The results of the presidential order to convert oil-fueled electric-generating plants to coal suggest the problems. Fifty specific conversion orders were issued by the government, but several years later none of the utilities named in the orders had actually converted to coal.[8] The utilities had many reasons to delay, but surely one factor is that they had just been required to convert *from* coal to reduce air pollution.

The environmental questions about coal pervade the whole process from mining through final consumption. The main questions about mining have already been mentioned, but by way of summary it can be suggested that the new reclamation laws, if they are enforced, are strict enough so that most of the environmental costs of mining will be internalized. The increases in safety, health, and environmental costs have driven up the price of coal, accounting for much of the increase in the steam coal price from 40 cents per million BTU in 1973 to $1.65 in 1982. Nevertheless, the still more rapid increases in prices of oil and natural gas have kept coal the cheapest fuel, even when the new laws are taken into account.

In the long run it is the environmental decisions at the point of consumption that will determine the future of coal. Unless air pollution requirements are made extremely strict, coal will provide the cheapest new electrical capacity in most parts of the United States for the foreseeable future. Utilities have not moved more quickly to consume coal for three reasons. First, the clear cost advantage of coal over nuclear power is a recent phenomenon and power plants are many years in the planning stage. Second, coal has been beset by political indecision on the important environmental questions. Third, higher prices for electricity in recent years have discouraged the growth of consumption to such an extent that utilities have been stretching out construction schedules and postponing plants planned long ago. Eventually, of course, such transition problems will be overcome and, if conditions remain unchanged, coal growth will accelerate. But this will give new emphasis to the questions about the degree of cleanliness expected in the mining and converting functions.

NOTES

1. The data exclude fuel wood because comprehensive and consistent data are hard to obtain. Including wood might lower the contribution of coal in 1900 from 89 percent to something in the vicinity of 67 percent. The discrepancy would gradually diminish to insignificance with the passage of time.

2. The technological history of the industry, as well as the physical performance and economic outcome, are discussed more fully in Bela Gold, William S. Peirce, Gerhard Rosegger, and Mark Perlman, *Technological Progress and Industrial Leadership: The Growth of the U.S. Steel Industry, 1900–1970* (Boston: D. C. Heath, 1984), chaps. 8, 9, and 10. The industry has been the subject of several extensive studies, including Carroll L. Wilson, *Coal—Bridge to the Future: Report of the World Coal Study* (Cambridge, Mass.: Ballinger, 1980); Richard L. Gordon, *U.S. Coal and the Electric Power Industry* (Washington, D.C.: Johns Hopkins University Press/Resources for the Future, 1975); and Martin B. Zimmerman, *The U.S. Coal Industry: The Economics of Policy Choice* (Cambridge, Mass.: MIT Press, 1981).
3. A power shovel has the bucket rigidly attached to a dipper stick extending down from the boom. A dragline has a bucket suspended by cables from the end of the boom.
4. Morton S. Baratz, *The Union and the Coal Industry* (New Haven, Conn.: Yale University Press, 1955) provides a discussion of the policy of the United Mine Workers toward mechanization. C. L. Christenson, *Economic Redevelopment in Bituminous Coal* (Cambridge, Mass.: Harvard University Press, 1962) gives a graphic account of the survival of old techniques into the modern era.
5. Data are available in the U.S. Department of Energy, Energy Information Administration, publications *Quarterly Coal Report* and *Annual Energy Review*. The *Minerals Yearbook* (annual) of the U.S. Bureau of Mines is the best source of earlier data. Information on the twenty-six largest energy companies is published by the Energy Information Administration annually in *Performance Profiles of Major Energy Producers*.
6. U.S. Department of Energy, Energy Information Administration, *Performance Profiles of Major Energy Producers* (1981), p. 69; *Quarterly Coal Report* (October–December 1983), p. xv.
7. For a cogent dissent from the whole idea of material progress, see E. J. Mishan, *Technology and Growth: The Price We Pay* (New York: Praeger, 1970).
8. U.S. General Accounting Office, "Financial and Regulatory Aspects of Converting Oil-Fired Utility Boilers to Coal," EMD-81-31, November 21, 1980.

CHAPTER NINE

The Oil Industry

INTRODUCTION

During most of its history, the oil industry has been dominated by the specter of overcapacity and the consequent efforts of strong parties in the industry to control its effects. Occasional episodes of scarcity of crude oil or refining capacity have triggered bursts of investment that have again produced excess capacity. The result within firms has been a strong emphasis on marketing, as large integrated firms have attempted to sell large quantities of product without letting prices fall.

In their efforts to maintain both output and prices, the firms tried whatever techniques were common at the time. These included (1) the early attempts at monopolization by Standard Oil, (2) the standard oligopolistic techniques of product differentiation, customer service, and promotion of the industry, and (3) efforts to enlist the aid of government to limit the impact of excess supplies of crude oil.

When viewed in this context, the behavior of the OPEC nations in the 1970s does not depart very far from the industry tradition. The main distinction is that the dominant OPEC nations acted in the interests of crude oil producers, rather than of particular firms or the industry as a whole. The OPEC nations also were temporarily more successful than the firms in their efforts to extract more from the consumers; the reasons for the difference in effectiveness are developed later in this chapter.

Figure 9.1 highlights the growth of the petroleum industry. Consumption grew almost without interruption, increasing tenfold between 1900 and 1919 and again by a factor of 10 between 1919 and 1964. In 1978 consumption of petroleum in the United States peaked at 38 quads per year.

Since growth in the first half of this century was much more rapid than the growth of total energy consumption, oil increased its share of the energy consumed in the United States as indicated in Figure 9.2. By the early 1950s oil accounted for more than 40 percent of energy consumption, but its relative importance showed no upward trend after that time.

When Edwin L. Drake drilled the first oil well in 1859, he was trying to find the raw material for lamp fuel. Whale oil was becoming increasingly expensive and the substitutes obtained from coal and oil shale were not very satisfactory. Not only did Drake strike oil but also (and even more important) the kerosene distilled from the Pennsylvania crude was accepted by the market, ending the whale oil crisis and the whaling industry. Drillers flocked into the hills of western Pennsylvania, while the price of crude at the wellhead fluctuated wildly from a high of $20 per barrel to a low of 10 cents. By 1900, however, the industry had become much more orderly under the domination of John D. Rockefeller's Standard Oil Trust.

The price of crude oil fluctuated but generally stayed below $1 per barrel from the 1890s until the boom years of World War I. Prices again dropped below $1 during some years of the Great Depression.

Figure 9.1 *U.S. Oil Consumption and Ratio of Reserves to Consumption, 1900–1982*

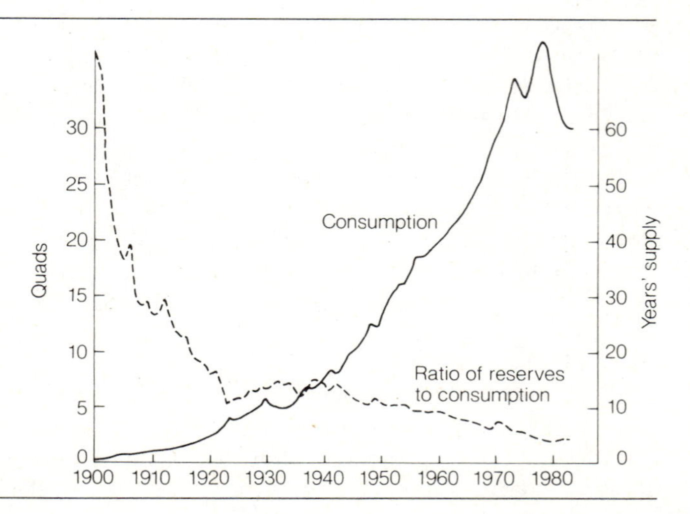

Sources: U.S. Historical Statistics *(1975), pp. 592–593; U.S. Department of Energy,* Annual Energy Review *(1983), pp. 71, 79.*

After World War II the situation became more complicated. The domestic price, shown in Figure 9.3, was influenced by foreign prices as imports grew in importance. This period is examined in more detail later.

Reserves, as might be expected, increased with production during the early years. The increases in domestic reserves ended in the 1950s as the immense magnitude and low production cost of Middle Eastern reserves became obvious. After the Alaskan discoveries kicked reserves upward in 1970, the decline set in again until renewed exploration finally yielded results at the end of the decade. Nevertheless, the ratio of domestic reserves to domestic consumption (Figure 9.1) remains low by historical standards.

DEVELOPMENT OF THE OIL INDUSTRY

Although the oil industry of 1900 had achieved a measure of stability, it was on the verge of a series of far-reaching changes. Some of the specific advances in exploration, drilling, and refining are mentioned in this section, but the big change in markets underlies the technological history and inspired much of the innovation in refining. Electricity in 1900 was rapidly displacing kerosene from the lighting market. At the same time the automobile was just becoming a significant market for gasoline. During most of the twentieth century, the oil industry has focused on supplying gasoline, the highest valued of the principal joint products of crude oil. The remaining products—most notably the heavy residual oil—have been sold at whatever price would compete with alternative fuels. Thus the pressures to find oil and refine it have come mainly from the gasoline market, which means the automobile.

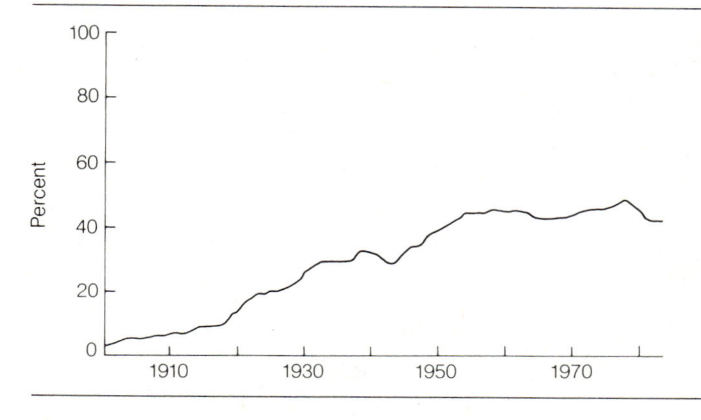

Figure 9.2 *Oil as a Percentage of Mineral and Electrical Energy Consumed in the United States, 1900–1983*

Sources: U.S. Historical Statistics *(1975), p. 588, M 83–90;* U.S. Statistical Abstract *(1981), pp. 577–578;* Monthly Energy Review, *December 1983, p. 4.*

Early oil wells were drilled in areas where oil seeped from the ground or had been detected in brine wells or streams. The first effort to bring scientific knowledge to bear on finding oil was to associate features of the earth's surface with the oil underground. What is relevant, obviously, is the underground geological structures, but early petroleum geologists had to guess at these from evidence visible from the surface. Every time oil is found in a place that seems unpromising by conventional geological standards, it forces a reexamination of the theories of oil formation and movement or of the relationship between surface phenomena and underground structures.

Drilling provides direct information about what is underground, but it is an expensive way to learn. Furthermore, the drill breaks the rocks into fragments with different characteristics that move up the drill hole to the surface at different speeds; thus interpretation of the cuttings requires considerable judgment. To supplement the information that can be inferred from the speed of drilling and the composition of the cuttings, the driller now lowers measuring devices ("sondes") to obtain such measures as spontaneous potential, resistivity, velocity of sound, and radioactivity. Recently, TV cameras have been lowered into the hole for a direct look.[1]

Since the 1930s geophysical techniques have been well enough developed to furnish additional clues before drilling. Magnetic surveys (by airplane since 1944) and gravity surveys are used for reconnaissance. Seismic studies, however, are the standard means used

Figure 9.3 *U.S. Average Price of Crude Oil at the Wellhead, 1900–1983*

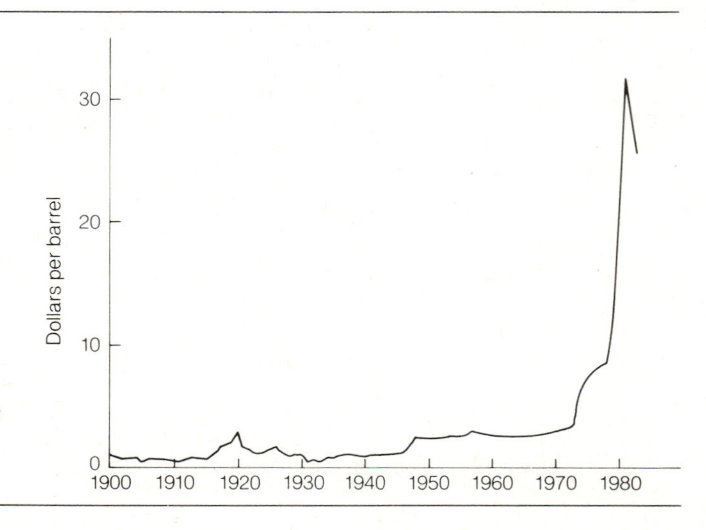

Sources: U.S. Historical Statistics *(1975), pp. 593–594; U.S. Department of Energy,* Annual Energy Review *(1983), p. 127.*

to obtain detailed information about subsurface structures. Because of the enormous cost of drilling, as new geophysical techniques—such as computer processing of data from seismic tests—became available, they were adopted rapidly.

Drilling

At the turn of the century, most oil wells were drilled with cable tool rigs. A drilling tool suspended from a cable was alternately dropped and hoisted a few feet to pound its way through the rock. A steam engine powered the drilling, as well as the hoisting operations that were necessary for removing the drill tool, lowering and raising the bailer to remove cuttings and returning the drill tool. The hoist was also used to lower the pipes into place when the well needed to be cased.[2]

The casing is a string of pipe as large as the diameter of the hole will admit, which is lowered in and cemented into place. After the completed portion of the well has been cased, drilling can proceed using a bit small enough to be lowered inside the casing.

By 1900 rotary drills similar to those used today had been tried in Texas. The rotary drill has a bit attached to the end of a pipe. As the pipe is rotated, the bit cuts through the rock. As drilling proceeds, more lengths of drill pipe are screwed on. When the bit is too dull, the whole drill string must be pulled out of the hole and the pipe sections unscrewed and stacked, then returned to the hole with a new bit. Otherwise, the process is continuous. While drilling proceeds, "mud" (a specially compounded fluid) pumped through the pipe comes through the bit, cooling it and flushing the cuttings out of the hole.

Initially, the rotary drills were used mainly in the soft formations that caved easily with the pounding of the cable tools. Under such conditions rotary drilling was fast, but the early bits did not last long in hard rock so the advantage of continuous operation was lost.

When Howard Hughes invented the rock bit in 1909, the range of applicability of the rotary drill was increased greatly. The basic design of conical cutters with beveled teeth has remained standard, although numerous refinements have been made and better materials have been developed. After the 1930s rotary rigs dominated oil drilling although cable rigs were also improved. The record depth for a cable rig was 11,145 feet completed in 1953, but rotary rigs had reached 15,000 feet by World War II and have exceeded 30,000 feet today. Rotary drills are much faster and are now the standard technique.

Sometimes a "mud motor," or hydraulic drill powered by the drilling fluid, is used. This technique has particular advantages when large deviations from the vertical direction are desired. Where the drilling platform is very expensive – the North Sea, for example – numerous wells are drilled from the same platform. Similarly, in the Arctic it may cost $100 million to construct a gravel island on which a drill rig can be stationed, so it is usual to drill several wells from the same site.

Refining

From 1861 until 1913 the refining process consisted of fractional distillation followed by various chemical treatments to clean and clarify the premium products. The heavier fractions could be sold only at low prices to compete with coal as a boiler fuel. Although various researchers (beginning with Benjamin Silliman in 1855) understood that heavy hydrocarbons could be cracked into lighter products by the application of heat, success on a commercial scale awaited the efforts of Dr. William Burton of Standard Oil of Indiana.[3]

Burton's four-year research effort yielded a highly profitable batch-type thermal cracking process that went on stream in 1913. It was particularly profitable for Standard Oil of Indiana because it extracted more gasoline from a barrel of crude at the very time when crude oil production in the Midwest was declining and gasoline consumption was increasing rapidly. Furthermore, the ready availability of cheap coal in the region meant that the heavier products of distillation had a very low market price. The particular firm was under a special pressure to increase gasoline yields, decrease crude requirements, and decrease production of residual oil, but these same motives continue to prevail in the industry today.

In the 1920s four continuous thermal cracking processes came on stream. As materials and equipment became available, operating temperatures and pressures could be increased and both the quantity and the quality (for example, octane rating) of product improved. Nevertheless, the profits earned by the more sophisticated processes did not match those that were generated by the Burton process.[4]

The next big change was the development of catalytic cracking by Eugene Houdry, who found that use of a catalyst results in a substantial improvement in the quality of the gasoline. The early process as operated in 1936 had to be stopped periodically to burn carbon off the catalyst. By 1942, however, continuous catalytic cracking was being used commercially and very quickly dominated the entire industry.

One of the continuing themes of the industry's technological history is the increase in the rate of output of refineries. This has been accomplished by building bigger units, moving toward continuous processes, and increasing the length of time that a unit could operate before being shut down for maintenance. Although these trends help to reduce unit costs, they also put small refineries at a disadvantage. There has been considerable concern among economists about the possibility of noncompetitive behavior in the oil industry, and certainly the economies of scale in refining would seem to make entry difficult.

During the 1970s, however, the number of small refineries grew more rapidly than the number of large ones. This reflected peculiarities of federal regulation rather than economic factors, because the government provided small refiners with preferred access to domestic crude oil at the controlled price. By 1980, when the controls expired, the twenty largest companies in the United States operated 112 refineries with an average capacity of 122,000 barrels per day apiece, whereas the 147 smaller companies operated 185 refineries with an average capacity of 22,000 barrels per day. In the absence of government preferences, a refinery would probably have to exceed 50,000 barrels per day to have reasonable costs. By 1984, 29 percent of the small refineries and 21 percent of the large ones had shut down.[5]

As significant as cost is the restricted flexibility of the small units. The big new refineries are technically sophisticated enough to process difficult grades of crude and to obtain high yields of gasoline. Indeed, a continuing theme in the growth of the industry is the ability to process a wider range of inputs. Just as exploration and drilling can exploit less accessible resources, so refining equipment and methods make it possible to produce high-valued outputs from crude oils that are heavy or high in sulfur content. Thus the general trend observed with coal toward the reduction of economic rents by technological change is also evident in the oil industry. Like improvements in transportation, such changes reduce the margin of superiority of the best fields and hence serve to reduce rents. In view of this trend, the subsidy that encouraged construction of small, primitive refineries seems anomalous.

OIL AND THE THEORY OF THE MINE

One effect of the innovations in refining has been a decrease in the input of crude per gallon of gasoline, but crude oil has generally been in ample supply because of excess capacity in the producing industry. Firms have invested in the new refining techniques to sell a larger proportion of output as expensive gasoline rather than as cheap boiler fuel. The observation that crude oil has been plentiful is not exactly consistent with the standard theory of the mine as developed in Chapter 7. In particular, in the simple version of the theory, the owner of the mineral rights will exploit the mineral at the moment that maximizes the present value of economic rent. Overproduction is impossible because as soon as price is driven down, production will be decreased to save the mineral for a future time when its price is expected to be much higher. If the theory is working out, the complaint of overproduction may be just the seller's perennial lament that prices should be higher.

Three specific features of the industry that intrude on the simple theory, however, furnish a stronger basis for such lamentation: (1) the fugitive nature of oil, (2) the uncertainty in the timing of discovery, and (3) the uncertain tenure of the right to receive rents. These are considered in the following paragraphs, but the net impact is to speed the rate of extraction beyond what the simple theory predicts. Different institutional arrangements, such as taxes and royalty agreements, can also influence the rate of exploration and extraction.

Fugitive Resources

The fugitive resource problem is easy enough to describe, although difficult to solve in practice. Oil, as well as gas and water, will flow underground in response to differences in pressure. Hence, if someone begins to extract oil from one well in a field, oil in the rest of the field will flow toward that well. If ownership of the field is divided among several people, the individual may face the choice between extracting oil today or watching the neighbors extract it today, rather than the classical choice of the theory between mining it today and mining it next year. State laws have traditionally applied to oil the "law of capture" developed during centuries of cases dealing with wild animals. Under this doctrine, people could obtain title to oil by capturing it through a well on their property, regardless of where it had flowed previously. Obviously, this distorts the timing of extraction as each landowner in the oil field rushes to capture as much oil as possible.[6]

The bad effects are not limited to the rate of extraction, because the total amount recovered from the field can be reduced by the excessive recovery rate and furthermore the total investment in field development will be excessive. Recovery is reduced for three reasons: (1) The geological facts are such that trying to speed the rate of recovery decreases the total amount that can be recovered. (2) Fields eventually require some form of treatment beyond merely pumping up the oil if recovery is to exceed about one-third of the oil in the field. Secondary techniques such as pumping in gas or water to maintain pressure are commonly used, and more advanced tertiary techniques that involve injecting chemicals or applying heat to the formation are being adopted. Obviously, the landowner will not inject materials in her well that will cause the oil to flow into her neighbor's well. (3) If a field is expected to produce for only a few years, it will not be worthwhile to build such related equipment as pipelines for transporting the oil to a refinery. Since a pipeline requires a large amount of capital to build, but then has a low operating cost, it is justified only when it can be used for a long time. Once it is in place, production can continue even after the extraction costs rise substantially, because the marginal cost of operating the pipeline is so low once it is built. If, however, it were desired to extract all the oil rapidly, a pipeline would not have time to pay off. Hence a technique using less capital but having higher operating costs (such as trucks) would be adopted. The high operating costs, however, would lead to earlier shutdown.

Although more rapid extraction will be associated with a higher ratio of operating cost to capital cost, it will also be associated with excessive investment in production, storage, and transportation facilities, which depend on the peak rate of output. If you are producing 1 barrel a year for a million years, you can produce from one well and carry the output in a 5-gallon can. If you want to produce 1 million barrels in one year, you will need many wells and substantial storage and transportation facilities. With the speeding of extraction that accompanies fragmented ownership, the investment both in wells and in the associated equipment will be excessive.

All of this suggests that fields should be operated as one unit; indeed, many are. State laws have been written to encourage unitary leasing or development, but the practice is by no means universal. If all participants in the earnings of a field knew exactly what the returns for individual development and for unitization would be, they could see that gains would accrue from unitization. It is very much like a city block that has been developed as separate house lots but that is now ripe for conversion into a more intensive use of the land as a result of urban growth. The ten houses separately might be worth $100,000 apiece or $1 million total, whereas the entire block might be worth $3 million as a piece of land suitable for a high-rise office or apartment building. The profits from assembling the lots into one unit are substantial, but even so, developers may have difficulty in buying out the last holdout. After all, that person may understand the numbers too and try to extract the last dollar of profit. The person might also be emotionally committed to living in that house forever and unwilling to sell at any price.

The situation in the oil fields, although similar in respect to the profits from unitization, is more complex because of the great uncertainties about the value of the oil and the extent of the field. The standard way to handle the problem of uncertainty is to use a royalty as the landowner's compensation. Having standard conventional royalties (for example, the landowner receives one-eighth of all oil produced) simplifies the bargaining process, but in a field with multiple owners it still leaves the problem of negotiating the fraction of total royalties to be received by each of the property owners. Furthermore, the mineral rights may have been subdivided and subject to a whole complex of overriding royalties and annual lease payments, which may complicate the bargaining process.

Because unitization is not universal, state conservation laws have been written, but these are a poor substitute. In the interest of preventing a wild scramble by all landowners to pump out as much oil as possible as quickly as possible, states have legislated such matters as the minimum number of acres per well, distance of wells from property lines, and maximum amount of oil that can be taken from a well. These are the famous "allowables" of the Texas law—the maximum daily output allowed by law to a particular well. The existing conservation laws were developed during an era of excess supply of domestic petroleum and should be reexamined now that conditions have changed.

Once the allowables were established for legitimate reasons of conservation, it was but a short step to "market demand prorationing," also performed by the Texas State Railway Commission. Under this system the Texas Railway Commission studied the market conditions for crude oil, then ruled on the number of days that wells in the state of Texas would be allowed to operate at the allowable daily rate. This device was used to maintain a high price for crude oil, rather than for conservation. One result, of course, was to maintain an emergency reserve of productive capacity. Since scarcity and OPEC raised crude oil prices, the wells have been permitted to operate at 100 percent of the allowable—the reserve is gone.

Uncertainty of Discovery and Cost

In addition to the problems posed by the fugitive nature of the resources, the oil industry is noted for the uncertainty about the existence of reserves and the costs of exploiting them. Whereas rational development of resources calls for the cheapest to be exploited first, the uncertainty results in a more nearly random pattern of development. The pattern is not entirely random, because development and exploration of areas that are known to be very expensive will be delayed until the price is high enough to offer some prospect of gain. At the other extreme, however, some new discoveries will turn out to be much cheaper than many of the fields already in production. This happened with such large, cheap discoveries as the East Texas field in 1930 and the Middle Eastern developments after World War II. If the discovery is small relative to the total market, the output can be sold without any impact on market price. The rent recipients can just pocket the extra money.

Extraordinarily high returns in a particular field will encourage even more development of that field and a search for additional ones. It is unlikely, however, that the high-cost fields that are already in production will actually be shut down because, once the exploration and development costs have been borne, the variable costs of producing oil are small. If the field required major new investment, such as reworking of old wells or the introduction of secondary recovery techniques, then it might be permitted to fade out of production. Until that time it will remain on the market as additional supply serving to depress the price of crude.

Uncertain Tenure

Although the fugitive resource problem is the traditional illustration of imperfect property rights, the fear of expropriation has served as an equivalent imperfection in the mineral industries in recent years. The oil company may suspect that its rights to produce oil or to retain a large share of the economic rents from production will expire with the next change in regime or the whims of international politics. Under these circumstances, as with the right to capture a fugitive resource, the producer will have an incentive to pump oil quickly. How quickly? That depends on the estimate of the time during which the firm will be permitted to continue production. The time horizon may be essentially infinite; for example, the firm may purchase a field outright in the United States and not worry about expropriation, which would require a political change so drastic as to be beyond the concern of company management.

At the other extreme, the firm might have favorable concessions in a country governed by a senile hereditary monarch whose profligate children denounce the impending revolution from opulent homes in London and Paris. If the revolutionary movement has succeeded in convincing the people that their poverty is the result of exploitation by the imperialist oil company working with the hereditary oppressors from the tottering monarchy, the firm will pump out oil as fast as it can in the few months before the shooting begins. Will the government attempt to conserve the natural resource by limiting production to the maximum efficient rate? Obviously not. The royal family will want to convert as much of that oil in the ground into dollars and gold as quickly as possible. They may negotiate about the division of the royalty with the company or try to have more of it flow directly into family accounts in Switzerland as distinct from the national treasury, but they will not want to slow down production. Even the funds that cannot be diverted from the national treasury may be useful in prolonging the life of the regime by purchasing weapons and buying off opponents.

Most cases in the real world lie somewhere between the extremes of infinite land tenure and impending confiscation. In most countries a producer of natural resources has to expect that a government supporting nationalization will take office at least once during the several decades before a field is exhausted. It would be self-defeating, however, to strip the field and run. This type of predatory behavior would certainly hasten the expropriation of the property. It will be more profitable to create a reputation for honesty, efficiency, and concern for national interests and the company's employees. The company will try to downplay its foreign origins and put citizens of the host country into responsible positions as quickly as possible. By acting responsibly and cultivating good relations with the public and as many potential leaders as possible, the firm can hope to postpone expropriation, improve the terms of compensation when it does occur, and maintain preferential access to the oil even after it no longer receives the rent from its production. In new mineral ventures today in underdeveloped countries, the outside firm often goes in as a manager and operator rather than as an owner, eliminating some of the sources of conflict from the start.

The firm has an incentive under all of these intermediate situations to produce steadily at full capacity. Steady production rates improve relations with the host country, and operating at full capacity increases the chances of recovering the initial investment before political conditions deteriorate. For all of these reasons, we should not expect the fine calculation of the exact moment at which sales are most profitable to have much influence on the output of existing wells. New wells, of course, are a different story. It is clear that drilling activity does respond to market conditions.

The basic choice in the simplest version of the theory of the mine should be rephrased to account for uncertainty and imperfect property rights. The modification is that the firm can choose to sell now or sell next year at an unknown price, provided no one else takes it first. It is clear that any shadow of imperfection in property rights, either from fugitive resources or from the threat of expropriation, makes earlier extraction more profitable. Uncertainty about future prices can work either way, of course, but if the uncertainty results from the prospect that large inexpensive fields will be discovered, then it too will encourage exploitation.

COMPETITION AND PRICES

As Figure 9.3 indicates, the price of crude oil at the wellhead in the early 1940s was about the same as it had been in 1900. After World War II the price rose from the vicinity of $1.20 per barrel to a range of $2.50–$3 per barrel, where it remained during most of the 1950s and 1960s. The rapid growth to a price exceeding $30 in 1981 marks the 1970s as a totally different era. The overall stability of oil prices before 1970 is particularly remarkable in view of the increase in other prices over the same period and the common expectation that a natural resource industry will encounter diminishing quality of reserves as output proceeds. Since the behavior of prices changed so drastically during the 1970s, the era of OPEC, it will be necessary to consider whether OPEC or other factors were responsible for the difference.

The two main deterrents to the formation of cartels in the domestic oil industry were opposition by government and the incentive for each firm to undercut an established price to capture more business. The most nearly successful domestic combination was the old Standard Oil Trust, which accounted for much of domestic production and more than 80 percent of refining in 1906. Because of its control of transportation, it played a more significant role for many sellers of crude or buyers of products than even that figure suggests. Standard Oil was dismembered by the federal government in 1911, however, and from that time on the industry has had a reasonably competitive structure.

There have been times when the government supported attempts to maintain prices, but these eventually failed because of the divergent interests of domestic companies. There have also been accusations that the foreign operations of the seven largest international firms (Exxon, Gulf, Mobil, Standard Oil of California, Texaco, BP, and Shell) have been jointly determined. It is easier to believe in an international cartel than a domestic one because of the small number of firms involved, the common interests in joint projects, and the lack of a single national government to exert control over the same group. But it is not really necessary to believe in conspiracy to suggest that a few giant firms would act in many ways like a cartel. After all, many of the international operations required extensive negotiations by companies jointly with host governments and often direct negotiations between governments of consuming and producing countries. These are the circumstances under which vigorous market competition is most difficult to maintain.

The behavior of the firms in international markets scarcely affected the United States market during the first half of this century. Imports were insignificant; the United States was a net exporter of petroleum until 1944. Even in the 1950s and 1960s the most significant factor for consumers was the restriction of imports imposed by the United States government under pressure from domestic producers.

As noted earlier, some state governments adopted output restrictions in the name of conservation, but then extended them to try to restrict output enough to maintain prices. The federal government cooperated in this effort to benefit producers at the expense of the consumer by prohibiting interstate shipment of oil produced in excess of state regulations (Connally "Hot Oil" Act of 1935). Thus a good deal of government effort was devoted to keeping prices high, although the federal government also adopted measures that encouraged investment in oil production, which presumably served to increase supply and reduce prices to consumers. The best known of these special advantages were the tax provisions for percentage depletion and expensing of exploration and development costs.[7] The net result of government actions on consumer prices was not fully certain because no coherent policy was involved, but merely a number of separate responses to varying pressures from different sectors of the industry at different times.

The companies, of course, had a clear incentive to keep prices high, but they were not able to. Since the elasticity of demand for petroleum products was quite low, a higher price would have resulted in higher profits. In an industry as fragmented as petroleum, however, it proved impossible to maintain high prices once the government prohibited outright monopoly. Oil production and the retailing of products are highly competitive in structure; there are more than 10,000 producing companies and 200,000 gasoline stations.

The refining stage is the easiest to imagine as oligopolistic, since a modern, efficient refinery is a huge investment with enormous capacity.[8] To build a fully integrated refinery with a capacity of 250,000 barrels per day would have cost $1 billion in 1980. This is certainly not an industry you enter with the savings from your sugar bowl. The United States, however, offers a vast market for petroleum products, so a total of 167 firms owning 287 refineries were in the business in 1980. Nor is entry totally closed; fourteen new firms entered the refining business during 1979.

As is often the case in industries that at first glance seem impossible to enter, the oil industry includes numerous firms that, with a few mergers and some investment, could become fully integrated firms capable of competing against the giants. A typical path is hard to specify, but it includes such possibilities as a marketing and refining firm (Sohio) integrating backward into production by developing Alaskan North Slope oil or a marketer (Occidental Petroleum) acquiring refining capacity and then lining up long-term crude supplies. With this type of market structure, it is no wonder that sporadic gasoline price wars and competition from discount stations kept prices of petroleum products low, despite the efforts of the major firms to compete only on the basis of advertising slogans, quality of service, or the other favorite techniques of the oligopolist.

THE ORGANIZATION OF PETROLEUM EXPORTING COUNTRIES (OPEC)

History

Although events in the international marketplace seemed quite remote from the American consumer during the first half of the century, for the next thirty years they were very important indeed. The most important event was the shift of the United States from net exporter to net importer of petroleum, occurring in the 1940s. By

1970 the United States was importing almost 10 quads of oil per year, or one-third of its total oil consumption. This in itself was not a problem but simply a response to the relative cheapness of imports, especially those from the Middle East where production costs of less than $1 per barrel were common.

The continuing discoveries of cheaper oil, however, were creating two destabilizing trends. In the first place, the United States was becoming ever more dependent on imports; and in the second, the continuing downward pressure on prices as new reserves were discovered and developed was a constant source of irritation to exporting countries, where oil was expected to finance economic growth as well as imports of expensive manufactured goods.

Although OPEC was formed in 1960, it had no noticeable power until 1970. The change during the decade was not really an abrupt reversal but merely a continuation of existing trends. In particular, the continuing rapid growth in consumption of petroleum worldwide and the feeling that other major producers, especially the United States, had begun their final decline combined to put the Middle Eastern members of OPEC in a very strong position.[9]

In retrospect it is easy to say that much of the apparent power of OPEC stemmed from a concatenation of unique circumstances. Energy prices in real terms had long been declining; although it was unreasonable to expect this to continue, many analysts did. Unreasonable or not, the assumption of cheap energy was built into all kinds of hardware including fuel-hungry automobiles, sprawling buildings, and industrial processes that substituted heat and power for labor, capital, and materials.

Meanwhile, the long downtrend in coal prices was ending as the traditional surplus of coal miners disappeared, wage increases accelerated, and the new safety requirements and environmental restrictions ended the rapid improvements in productivity that had offset higher wage rates since World War II.

The natural gas industry was facing as big a change for different reasons. By controlling the wellhead price of natural gas at low levels, the Federal Power Commission had eliminated the incentive for anyone to look for more gas and had increased the quantity of gas demanded by firms and households. By 1970 the era of cheap and plentiful natural gas was drawing to a close, as we discuss in the next chapter.

The electric power industry was also moving into a period of crisis. With natural gas becoming scarce at the controlled price, utilities could not count on burning that. Coal-fired plants aroused opposition because of increasingly stringent environmental standards. The nuclear alternative that had seemed so promising during the previous decade was encountering increasingly serious opposition from a vocal fraction of the public and, even more worrisome, began to seem increasingly expensive and unreliable.

This thumbnail sketch has stressed conditions in the United States, but it is essential to realize that the same basic picture of a rapidly growing demand for energy that focused increasingly on oil as a preferred source characterized most of the world. It was the perfect time for sellers to combine.

Theory of Cartels

As OPEC achieved its initial successes in the early 1970s, economists turned to their models of the behavior of cartels, and other commentators tried to supplement such economic models with political or other explanations. Let us therefore first consider the assumptions on which our analysis of cartels is based. The basic economic postulate is that each individual tries to maximize his own utility. Since utility cannot be measured, however, the analyst usually works with some imperfect substitute. It is convenient to use the assumption that individuals try to maximize income or wealth, although we know that is an inadequate description of the complex motives of most people.

The distinctive feature of the economic postulate is that it refers to the goals and behavior of individuals, not of firms, cartels, nations, or groups.[10] This focus on the goals of individuals is the key to unraveling the complexities of political behavior, which is often assumed to be something more mysterious and less rational than economic behavior. In this study I maintain the view that individuals will pursue their own interests within a group or bureaucracy or throughout political negotiations. Accordingly, we have to be cautious in ascribing goals to a firm or government or nation or especially to OPEC, since the relevant goals are those of the people who happen to be making the decisions. If the goals of individual decision makers are not identical with the objectives of the organization, the actions of the organization are likely to be inconsistent, reflecting the shifting alliances and compromises of the individuals involved.

When OPEC began to have an apparent impact during the early 1970s, it was natural for economists to explain the phenomenon with the aid of the simplest model of pricing under competition and monopoly. This is shown in Figure 9.4, which indicates a competitive market equilibrium price at *OC*, where the supply curve for the industry intersects the demand curve for the industry. If the competitive producers can bring themselves to act as a unified whole (that is, as a monopolist), then they could cut back output from *OA* to *OB* where marginal cost equals marginal revenue, raise the price to *OD*, and enjoy vastly increased profits. This analysis was developed to describe the pricing situation for an ordinary manufactured item where the competing firms are bought up and put under single ownership and management. It has numerous weaknesses when applied to the complex oil industry, but may nevertheless have been useful in describing the behavior of OPEC as it first began to use its power and grope toward the profit-maximizing price without knowing the exact shape of the demand curve.

The fundamental problem with using the simple model of monopoly is that OPEC is a negotiating committee of countries in which about half of the world's oil was produced, not a unitary decision-making monopolist that can easily choose its rate of output. Under the peculiar circumstances of the early 1970s, especially with the demand for oil growing rapidly and the extreme short-run inelasticity of demand for oil attributable to its very low price, it was possible for OPEC to raise prices substantially without forcing any producer to accept a cut in production. The OPEC ministers were also able to come up with a formula that transferred the monopoly profits and economic rents to the governments, rather than leaving them in the hands of the crude oil producers.

Once the price of crude oil was increased from the vicinity of $1.40 per barrel in 1970 to more than $8 per barrel in 1974, many economists began forecasting the breakup of the cartel. The reasoning stemmed directly from the nature of a cartel, as well as ample experience with manufacturing cartels. The basic point is that a

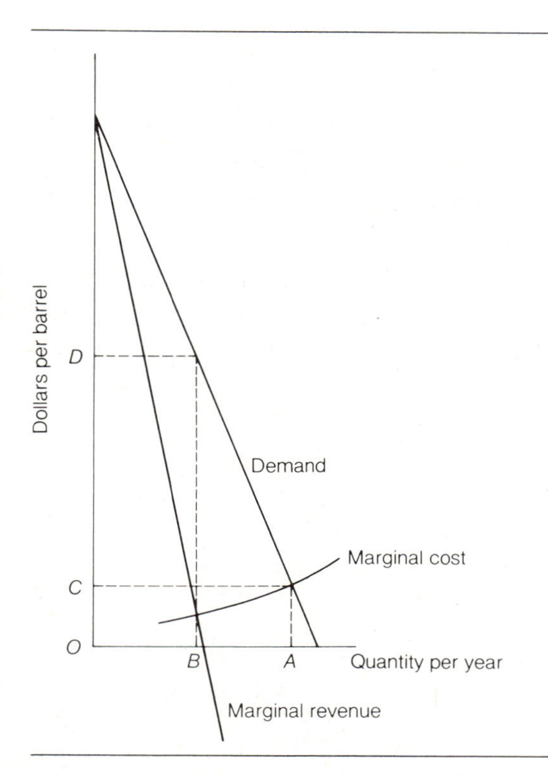

Figure 9.4 *Competitive and Monopolistic Pricing*

cartel is not a single seller but rather a collection of producers. Once the group acts together to raise the price by restricting output, conditions are ideal for each cartel member to "cheat" by shading the established price just enough so that it can sell all of its potential output. Since each participant has the same incentive as long as the revenue it obtains for additional output exceeds the cost of producing it, the cartel will soon cease to control output or price and the situation will revert to something not much different from competition or perhaps oligopoly. Why did the cartel not suffer this fate in the latter half of the 1970s?

The four main economic factors that made OPEC different from a standard manufacturing cartel were the natural resource base, the limited absorptive capacity of some members, the exogenous reductions in production, and the incredible stupidity of United States policy.[11]

The fact that oil reserves are limited can change the economic behavior of some producing countries. The cost of the oil that is sold today is not just the production cost of that oil but also the present discounted value of the economic rent that will not be received when the reserves of oil are gone. If reserves are adequate for a century or more, this amount is negligible. If reserves will be gone within the decade, this can be an important consideration. A manufacturing industry does not usually face such a problem, since it is ordinarily purchasing standard materials, rather than exhausting a finite resource.

Not only does the threat of exhaustion predispose nations to restrict output more than instantaneous profit maximization would suggest but it also sets up strains within the cartel, since different nations are differently endowed with resources. In particular, Saudi Arabia has resources that are unlimited in an economic sense, whereas Iran and Nigeria have to take account of the possibility of exhaustion. Accordingly, we would expect the latter countries to press for an extremely high price to earn as much as possible now and to slow the rate of consumption of the limited resource. Countries with essentially unlimited reserves will prefer to keep prices low enough to avoid encouraging the development of substitutes such as solar energy or oil shale.[12]

The factor that complicates this analysis is that decisions are not made by "Iran" or by "Saudi Arabia," but rather by individual people negotiating, discussing, maneuvering, and perhaps voting. It is not at all clear that the rulers of a country with oil reserves that will last for a century will have that as a time horizon. They may sense that events will force them out of office within a decade. As in the case of a private firm that expects its reserves to be expropriated by a host government, uncertainty of tenure (imperfection in property rights) hastens the optimum rate of extraction. Not only does Iran have limited reserves by standards of the Middle East but it also seemed to some observers that the Shah was in a great hurry to extract them. In some quarters it was also alleged that some of the revenue found its way into family accounts in Europe and the United States. If indeed such things occurred, could they be considered a sign of irrationality or poor planning, in view of the reluctance of former friends and allies to give anything to the Shah after he was displaced by the revolution?

Absorptive Capacity

This entire question blends into the issue of absorptive capacity. Most economies have to struggle to obtain the savings necessary for investment and growth. The countries with vast oil reserves, small population, and rudimentary economies, in contrast, were faced with the difficult question of what to do with the torrent of money pouring in.

In the simplest economic theory, the idea of absorptive capacity can be illustrated by a marginal efficiency of investment (MEI) curve (Figure 9.5). Each point on the MEI curve shows the highest rate of return (on the vertical axis) that the firm can obtain on the last dollar invested (shown on the horizontal axis). This slopes downward because all of the investment possibilities were supposed to be ranked that way—the best first, then proceeding to progressively less remunerative projects. Even for a tiny firm in a large economy, this is downward sloping because the firm would lack the managerial capacity to do many projects well at one time.

Figure 9.5 *Possibilities for Investing Oil Revenues*

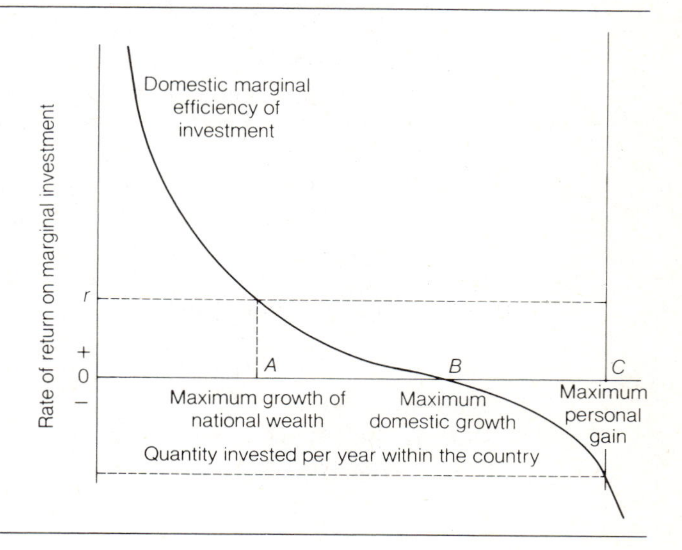

The rational firm will stop investing when the best return it can get on its own projects is just equal to the financial rate, that is, the rate at which it must borrow funds or the rate it can obtain by lending out its surplus funds.

For the entire economy the picture is just about the same, except that the amount of investing that can be done effectively is more rigidly constrained. In addition to the shortage of managerial talent that limits growth by one firm, all forms of labor may be scarce, materials may be unavailable, transportation facilities can be overburdened. Many scarce factors including labor can be imported, but often only at an increasingly exorbitant cost. Congestion in port facilities can prevent the import even of materials and equipment that routinely move in international trade.

Most of these constraints refer to the rate at which investment can occur. Projects that are too expensive this year can be carried out profitably next year, and many of the constraints on rapid investment will be eased from year to year as managers are trained, transportation facilities are built, and inventories of spare parts are accumulated. Nevertheless, the picture in Figure 9.5 still applies. As the rate of investment increases, the marginal return is driven down and finally becomes less than the rate obtainable by investing oil revenues in international financial markets. Beyond this point the rate of return goes to zero and then becomes negative.

What is the best policy for the ruler of a country with huge oil revenues? If the curve is properly drawn, the proper procedure for maximizing the growth of the nation's economic wealth is to invest in physical capital such as factories and roads within the country until the MEI equals the rate of return available in world markets. Any additional oil revenues can be invested in other countries either directly (buy office buildings in London or factories in Texas) or in financial instruments. Here the return on U.S. Treasury bills, r, symbolizes the best alternative investment outside the country. Oil should be sold as long as the marginal rate of return from the proceeds is greater than the expected return from leaving oil in the ground.

Whereas the total domestic plus foreign wealth of the country is maximized at Point A, the country's internal wealth can be increased by investing up to Point B. As long as any added investment yields a positive rate of return, it will make the wealth of the country grow. The serious question is whether the country should view investments in other countries as being equivalent to domestic ones, in which case A is better, or whether the growth of the domestic economy is to be increased, even at some cost in total wealth. Individual firms often are faced with equivalent choices.

The ruler might want to depart from either wealth-maximizing rate to improve his own welfare. On the one hand, he might consider a slower rate of economic growth (and hence of social change) to be more conducive to the retention of national traditions including his own rule over the country. Hence he would want to decrease the rate of investment within the country below the apparent optimum. It might be politically difficult to do this by investing more abroad, however. The popular appeal of leaving the oil in the ground to benefit future generations is greater than that of pumping it out and selling it to buy U.S. Treasury bills.

Of course, the ruler may have a conflicting motive. It may be easier for him to arrange the transfer of large amounts of money into his pockets and those of his family under the chaotic conditions of excessively rapid development. The MEI schedule, moreover, may be totally irrelevant for a nation where government is taking a substantial role in the economy since the criterion for investment is then political, rather than economic, benefit. The most realistic approach, although slightly cynical, may be to think of the ruler as desiring to maximize the foreign wealth of his family. This can be increased by increasing the rate of expenditure on imports, construction, and government purchases (those expenditures from which the family can most easily skim a portion) and by increasing his duration in office.

Unfortunately for the ruler, he does not know how to arrange for rapid development without shortening his tenure in office. This implies that the ruler should fire his staff of economic planners (perhaps transferring one economist to his Swiss office to assist in managing his personal portfolio) and look for a competent sociologist and political scientist to help him retain power. In any event, the socially disruptive effects of rapid growth have probably been partly responsible for decreases in oil output in countries such as Iran and Libya.

Vulnerability of International Trade to Disruption

Reluctance of producing countries to expand exports is one kind of problem, but interruption of established patterns of exports is a much more dramatic one. In addition to the embargo of 1973 that heralded the new power of OPEC, two such disruptions helped to shore up the cartel. The first of these was the revolution in Iran, which eliminated that country from the ranks of the major exporters in 1979. The second was the war between Iraq and Iran, which reduced the exports of the former country as well in 1980.

One measure of the vulnerability of the world to such disruptions of output is the concentration of supply in a few locations. As Table 9.1 indicates, the dependence of the world on Middle Eastern and OPEC oil decreased during the decade following the embargo of

1973. The Arab members of OPEC, which had supplied 32.3 percent of the world's oil in 1973, were below 20 percent by 1983. Over the same period, however, the U.S.S.R. had enhanced its position, moving from 15.2 percent to 22.7 percent of world production. It is not entirely obvious that the vulnerability of the world to disruptions of supply had been reduced at all, although the risks had been spread away from the Persian Gulf. The rise of production in the United Kingdom, Mexico, and the miscellaneous "other" countries helped limit OPEC power.

Although many of the OPEC countries reduced production to comply with OPEC policy, the decline of Iran and Iraq as major exporters was directly related to revolution and war. Significantly, none of the reductions in output was forced by a lack of reserves.

Table 9.1 *Principal Oil Producers, 1973 and 1983*

	Production (thousands of barrels per day)		Share of world output (percent)	
	1973	1983	1973	1983
Algeria	1,097	675	2.0	1.3
Iraq	2,018	1,005	3.6	1.9
Kuwait	3,020	1,064	5.4	2.0
Libya	2,175	1,076	3.9	2.0
Qatar	570	295	1.0	0.6
Saudi Arabia	7,596	5,086	13.6	9.6
United Arab Emirates	1,533	1,147	2.8	2.2
Total Arab OPEC	(18,009)	(10,348)	(32.3)	(19.5)
Indonesia	1,339	1,385	2.4	2.6
Iran	5,861	2,426	10.5	4.6
Nigeria	2,054	1,241	3.7	2.3
Venezuela	3,366	1,768	6.0	3.3
Total OPEC[a]	(30,989)	(17,562)	(55.7)	(33.1)
Canada	1,800	1,450	3.2	2.7
Mexico	465	2,686	0.8	5.1
United Kingdom	2	2,291	0.0	4.3
United States	9,208	8,688	16.5	16.4
China	1,090	2,120	2.0	4.0
U.S.S.R.	8,465	12,034	15.2	22.7
Other	3,655	6,150	6.6	11.6
World	(55,674)	(52,981)	(100)	(100)

Source: Monthly Energy Review *(September 1984), pp. 108–109.*

[a] *Includes Ecuador and Gabon.*

It seems evident that in the absence of the reductions forced by revolution and war, it would have been difficult for the cartel to hold together. Saudi Arabia, of course, could reduce output enough to maintain current price levels for some time to come, but it is by no means clear that it is in the interest of the Saudi royal family to do so. The potential for painless reductions by other members of OPEC has been largely achieved, so the future of OPEC will be determined by the behavior of Saudi Arabia, which increased its share of total OPEC production from 24 percent in 1973 to 29 percent in 1983.

The Future of OPEC

What does all this imply for the future of OPEC and oil prices? The literature of the 1970s is littered with forecasts that look ridiculous from the vantage point of today. The item that is most difficult to predict is the stability of Saudi Arabia as a producer. After the experience in Iran, it would certainly be rash to predict that a successor to the current Saudi regime would choose to sell about as much oil. If the Middle East does remain reasonably stable, then prices should not rise much above the current level except to keep pace with increases in the prices of such substitutes as coal and nuclear electric power or possibly oil from shale. Now that oil has become the highest-priced fuel, purchased mainly for its special advantages in mobile uses and in stationary applications that have not yet been converted, the elasticity of demand is far higher than it was in the days when it was still a bargain even when prices doubled. The long-run cohesion of OPEC is by no means a certainty if the price of crude remains stable or begins to decline, because several producers that are currently exporting very little, notably Iran, Iraq, Kuwait, Saudi Arabia, and the United Arab Emirates, have huge reserves that they may want to exploit if it becomes clear that their value in the ground is no longer increasing.

U.S. POLICY

The other feature of the 1970s that preserved the strength of OPEC was the sheer stupidity of policy in the United States. This is not to imply that the policy was made by stupid people, but only that the net result of the compromises and political opportunism was a policy that did more to advance the interests of OPEC than of the consuming nations. The main themes of energy policy during the 1970s were (1) to prevent the substitution of other fuels for oil, (2) to keep prices low to consumers so they would not conserve oil, (3) to inhibit domestic production and refining, and (4) to prevent domestic producers of oil and gas from making large profits. These were not necessarily the expressed goals of policy, but if we look at the legislation and regulations, these would seem to be the directions in which they would push the economy. By the end of the decade, many of the worst aspects of government behavior had been alleviated so that prices to the consumers of energy had begun to reflect free market transactions. Application of the windfall profits tax to oil and the continuation of some price controls on natural gas had dulled the economic incentives somewhat, but the economy was finally being given signals to move in the right direction and was beginning to respond.

Although most of the history of energy prices during the 1970s was written in the Persian Gulf, the role of the United States government was substantial and may become more so. The most disastrous policies were attempting to keep prices down and hobbling the substitutes for oil, but the federal government also controls much of the land still suitable for oil drilling in the United States. In leasing policy, as in other aspects of energy policy, it has been difficult for the federal government to come to grips with the question of what it is trying to achieve. In particular, the policies may be oriented, as are those of OPEC, toward extracting the maximum amount of economic rent for the government. Another possibility is to try to speed development as much as possible. In practice, the U.S. government's motives have included a mixture of these plus numerous stringent environmental constraints.

Leasing Policy

These compromises first arise in connection with the amount and location of the federal lands that will be made available to the oil companies. It is here that the environmental constraints have been most binding in slowing the release of land on the outer continental shelf (OCS) and in locking up as wilderness large parts of Alaska and promising areas of the Rockies.

The particular terms on which properties are leased is not as significant as whether they are leased at all; but because it is a nice problem in microeconomic theory, analysis of lease terms has received a lot of attention. The conventional approach is to make tracts available subject to a standard royalty of one-sixth of gross output for the government. The tracts are auctioned by competitive bidding to the company or group of companies offering the largest immediate sum as a bonus bid. The tracts are also subject to a small annual rental of a fixed amount. The main objection to this standard procedure is that the capital requirements are so high that small firms may be squeezed out of the market. Accordingly, various alternative leasing systems have been tried.

Experiments have dealt with possibilities such as the following: (a) higher royalty rates combined with conventional bonus bids, (b) fixed bonus payments combined with bidding on the royalty rate, (c) bonus bid plus fixed share of profits, (d) fixed bonus plus bidding on share of profits, or (e) payments per unit of time. The problem that arose with several of the variants is that by making the extra cost of exploitation so high and the initial acquisition price so low, they encouraged firms to buy up leases that they had no intention of using right away. As far as the main objection was concerned, it did not appear from an analysis of the experimental results that the changed lease arrangements made the industry more competitive. Firms that risk several million dollars on a dry hole are not poor enough or so averse to risk that they will let another few million dollars of "up front" money stand in their way.[13]

Economic Rents

Within the historical context of the industry, various lease arrangements can be considered as devices by which governments try to extract a larger share of the economic rent. When the government owns the land, of course, economic theory holds that as the landowner it can be expected to receive the entire economic rent, leaving the oil producer with only a normal return on its investment. The main practical difficulty with this is the enormous uncertainty about the productiveness of the field and the price of the oil at the time that royalty agreements are signed. The government can try to share net income (profit, in the accountant's sense), rather than gross as in the typical royalty arrangement. Yet until the profit is received, it is impossible to know what share of it would compensate the producer for its investment. Furthermore, once the government begins to take a substantial share of net income, it has to monitor the business very closely to make sure that the producing firm does not either (1) convert economic rents into the costs that make life more pleasant for the manager or (2) transfer them to a less heavily taxed time or stage of processing. The simplest such device is to sell the crude at cost to the refinery, reducing net income—and the government's take—to zero. As the government catches on to each scheme, more complex variations can be devised to stay ahead of the increasingly complex rules.

Governments also try to extract economic rent when they do not own the property. The most recent effort in the United States is the windfall profits tax. The fundamental situation that inspired the tax is depicted in Figure 9.6. Any domestic oil wells that were earning a reasonable profit in 1970 (with a domestic crude oil price of $3.18 per barrel) must have been doing very well by 1980 (when the import price of about $30 per barrel determined domestic prices). The large area labeled windfall profit in Figure 9.6 attracts the attention of governments. But devising a way to extract the windfall profit without damaging incentives is difficult. The technique enshrined in the windfall profits tax is to assume that the production cost of a well is revealed by the year it began production. Thus wells are assigned to classes according to when production began. Then the law permits wells of each class to deduct a specified assumed production cost for that class from the selling price of the oil. The difference is then taxed.

If the world were perfectly orderly and predictable and firms operated with perfect information, the tax would come close to working properly. The main problem is that wells of different cost are drilled simultaneously because no one knows costs exactly until production is under way. So the time when a well was drilled does not convey much information about cost. Furthermore, output and cost change over time. The law recognizes the extreme forms of change by providing special exemptions for wells requiring expensive treatment to enhance recovery. All such devices are very blunt, however, and tend to set up incentive structures that divert the effort of producers from finding and pumping more oil to devising ways to profit from the regulations.

REFINING

It should not be supposed that the only problems the industry faces are finding crude oil and contending with government regulations. The rapid changes of the 1970s have complicated some of the traditional questions about refineries as well. A refinery typically was designed to process one type of crude into a variety of different outputs. With the rapid changes in prices and sources of supply, however, flexibility in adapting to different inputs has been very

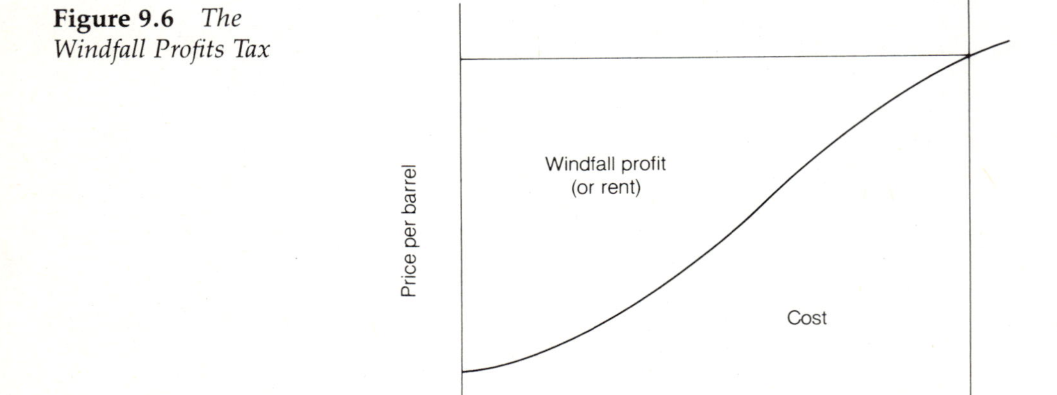

Figure 9.6 *The Windfall Profits Tax*

valuable. Much of the crude oil that has been relatively cheap and available during recent periods of shortage has been heavy oil of high sulfur content. The light, low-sulfur crudes have commanded a premium in the market, which has furnished an incentive to rework the refineries to process inferior inputs.

Similarly, at the output end the profit-maximizing mix may be changing as well (Figure 9.7). In particular, residual oil must compete against coal as a fuel for large industrial boilers and electric power generation. When crude oil was cheap and plentiful, residual oil could undersell coal (with allowance for the greater convenience and cleanliness of oil and the smaller boilers required) in many locations while still yielding a respectable price to refiners. Despite the restrictions on air pollution that have made the use of coal difficult, however, the situation is now changing. For example, in 1973 coal was delivered to electric power plants at an average cost of 40.5 cents per MMBTU. By 1981 this had nearly quadrupled to 156.5 cents. In the meantime, residual oil had gone from 78.8 cents per MMBTU to 502.4 cents. The price increased at a faster rate, and also the price differential between the two had widened enough to compensate more users for the greater handling cost of coal.

Not only was residual oil losing markets to coal because of its high relative price but it had also not been possible to raise the price of residual to cover the increases in crude oil prices. The residual fraction of output was being sold for less than the crude oil to make it. This did not imply disaster for refiners because residual had always been a low-valued by-product of producing the premium products such as gasoline and home heating oil. What is significant is the total profit on the products refined from the barrel of crude, not the margin on any one product. Nevertheless, with residual oil

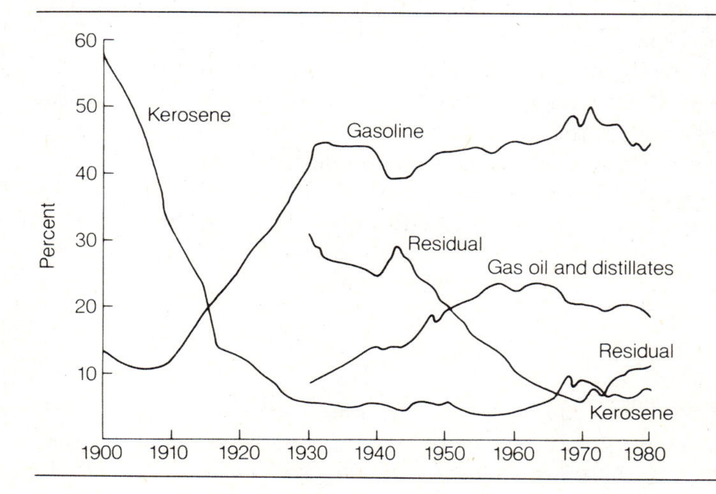

Figure 9.7 *U.S. Distribution of Refinery Output, 1900–1980*

Sources: U.S. Historical Statistics *(1975), p. 596;* U.S. Statistical Abstract *(1981), p. 735.*

selling for less than crude oil, the desirability of shifting the refining mix toward lighter products is reinforced.

One approach to this has been to use the residual oil as a very low-quality feedstock to substitute for additional crude. Residual oil requires a substantial amount of processing because it contains such contaminants as sulfur and heavy metals and requires the addition of hydrogen to produce the lighter products from the heavy base. The United States has traditionally imported residual oil because of the very high proportion of gasoline and distillate fuel oil produced here compared with the rest of the world.

Nevertheless, the trend for the future would certainly appear to be the substitution of other sources of energy for petroleum in stationary applications. Even residual oil is so expensive now relative to the other possibilities that announcements are commonplace of utilities and large plants converting to coal-fired boilers and of paper mills and sawmills reverting to the practice of burning wood waste—a practice abandoned earlier because of cost, inconvenience, and air pollution regulations. Similarly, home heating installations stress electricity, wood, passive solar design, or natural gas.

Even when oil has no cheaper substitute, the high price has prompted great effort and investment in conservation techniques. This leaves the mobile applications as the main stronghold of petroleum. In highway transportation, gasoline may give ground to diesel, but nothing else seems likely to capture much of the market soon. Airplanes will be petroleum powered for the foreseeable future. These all use light products. In the transportation sector only large ships have typically used residual or crude oil, but even there diesel engines are increasing fuel efficiency compared with steam turbines. For the very largest ships, of course, coal and nuclear power are technically feasible substitutes for oil.

The trend is clear, and indeed it is just a continuation of the long-term trend toward extracting more light products from crude. The changes are that gasoline may not account for 45 percent of the product in the future, the market for home heating oil will not expand, and the crudes that become available may be heavier and dirtier. This raises some difficult problems for the industry in devising the kind of flexibility in refineries to deal with changes in product mix and rapid changes in sources of crude oil (and hence its refining characteristics) as international events influence the oil industry.

THE FUTURE OF OIL

The tone of this analysis is that oil shortages are a thing of the past, not because oil has been discovered in huge quantities, but because the market has finally been set free. Prices have risen enough to lead consumers to substitute other fuels and conservation. If this is a good description of the world, what does it suggest for policy?

Any policy other than laissez-faire has two serious defects. In the first place, the political process is so caught up in questions of the distribution of income and wealth that it cannot come to grips with trivia like efficiency or even survival. The great political debate over energy policy in the United States during the 1970s was reminiscent of two ten-year-olds squabbling about sharing an ice cream cone while the ice cream melted onto the ground. This behavior is inevitable since the distributional issues are very obvious (Will producers be permitted to charge the consumers more?), whereas the implications for efficiency and welfare require some foresight or analysis to comprehend (Will consumers be able to buy enough if the price is held to a level that would have seemed very high two years ago?).

The second flaw in the political process is its slowness. The market begins to respond to impending shortages or disruptions before the general public is even aware that a problem may develop. The political process, in contrast, lags somewhat behind the newspaper headlines and often does not really come to grips with an issue until after it has been solved. Even after an issue is on the political agenda, the process of hearings, debate, and introduction of legislation repeated session after congressional session can consume years before yielding some compromise solution. The process can deal with some issues but is not effective in solving economic problems in a fast-changing environment.

The cumbersome nature of the political process guarantees that it cannot deal with many issues at one time. Waiting for a political solution to a politically complex issue such as energy pricing may well mean a decade of inaction. In the meantime, more pressing issues of national policy may be shoved aside by interminable debates on the windfall profits from increases in energy prices.

What is the appropriate policy for the United States in an era when OPEC shows signs of weakening? Some economists have suggested that the United States follow a hard-line policy of trying to break the cartel. One suggestion is that some United States government agency be given a monopoly over oil imports. The agency would then solicit secret bids from exporting countries and award the contracts to the lowest bidders. The idea is that the monopsony power of such an agency could offset the monopoly power of the cartel. Is it not equally likely, however, that having such an official government agency in the market would produce severe diplomatic strains forcing producing nations to decide explicitly whether they were cooperating with the United States or loyal to the leading powers in OPEC? The cartel would not be likely to crack as a result of secret bidding unless secret delivery could also be arranged. The advantage of the present arrangement is that it pits cartel members against each other, with the government of the United States as a bystander, rather than pitting the United States government against countries that are still struggling to demonstrate their independence from colonialism and Western paternalism.

A less extreme proposal is that the United States impose a tariff on imported oil. It might be made equal to some calculated external cost to the economy of the marginal barrel of imports. Estimates of external cost have ranged from negligible to about $100 per BBL, which would be a very extreme tariff indeed.[14] If the tariff were set at such a level, it would provoke political pressure for spreading the burden among all refiners, rather than concentrating it on those most dependent on foreign oil.

The notorious "entitlement" system of the 1970s did that, with the result that the pressure to diminish imports was substantially lessened. The entitlement system originated during the embargo by Arab producers on shipments of oil to the United States during the winter of 1973–74. Since import-dependent firms were faced with severe difficulties (which could translate into severe burdens for particular geographic areas served by the hard-pressed firms), the government imposed the requirement that the reductions in foreign supplies be shared equally among firms. This was achieved by the physical movement of oil from firm to firm during the embargo. After the embargo physical movement of oil ended, but since domestic crude was controlled at a cheaper price than imported, refiners of domestic would have enjoyed a great cost advantage over refiners of imported oil. To equalize costs any refiner intending to use domestic crude was forced to buy an entitlement to do so from refiners of foreign crude. The device satisfied some notion of equity by giving all refiners nearly equal crude costs. It also could be interpreted as a subsidy for the use of imported crude or a penalty on those refiners who were mindful enough of national security to develop domestic sources of supply. With the end of price controls on domestic crude oil, the entitlement program also lapsed, but any proposal to levy a high tariff on imported oil would bring about the same disparities among refiners that inspired the original entitlement program, unless the price of domestic oil were permitted to rise without limit.

A more modest tariff proposal on imported oil calls for the United States government to match every increase in OPEC price beyond a set amount by a tariff that is related to the price increase. If, for example, the base price is set at $30, the actual OPEC price is $35, and the matching rate is 1, then the United States would levy a tariff of $5 on all imports. This could even be made an excise tax on all oil to avoid putting certain refiners at a disadvantage. The net effect would be to raise the price to the consumer more than OPEC alone would. The consumer would have a greater incentive to curtail consumption and the elasticity of demand perceived by OPEC would increase. It is hard to imagine such a plan being politically acceptable.

Although predictions about the future are difficult, it is clear that higher energy prices of the 1970s, when permitted to have an impact on producers and consumers, resulted not only in a decline in the use of petroleum products but also in a vast surge in drilling. The data are shown in the next chapter. The number of active rotary drilling rigs increased from fewer than 1,200 in 1973 to more than 4,300 by the end of 1981. The number of wells completed grew from 26,000 in 1973 to 60,000 in 1980. That this spurt of drilling would pay off was doubted by numerous skeptics. Nevertheless, by 1980 the decline in oil and gas reserves in the United States had been arrested and it appeared as though reserves might even increase despite the higher production rates as the United States moved closer to self-sufficiency in energy.

The new reserves are, on the average, more expensive, which is hardly surprising in view of the very definition of reserves as resources that are known to exist and can be exploited at prevailing prices. As market prices have risen, the boundary line between what is worth exploiting and what is submarginal has also shifted. In addition to the purely definitional change, however, the vast increase in drilling activity has included searches in areas that were not considered particularly promising in the past.

It would be a mistake to think that exploration and drilling are proceeding in remote and difficult areas only because prices have risen. In fact, the costs of locating and producing oil in such places also have fallen as a result of continuing improvements in geological knowledge, techniques for obtaining information about what is underground, and techniques for drilling. In particular, although the cost per foot of depth in a well still increases rapidly as depths exceed 10,000 feet, wells are routinely being drilled to depths approaching 20,000 feet, and greater depths are not unheard of. Drilling in the North Sea and on the Outer Continental Shelf has required a series of advances in materials and design.

The fear has been expressed that production of oil might soon require more energy than it produces.[15] In a well-functioning market economy, the fear would be irrelevant. Firms would not start projects unless they guessed that the proceeds would cover all costs and leave a profit. If all energy sources cost the same amount per BTU, that would be enough to guarantee that, on average, successful firms would not embark on energy-losing projects. Because the value of the energy output must exceed the value of all direct and indirect energy inputs, under these assumed conditions the net contribution of any well to the energy supply would be positive.

In the real world, however, different forms of energy are not necessarily equal. In particular, the easy transportability of liquid fuels is worth a premium in many uses. We could imagine a situation in which coal containing 100 BTU of energy is used to make steam and electricity to extract oil containing 50 BTU of energy. Consumers might still be willing to buy gasoline, even if it were selling at a price per BTU several times that of coal (as it is today). Such an outcome would be neither surprising nor harmful to consumers or the economy.

More serious is the possibility that government regulation could distort economic activity into socially undesirable channels. The general formula by which the United States has regulated energy prices is to make the price to consumers equal to the average cost of production. This is common in the regulation of electric rates and natural gas prices and was a criterion for the control of domestic crude oil prices during the 1970s. At the same time, producers are generally allowed to recapture the cost of new facilities; that is, they receive the marginal cost (or sometimes even more than that) as an incentive to increase supply. One result is that the energy used to produce new energy is priced much lower than the new energy that is produced. As a result, the fear that energy production could absorb more energy than it provides must be taken seriously whenever prices are controlled.

NOTES

1. For a good discussion of exploration techniques, see *Our Industry Petroleum*, 5th ed. (London: British Petroleum Company Limited, 1977), chap. 5. The Seventy-fifth Anniversary issue of *Oil and Gas Journal* (August 1977) provides a good survey of technology.
2. *Our Industry Petroleum*, chap. 6.
3. John Lawrence Enos, *Petroleum Progress and Profits: A History of Process Innovation* (Cambridge, Mass.: MIT Press, 1962).
4. John L. Enos, "Invention and Innovation in the Petroleum Refining Industry," in *The Rate and Direction of Inventive Activity*, NBER (Princeton, N.J.: Princeton University Press, 1962), pp. 299–321.
5. *Oil and Gas Journal*, March 24, 1980, p. 77; and March 26, 1984, p. 73.
6. Stephen L. McDonald, *Petroleum Conservation in the United States: An Economic Analysis* (Baltimore, Md.: Johns Hopkins University Press/Resources for the Future, 1971).
7. Under the tax law that prevailed during much of the twentieth century, oil producers could deduct 27.5 percent of gross income (from sale of oil at the wellhead) from net income in calculating taxable income. (Net income could not be reduced by more than 50 percent, however.) The applicability of the provision was narrowed substantially during the 1970s, so now it is a special tax break for small producers.

8. For a detailed discussion, see Thomas D. Duchesneau, *Competition in the U.S. Energy Industry* (Cambridge, Mass.: Ballinger, 1975). A more recent introduction to the structure of the industry is Walter S. Measday, "The Petroleum Industry," in *The Structure of American Industry*, 6th ed., ed. Walter Adams (New York: Macmillan, 1982).
9. For detailed background on the origin of OPEC and its success in 1973, see Raymond Vernon, ed., *The Oil Crisis* (New York: W. W. Norton, 1976). The most accessible guide to the economic literature dealing with OPEC is Dermot Gately, "A Ten-Year Retrospective: OPEC and the World Oil Market," *Journal of Economic Literature* 22 (September 1984), pp. 1100–1114.
10. The literature of "public choice" explores the consequences for organizations of individuals' pursuit of their own goals. See, for example, Mancur Olson, *The Logic of Collective Action* (Cambridge, Mass.: Harvard University Press, 1965) and Gordon Tullock, *The Politics of Bureaucracy* (Washington, D.C.: Public Affairs Press, 1965).
11. For a detailed analysis of an important aspect of U.S. policy, see Joseph P. Kalt, *The Economics and Politics of Oil Price Regulation: Federal Policy in the Post-Embargo Era* (Cambridge, Mass.: MIT Press, 1981).
12. For a discussion of behavior rational for the country, see Harry D. Saunders, "Optimal Oil Producer Behavior Considering Macrofeedbacks," *Energy Journal* 4, no. 4 (October 1983), pp. 1–27.
13. See William R. Moffat, "Federal Energy Proprietorship: Leasing and Its Critics," in *Options for U.S. Energy Policy* (San Francisco: Institute for Contemporary Studies, 1977); and "Impact of Making the Onshore Oil and Gas Leasing System More Competitive," U.S. General Accounting Office, EMD-80-60, March 14, 1980.
14. Robert Stobaugh and Daniel Yergin, *Energy Future* (New York: Ballantine Books, 1979), p. 293 provide the estimate of $100/barrel. For a rigorous analysis with numerous references, see Elena Folkerts-Landau, "The Social Cost of Imported Oil," *Energy Journal* 5, no. 3 (July 1984), pp. 41–58.
15. Charles A. S. Hall and Cutler J. Cleveland, "Petroleum Drilling and Production in the United States: Yield per Effort and Net Energy Analysis," *Science*, February 6, 1981, pp. 576–579.

CHAPTER TEN

The Natural Gas Industry

INTRODUCTION

Consumption of natural gas is second only to that of oil among energy sources in the United States, accounting for about 20 quads of the total consumption of 75 quads per year (Figure 10.1). Since relatively little is imported, natural gas provided a slightly larger share of total domestic energy production than did oil in 1981, although oil, gas, and coal were each in the vicinity of 20 quads. As a clean and convenient source of energy for many stationary applications, natural gas could play an even bigger role in the future if lingering doubts about its availability and cost could be stilled.

In addition to the quantitative importance of natural gas, the industry is of considerable economic interest because its three segments–production, transmission, and distribution–are so different. The production segment must be analyzed jointly with oil because the exploration and extraction techniques for oil and gas are essentially identical and indeed gas is often a product of the search for oil. Transmission, the transportation of natural gas, nearly always involves pipelines, which are characterized by their own peculiar economics. At the distribution end, the natural gas industry is a regulated public utility that has some features in common with the local distribution of electric power.

The overriding fact that has colored the development of the industry is that natural gas is difficult or expensive to transport by any means other than pipeline. One million BTU of natural gas occupies a space of 1,000 cubic feet (1 mcf) at room temperature and atmospheric pressure. About 7½ gallons of gasoline has the same energy content. Alternatively, 1 mcf of stacked, dry firewood contains about 150 million BTU. The gas can be compressed, but this imposes requirements for strong, leakproof containers. Not until the 1970s did the movement of liquified natural gas (LNG) by large oceangoing ships become common enough so that technical feasibility could be assumed, and it is still an expensive proposition.

In the absence of long-distance pipelines, gas at the wellhead has very low value. In remote parts of the world even today, much

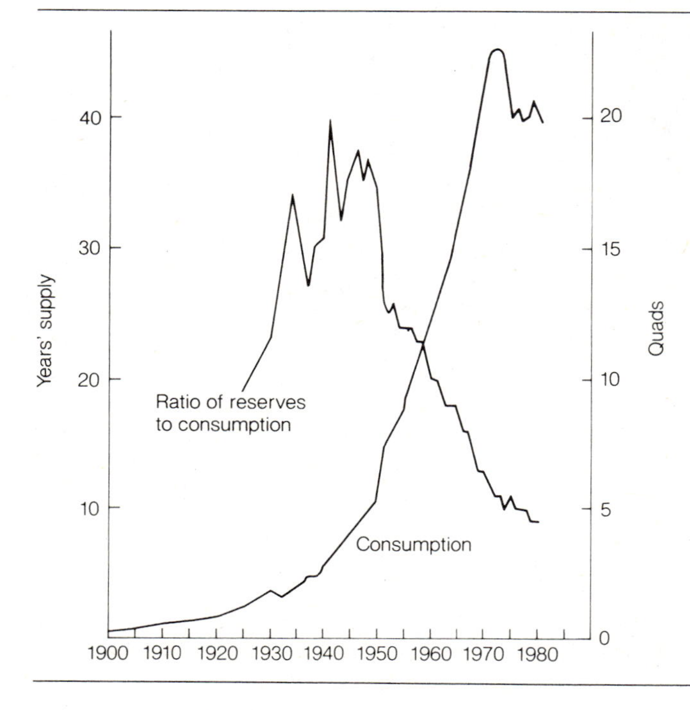

Figure 10.1 *U.S. Natural Gas Consumption and Ratio of Reserves to Consumption, 1900–1983*

Sources: U.S. Historical Statistics *(1975), p. 595; U.S. Department of Energy,* Annual Energy Review *(1982), pp. 43, 105.*

gas is flared (burned as waste) at the well because no transportation link to a consuming center has been built. The impact of transportation cost on the value of the mineral at the mine is a common element in the theory of the mine, but the effect is strongest in the case of natural gas because of the absence of alternatives to pipelines. Crude oil too is generally moved by pipeline because it is the cheapest mode (except for ocean routes over which very large tankers are cheaper), but other techniques are readily available.

When natural gas has little value at the wellhead, no one expends great resources in looking for it. Often gas is the sole product of wells that are drilled to obtain oil, however, and often it is a joint product of wells producing oil. In 1981 about one-fifth of the gas produced in the United States came from oil wells. If the gas is available when pipelines are not, it can be flared off, reinjected into the oil formation to maintain pressure, used to power oil field equipment, or sold on the local market. The oil-producing areas of the Southwest developed industries using cheap natural gas for such purposes as petrochemical feedstock or to produce carbon black or as an input for steel production. Natural gas was also a cheap and convenient fuel for generating electrical power, so industries locating near cheaper sources of electricity were being attracted indirectly to the natural gas. This same process is under way in the oil-exporting countries now as they try to find uses for the natural gas associated with oil production.

The first gas well in the United States was drilled in 1825, long before the first oil well. It was connected by pipeline to users in Fredonia, New York. By 1886 an 8-inch gas line had been run eighty-seven miles from Kane, Pennsylvania, to Buffalo, New York. The gas utilities that developed during the nineteenth century to supply lighting for streets and homes had to rely either on such nearby sources of natural gas or on manufactured gas.[1] Cleveland, for example, was close enough to the producing areas in eastern Ohio and West Virginia to obtain natural gas, but the cities along the East Coast had to rely on gas manufactured from coal and other fuels. The manufactured product had a high carbon monoxide content, which made it something less than ideal for a household fuel. The low BTU value also meant that transmission was expensive relative to heat content, and this confined gas utilities to densely populated areas.

In the 1930s the techniques available for building high-pressure, long-distance transmission lines had improved enough so that it was possible to think about connecting the population centers of the Middle West and East with the producing areas of the Southwest. The real boom in construction of gas transmission lines occurred in the fifteen years after World War II. Concurrently, the consumption of manufactured gas declined from 260 trillion BTU in 1945 to 23 trillion BTU in 1965.[2]

Pipelines

With the development of the long-distance transmission lines, conditions began to change. The difference between retail prices in producing and consuming areas narrowed, as always happens when transportation is cheapened. The dependence on pipelines, however, imposes a variety of constraints. First of all, a pipeline is a fixed link between stationary producers and stationary consumers. That means it is not worth building unless the gas is expected to be available for many years and the customers are expected to keep buying it.

Pipelines, moreover, are natural monopolies at the gathering and distribution ends and often for long-distance transmission as well. This brings all of the usual problems of economic regulation. In a sense, the monopoly of a gas utility is not as complete as that of the phone company or the electric company, because homes can be heated with other fuels and electricity can substitute for gas in all household uses. Nevertheless, the homeowner installs equipment that will last ten or twenty years or more and hence will probably want the assurance of regulation. Gas distributors probably feel themselves at the mercy of their customers as much as customers do of them, because the pipes in the ground tie distributors to a fixed group of customers, just as they restrict the customers to one source of supply. In any event, regulation at the retail level is a long-standing tradition, the implications of which must be examined.

Fixed Costs

Not only are pipelines embedded in particular locations but they also inherently involve very heavy fixed costs and low operating costs. Heavy fixed costs always impose pressure for uniform operating rates. If output rates are cut in half and fixed costs comprise all of total costs, then total cost per unit of output doubles. Conversely, firms with heavy fixed costs will be reluctant to invest in capacity that is used only sporadically. This poses a particular problem in normal times for the gas industry because sales to households, dominated as they are by home heating, are strongly concentrated in the winter. At the same time, gas because of its nature is expensive to store. In addition to this normal seasonal fluctuation, the industry now has to contend with instabilities in the quantity demanded stemming both from changes in price and from changes in government regulations.

Seasonal Demand

The traditional means for dealing with seasonal fluctuations were storage, special rates, and promotion of off-season uses. Storage near point of consumption is sometimes arranged in old gas wells or underground caverns. This adds a cost in the vicinity of $1 per thousand cubic feet in areas where the appropriate geological features are present. The pipeline itself provides some storage capacity because the pressure can be built up during the slack months and drawn down during the coldest weather.

Special rates and other promotion of off-peak uses pose some more interesting questions. Traditionally, the gas distributor guaranteed delivery to residential customers but charged a relatively high price. Industrial users were able to buy gas very cheaply if they agreed to an interruptible service. If supplies grew tight in cold weather, the gas company could stop delivery to its interruptible users, who could then shut down or switch to more expensive oil or less convenient coal. This is a very reasonable approach when the only problem is the seasonal nature of the premium market combined with steady delivery rates. If, however, we are concerned also with the long-run adequacy of supplies, as some critics are, then the wisdom of the low-priced sales can be questioned. The Natural Gas Policy Act of 1978 (NGPA) in fact moves toward the position that gas should not be used as a boiler fuel by large users. They after all are capable of using coal, which is plentiful, and of installing the equipment to burn it cleanly.

This takes the analysis back to the high fixed costs of pipelines and the difficulty of storage. If the total flow through a pipeline is decreased by the elimination of load-balancing industrial consumption, then fixed cost per unit rises and rates to residential consumers must be increased. This will inspire consumers to conserve more, with the result that overhead costs per unit and prices rise even higher. In a growing market such responses by individual consumers are not troublesome, but overall consumption of natural gas has stopped growing in recent years, as is shown in Figure 10.1. Individual users of gas have decreased their consumption as prices have risen, and this has offset the new business generated by the formation of new households and by conversions from oil.

Economies of Scale

Not only do pipelines have high fixed costs, which predispose toward high output rates once they are built, but they also have significant economies of scale when they are still in the planning stage. This is indicated in Figure 10.2, which relates pipeline costs to throughput (amount put through) for an oil pipeline. Bigger pipelines are more expensive to build than are small ones, but the cost does not increase as rapidly as does the capacity. If the market is growing, it is better to build some excess capacity when the first pipe is laid, rather than to add an additional pipeline soon.

Capacity of an existing pipeline can be increased by adding more pumping stations. This adds some flexibility to the design because a pipeline can be built larger than necessary without incurring very much extra capital cost. Capacity can then be raised in stages by adding compressor stations.[3]

One unfortunate implication of the economies of scale in pipelines is that there is no uniquely correct way of allocating the costs among consumers; that is, a pipeline from Texas to Chicago and going on to Cleveland will supply cheaper gas for both Chicago and Cleveland than will separate lines to each city. The savings could very well be used up in arguments over who should receive them in the absence of some rules of thumb. Natural gas pipelines have been regulated since 1938, however, and the one thing that regulators are good at is developing the requisite rules of thumb.

RESERVES OF NATURAL GAS

"Seek and ye shall find" is not a perfect description of minerals exploration: If it is not there, you will not find it no matter how hard you look. It is more nearly accurate to say, "If you do not seek it, you will not find much." When only a small amount of natural gas could be sold, the amount that was a by-product of oil exploration was sufficient. That has ceased to be so for several decades. Figure 10.3 shows the upsurge in recent drilling activity that has finally ended the decline in reserves, as shown in Figure 10.4.

The increased drilling has added to reserves, but much attention during recent years has been devoted to an apparent deterioration of the relationship between wells drilled and reserves discovered. Of course, if a producer has already counted the reserves

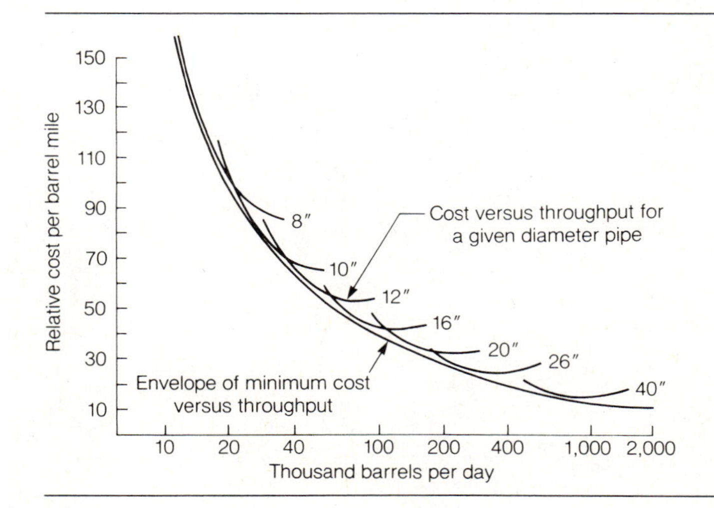

Figure 10.2 *Relative Pipeline Costs versus System Throughput (based on 1975 data)*

Source: *Exxon Company, U.S.A.,* Competition in the Petroleum Industry *(1975), p. 30. Reprinted by permission.*

that are underground and just drills a well to develop them, additional discoveries would not be expected. It has been suggested that the drilling of the 1970s was directed largely at developing known reserves to satisfy the shortages that existed or impended.

Just as in the case of oil, the gas-pricing regulations attempted to keep the price of old gas down while permitting the prices to be higher for new wells to encourage more drilling. The flaw in this approach is that it creates an incentive for producers to drill new wells to extract gas that would still come from an old well. The regulations attempted to deal with this by specifying a minimum distance from an existing well before one newly drilled could be counted as new. Nevertheless, we can assume that some drilling activity is misdirected by any such regulation.

In any event, by 1980 the drilling activity was beginning to locate large quantities of reserves. The long-term decline in reserves was arrested and, more significant for the future, drillers were increasingly successful in tapping the complex geology of the Western overthrust belt and in discovering deep gas in such places as the Anadarko basin (Texas and Oklahoma) and the Tuscaloosa trend (Louisiana). The truly exciting possibilities implicit in geopressured methane are examined shortly.

Whereas expanding productive capacity to meet immediate requirements is important, discovery of new reserves is essential to the long-run health of the industry. In the absence of adequate reserves, long-distance transmission lines and seasonal storage facilities will not be built. Now that reserves of natural gas have stabilized, investment in distribution facilities can proceed.

The restrictions on new connections that forced potential consumers into inferior or expensive alternatives during the 1970s have been lifted. This is due in part to the increasingly successful search for new gas, but more significantly to the large reductions in use by

Figure 10.3
Exploratory Wells Drilled for Oil and Gas, 1952–1982

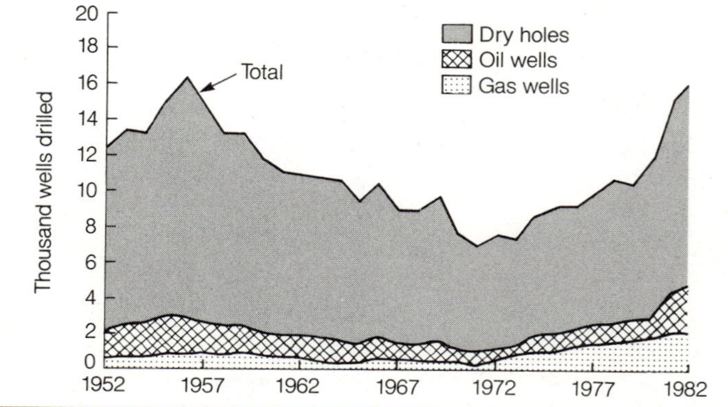

Source: *U.S. Department of Energy,* Annual Energy Review *(1982), p. 36.*

individual consumers in response to the higher prices. Many utilities were reporting declines (adjusted for temperature) in consumption averaging more than 20 percent per household between the winter of 1981–82 and that of two years earlier. Nor was this accomplished by a switch to oil or electric heating, since heating oil sales were down sharply and electricity consumption had leveled off. As the natural gas industry entered the 1980s, retail sales—at the regulated prices—were limited by demand, rather than by shortages and restrictions.

GOVERNMENT REGULATION

General Background

Local distributors of natural gas have long been regulated by states as public utilities. They were included within the "natural monopoly" concept, since it seemed inefficient to permit more than one company to lay gas mains under the same street. Once the transmission lines began to cross the country, however, Congress concluded that serious matters of interstate commerce were involved. The Natural Gas Act of 1938 gave the Federal Power Commission the authority to regulate the charges for transporting gas through the pipelines crossing state lines. The basic idea was to treat the pipelines like public utilities, with prices limited to costs plus a fair rate of return. The economies of scale inherent in pipelines and the

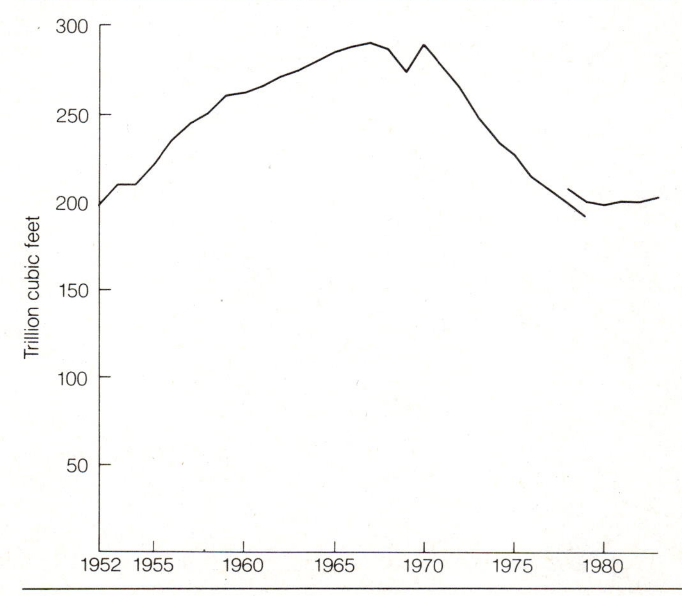

Figure 10.4 *U.S. Reserves of Natural Gas at Year End, 1952–1983*

Sources: *U.S. Energy Information Administration,* Annual Energy Review *(1982), p. 43;* Monthly Energy Review *(September 1983), p. i;* Oil and Gas Journal, *May 21, 1984, p. 94.*

Notes: *Data compiled by American Gas Association, 1952–1979, and U.S. Energy Information Administration, 1978–1983. Alaskan reserves added in 1970.*

lack of competition once the line is in place furnished plausible reasons for regulation of pipeline rates.

In 1954 the Supreme Court ruled in *Phillips Petroleum Co. v. State of Wisconsin* that the Natural Gas Act of 1938 had given the Federal Power Commission the task of regulating wellhead prices of natural gas, as well as subsequent transportation charges. The task was large, difficult, and unnecessary, and the agency performed it in such a way that it created the shortages of natural gas that recurred during the 1970s.

The full story of the regulatory disaster has been recounted so often that only a brief sketch is offered here.[4] The essential point is that the FPC set the wellhead price too low. This was no problem at first because reserves were ample, and indeed reserves continued to increase until the end of 1967 when they amounted to sixteen years' production (see Figure 10.1). Average intrastate prices, which at that time were unregulated, moved closer to the interstate rate (Figure 10.5) until 1972. Then the unregulated price moved quickly above the regulated, which meant that few producers were interested in selling to the interstate market. The decline in reserves also seemed alarming.

Economists were inclined under those circumstances to argue that more gas would be found and become available in the interstate

Figure 10.5 *Average Pipeline Acquisition Price of Natural Gas, 1963–1984*

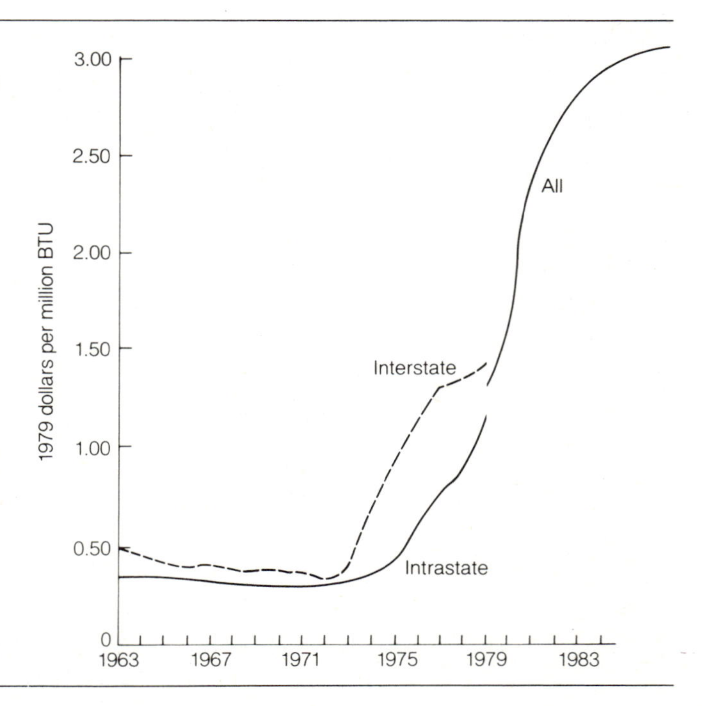

Source: Monthly Energy Review *(January 1982), p. ii;* Monthly Energy Review *(September 1984), p. 102.*

market if price controls were eliminated. Many learned arguments were presented, however, which purported to prove that supply would not respond much to price increases, so that the main effect of higher prices would be a massive transfer of income from consumers to producers. The political debate was complicated by a totally irrelevant discussion of "withholding." Some of the politicians and consumer advocates claimed that the producers had more gas but were withholding supplies to force a price increase. The producers angrily denied such behavior. The irony, of course, is that withholding supplies when prices are low is precisely the behavior that the economic theory of the mine assumes that the rational mineral owner will follow. Furthermore, it is precisely the behavior that maximizes public welfare (as well as producer profits) because the gas is conserved until such time as consumers will value it more highly. So the opponents of the gas producers were accusing the producers of serving the public interest, whereas the producers themselves vehemently denied that they were sufficiently rational to maximize profits.

The well-publicized shortages of the 1970s did include bans on new gas connections, mandatory cutbacks of supplies to interruptible customers, school closings, and exhortations from politicians to reduce consumption. It is not clear that these shortages reflected a lack of production capacity in the field, but without adequate reserves it made no sense for any firm to build pipelines to the consuming areas. Furthermore, cutbacks in industrial use and some closings of schools may be wiser than building sufficient storage and transmission capacity to heat everything during the coldest winter in a hundred years.

The shortages set off a policy scramble. The Federal Power Commission and its successor, the Federal Energy Regulatory Commission (FERC) allowed the price of newly found gas to rise during the 1970s. Other policies came and went with the winds of crisis. During the era of air cleaning, various utilities and industrial users had been ordered to reduce the air emissions from coal-burning plants. Many had converted to oil or natural gas. With the OPEC oil embargo in 1973 and subsequent oil price increases, utilities were ordered to stop using oil. But in many cases the coal-handling equipment had been dismantled, and in any case the air still had to be kept clean. The use of natural gas nevertheless was forbidden, leaving the electric utilities with no good options.

Similarly, to meet commitments to existing customers, gas utilities often refused to accept new customers. This left the owners of new homes with the choice of burning imported oil or relying on electric heat. The electricity in turn was generated by burning imported oil.

Yet even as Congress was grinding out the Natural Gas Policy Act of 1978 (NGPA), conditions were changing. The NGPA was based on the assumption of inherent scarcity of natural gas, a scarcity that should not be eliminated by permitting prices to rise. Hence it was decreed that natural gas should be reserved for household use because it is clean and efficient to use in small quantities (and perhaps because households include voters). Most large consumers would be required to convert to other fuels (generally coal) for firing large boilers. Even as the law was being passed, however, a surplus of natural gas was developing (the "gas bubble"). The Department of Energy was beginning to advocate temporary use of gas by industry to bring about an immediate reduction in oil imports, but always accompanied by the warning that industry must expect to rely on coal soon because gas would become very scarce. The question of the scarcity of gas must be deferred until after a more detailed discussion of the regulation of natural gas prices.

Price Controls and Misallocation

Differing judgments about the usefulness of price controls (to the extent that they employ rational analysis) are based on different estimates of the elasticities of supply and demand for the regulated commodity. Some statements on either side of the regulation question are purely ideological, of course, and are therefore not susceptible of proof or disproof through analytical or empirical methods. These include such statements as "Price controls are bad because they interfere with human freedom" and "People have a right to cheap fuel for necessary household uses." Most people, however, are probably more interested in the consequence of price control: Will price be prohibitively high in the absence of controls? Will I be able to buy what I want, when I want it, without any red tape or waiting?

The view of the world that makes controls seem most reasonable is the belief that quantity demanded and quantity supplied are both almost unresponsive to price; that is, people need a certain amount of gas and cannot be induced to use more or less. Similarly, producers can supply a set amount and this cannot be altered by economic incentives. If demand and supply do not match up, then some form of direct allocation must be used because (in this view) outrageous price increases would be required to alleviate the shortage. Accordingly, some administrative fiat ("The company is forbidden to take on new customers," "No industrial boiler burning more than 300 mcf per day of natural gas may be operated if gas becomes scarce") must restrict the quantity demanded so that it does not exceed the available supply. If we assume that raising the price would have no effect on the quantity supplied or demanded, then price just determines the relative wealth of consumers and producers, so each representative's vote on decontrol could be expected to reflect the composition of his or her district—does it have more producers or consumers?

If we believe that either supply or demand for natural gas is responsive to price, as most economists do, then the case for price controls is very difficult to make. Not only will an unrestrained price move to such a level that quantity supplied equals quantity demanded, it promotes efficiency in a variety of ways. Even the most heavy-handed controls cannot do as much to encourage conservation in all the uses of natural gas nor to encourage production from all potential sources. In particular, residential and commercial customers have an incentive to reduce total energy consumption or to switch partially or completely to other sources of energy, if natural gas becomes more expensive. Similarly, industrial consumers can make the same kinds of substitutions.

The factor that confused the empirical evidence about the elasticity of demand for natural gas is that at the time when the problem began to be discerned, the price of gas was very low relative to the prices of substitutes. Through an appreciable range of prices, the only feasible response of most consumers was to continue using gas but to try to decrease the quantity. After the initial response of turning down the thermostat, there was not much that either a household or a firm could do right away. The important responses, such as insulating, capturing more solar energy, installing better controls and more efficient combustion equipment, all took time, planning, and investment. By the end of the 1970s, such measures were beginning to have a significant impact. By that time too gas and oil prices to consuming firms and households had risen enough to make coal and wood seem inexpensive in certain applications.[5] It appears that the price of gas has gone up to the range where the quantity demanded is much more responsive to price changes.

A more immediate problem with the old controls on prices of gas moving interstate was that the interstate market was starved by the uncontrolled intrastate markets. Reserves committed to the interstate market plunged sharply after 1967. Ironically, the regulations that were supposed to protect interstate customers from high prices of gas forced them to consume the even higher-priced alternative fuels. Since intrastate prices were not controlled by the federal government before the NGPA, intrastate customers could bid more for gas than interstate customers were permitted to pay, even if the latter valued the gas more highly. Specifically, whereas consumers in the Northeast were forced to use expensive electric or oil heat, utilities and industries in the Southwest, which could have converted to coal, were using natural gas as a boiler fuel.

The solution adopted in the NGPA was not to free interstate prices but to control all prices. The decontrol of most natural gas prices did not happen until January 1, 1985. But the rapid changes in the industry left consumer prices far above the apparent marginal cost of gas. Large consumers in gas-producing areas still have the option of drilling their own wells or of buying gas from producers and paying pipelines to move it, so they can protect themselves.

The Natural Gas Policy Act

The Natural Gas Policy Act of 1978 (NGPA) is a typical piece of modern federal legislation: Congress faced a serious problem created largely by earlier government action. It worked long and hard to devise a complex solution when a simpler one was available. In the end it dealt mainly with the question of distributing economic rents, leaving a piece of legislation so woefully incomplete that much of the substance would have to be filled in by the bureaucracy as it wrote the implementing regulations.

The serious problem confronting Congress in this case was a real shortage of natural gas. The quantity demanded was growing, the quantity supplied was showing signs of impending decline, and it was becoming increasingly questionable each winter whether major shutdowns of industries and schools would be necessary to maintain residential service. The cause of the shortage was the very low price ceiling set by the Federal Power Commission, so the problem could have been solved by eliminating all government regulation of wellhead prices of natural gas.

The objection to the simple solution was that it did not come to grips with the main fact of political importance. Politicians are reelected, not for solving real economic problems, but for satisfying their constituents. The basic political problem was the distribution of economic rents. The economic rent that could be transferred from consumers to producers by freeing prices was estimated to be around $35 billion for 1982.[6] In a free market, the price at the wellhead would rise for all new contracts signed. As old contracts expired, the gas from old wells would also be priced higher. Although market imperfections would delay the process, producers would eventually capture most of the economic rents.

Consumers would have been hit with huge increases in natural gas prices during the early 1970s if the markets had been free. The result would have been enormous pressure to conserve fuel, a pressure that would have affected every consumer. This would have reduced energy consumption much more rapidly than the policy of maintaining low prices to existing consumers, while forcing occupants of new homes to use alternate fuels. It would also have encouraged more drilling in the most promising sites.

Congress, however, was not prepared to let the producers have the rents. In the case of oil, prices were freed from controls, but much of the economic rent was extracted by the government's windfall profits tax. The price increases for oil were necessary to reduce imports by reducing consumption. Very little gas is imported, however, so it would be difficult to blame foreign producers for the price increase. Congress therefore tried to transfer the economic rents to the consumers by keeping prices low. The approach was to try to keep all old wells selling at the price at which they had been selling, while allowing higher prices for new gas wells.

The two problems with this are that the gas is not being sold at marginal cost, which raises problems that are discussed later, and that the seemingly simple procedure is difficult to carry out in practice. The practical problems include the creation of an incentive to extract old gas from new holes. Conversely, if producers cannot find a way to obtain the higher prices, they will refrain from any action necessary to maintain production at a high level. We were treated to the anomaly of old wells at $2 per thousand cubic feet (and less) selling gas in the same markets with imported gas at almost $5 per thousand cubic feet and even some gas from deep wells at $9 per thousand cubic feet (Figure 10.6). The chances are that the very expensive marginal sources of supply do not yet need to be developed because (1) if all producers received the highest price paid to any, substantial amounts of cheaper gas would be produced, and (2) if all consumers paid a price equal to the marginal cost of producing

Figure 10.6 *Interstate Gas Purchase Price Distribution by NGPA Title I Category, Early 1981*

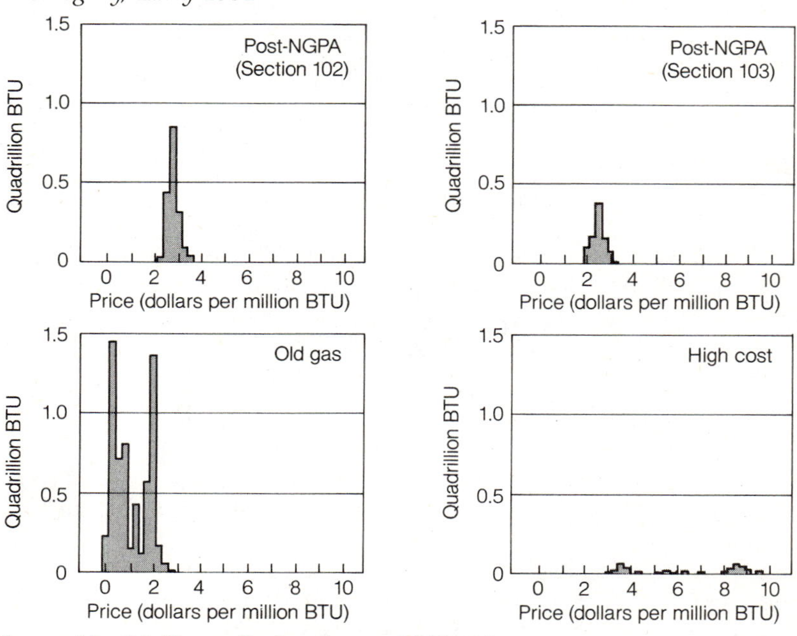

Source: Monthly Energy Review *(January 1982), p. iv.*

gas, they would conserve enough so that the $9 gas would not need to be produced yet.

By keeping most gas prices low (in the vicinity of $2 per thousand cubic feet), the government tried to reserve economic rents for the consumers. To satisfy consumer demands at such low prices, the industry turned to relatively expensive sources of additional supplies. Since the law exempted "deep" gas (wells more than 15,000 feet deep), much drilling effort was diverted to them. Contracts for such gas were signed at prices as high as $9 per thousand cubic feet during the periods of scarcity, but by mid-1982 prices in new contracts had drifted down to $5. Similarly, the gas from Mexico and Canada was available at $4.94 per thousand cubic feet but was in excess supply by the early 1980s. By 1984 the Canadian government permitted some exports to the United States at $2.89 per thousand cubic feet.[7] The years of scarcity had also prompted experiments with coal gasification, but these efforts too met with less enthusiasm as the uncontrolled prices dropped below $5. One plant went on stream in 1984 to the accompaniment of bickering about who would bear the high costs.[8]

Some liquified natural gas (LNG) was shipped into the United States from Algeria in the 1970s. When natural gas is liquified, its volume is only about $\frac{1}{600}$ of the volume of the gas under standard conditions. To liquify it at atmospheric pressure, however, it must be cooled to $-162°C$. At that temperature the hull of a steel ship would become very brittle and probably crack, so the cargo tanks of the LNG tanker must be carefully insulated. If a ship were to break open, many people think it would explode with more force than any other object of human origin except a large nuclear weapon. Even barring disasters, the process is not cheap. Estimates of the cost of liquifying natural gas, transporting it from Algeria to the United States, and feeding it into the domestic pipeline system are of the order of $3 per thousand cubic feet.

When Algeria attempted to drive too hard a bargain on the price, the equipment was shut down and the specially built ships were sold as ordinary bulk carriers.[9] Nevertheless, the technique is still used by other countries, and it is one option for making use of the natural gas available in Alaska.

To protect residential consumers from the burden of these costly sources, the NGPA adopted a complicated scheme known as incremental pricing. This is totally different from marginal cost pricing as the economist talks about it. To clarify these differences and their implications, let us look at Figures 10.7, 10.8, and 10.9. Figure 10.7 shows the conventional textbook model of the purely competitive firm. The central message of microeconomics is that the competitive industry will settle at an equilibrium such that each firm will be in the position of the one in the figure: To maximize profits the firm will produce at such an output that marginal revenue is equal to marginal cost. Because the industry is purely competitive, this occurs at the point where price is exactly equal to minimum average total cost.

The advantage of this condition is that it leads to the maximization of welfare in the entire society, under the conditions spelled out in textbooks on microeconomics and welfare economics. In particular, when price is equal to marginal cost, the consumer has the correct information about the cost to society of an additional unit of the good. If the gas company is paying $8 per thousand cubic feet for gas from deep wells and the consumer faces that same price, then the consumer will decrease her consumption until the last thousand cubic feet she uses is worth just $8 to her. Equivalently, for welfare to be maximized every consumer should face the same price, except for any differentials that reflect differences in the cost of providing service. It makes no sense to have one customer using gas that is worth only $4 to him, while another goes without gas she would gladly pay $6 for. Of course, off-season rates can be lower because they do not have to include storage costs and the entire overhead cost of pipelines to provide winter fuel. Similarly, interruptible rates can be lower because the other customers are paying for the capacity to meet the most severe weather conditions.

The distribution of natural gas, however, is not purely competitive. The regulation of the industry developed under the assumption that gas companies had substantial monopoly power. This common view may exaggerate the long-run power of the firms today because, given time, most users of natural gas can easily switch to other fuels. Indeed, nearly half of the population gets by without gas for heating, and for other residential uses substitution of electricity for gas is easy. In most industrial processes it is also possible to substitute other fuels.

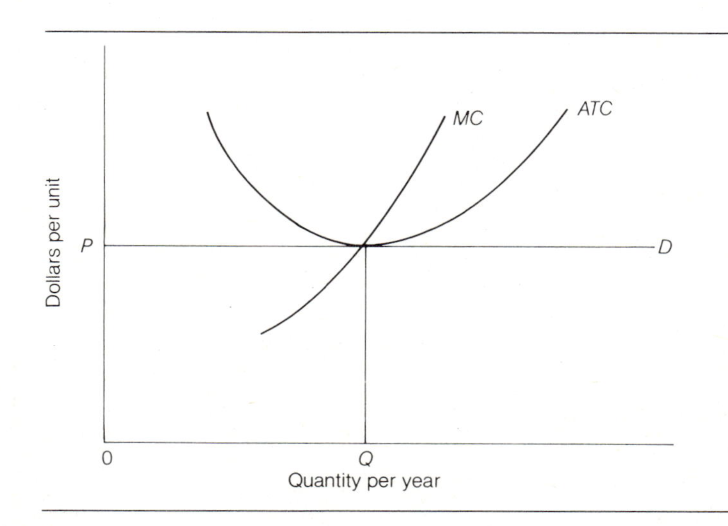

Figure 10.7 *Purely Competitive Firm*

In any event, gas companies have been regulated as natural monopolies by public utility commissions of states. The standard theory of regulation of decreasing cost industries seemed to fit natural gas very well for a long period of time. This basic notion of the industry is indicated in Figure 10.8. The two key differences from the competitive picture are the downward-sloping demand curve of the monopolist and the decreasing average total cost (*ATC*) curve derived from the immense fixed costs and the constant marginal costs (the cost of the gas to the utility).

An unregulated firm in such a situation can maximize profits by selling the quantity such that $MR = MC$ at the price given by the corresponding point on the demand curve. The result is not optimal for society because price exceeds marginal cost, which means that people do not consume as much of the commodity as society could provide at a cost that is less than their perceived benefit. This is the economic argument for regulation.

The political argument for regulation is that it prevents the monopoly from extracting excessive profits from the citizens. This starting point probably explains actual regulation far better than does the economic rationale. In any event, the simple economic criterion of letting price equal marginal cost cannot easily be followed in the classic decreasing cost industry of Figure 10.8. Either the firm would suffer losses or it would require a subsidy from general tax revenue, both of which have economic and perhaps

Figure 10.8 *Traditional Regulated Utility*

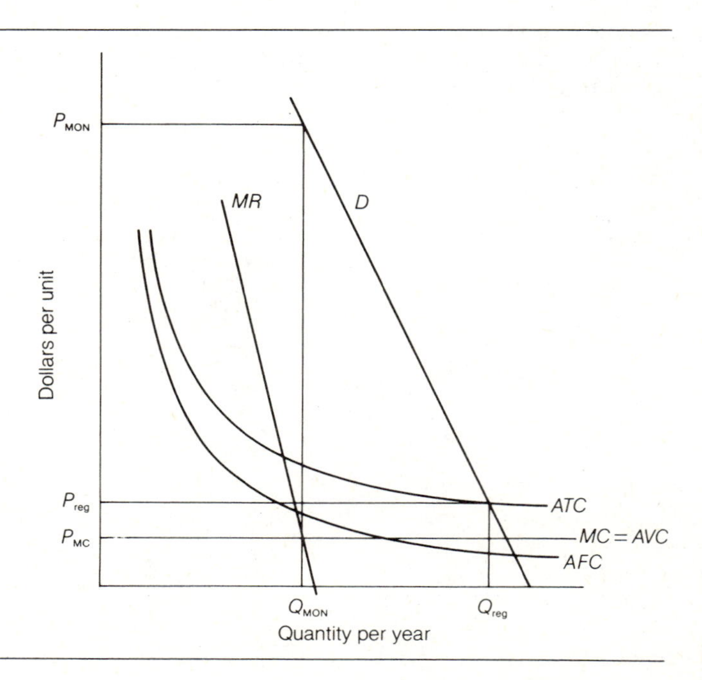

political disadvantages. The usual solution of regulatory commissions is to permit firms to recover full costs including a "fair rate of return" on investment. This is equivalent, if everyone uses the same assumptions, to setting the price equal to average total cost (*ATC*) on Figure 10.8.

This is a fairly good solution to a difficult problem, but it has some drawbacks (aside from problems of implementation). One drawback is that price exceeds marginal cost, so that both the firm and the consumers could benefit by somehow reducing rates for added consumption. In fact, it has been common for public utility rate schedules to offer declining prices per unit as consumption increases. Although such rate structures were often attacked by conservationists and spokesmen for the poor, they made a great deal of sense as long as the conditions resembled those depicted in Figure 10.8.

Conditions changed during the 1970s, becoming more like those depicted in Figure 10.9. Although fixed costs are still large, the marginal cost of gas is rising in the vicinity of current consumption levels. Since marginal cost exceeds *ATC*, it would be easy for the regulatory commissions to set price equal to marginal cost. During the 1970s the utilities could have made substantial profits by pricing at marginal cost.

The significant point is that the gas industry, like the electric utility industry, has changed so much that the old regulatory objectives are no longer valid. Marginal cost pricing, which has great advantages in making the economy more efficient, now will work. If the resulting costs to consumers (or industry profits) are deemed to

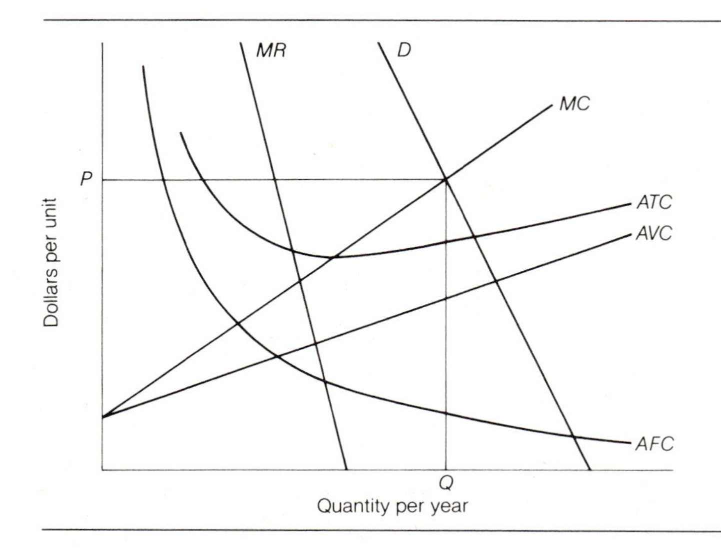

Figure 10.9 *Gas Distributor under Current Conditions*

be too high, then either a tax on the industry's profits or lower rates on the initial quantities consumed could be appropriate.[10] Both, however, are easier to propose than to implement seriously.

The NGPA provides for incremental pricing rather than marginal cost pricing as described here. Instead of pricing all gas to every customer at the cost of the highest-cost gas supplied to the system, an average cost approach derived from traditional regulation is maintained. Whereas a pure *ATC* would "roll in" the high-priced gas along with low-priced gas under old contracts to obtain an average applied to all, incremental pricing treats the high-priced gas as though it were reserved for low-priority customers. The low-priority users are large industrial boilers. The excess cost of the high-priced gas is passed on to such low-priority users, although their maximum price is kept below a ceiling based on the price of alternate fuels (no. 2 fuel oil and residual oil) to discourage them from switching to imported oil while gas is available. In practice, the competition from oil has prevented serious application of incremental pricing.

The political turmoil surrounding natural gas policy abated during 1983 as the markets began to function more smoothly and energy prices leveled off. Under NGPA all new gas was to be decontrolled on January 1, 1985, but gas from wells begun before January 1, 1977, was to remain controlled. It was feared by many analysts that a combination of contract terms would cause prices to soar after decontrol. A few contracts specify that after decontrol pipelines will pay producers 110 percent of the BTU equivalent price of no. 2 fuel oil (home heating oil). Many contracts state that the pipeline will pay the producer the highest price it pays anyone else. Combining these two kinds of contracts could lead to dramatic increases in average prices. With ample supplies of low-cost gas available, however, several pipelines broke contracts with suppliers and negotiated lower prices. Messy though this process is, it seems preferable to a new onslaught of federal intervention.

IS GAS SCARCE OR PLENTIFUL?

During the decade of the 1970s—the era of energy scarcity—natural gas was considered the premium source of energy but unfortunately very scarce. "How to Slice a Shrinking Pie" was the label Stobaugh and Yergin attached to their chapter on natural gas in *Energy Future*.[11] Nor was their view idiosyncratic. It appears to capture much of the sentiment behind the NGPA, the dominant attitude in the Department of Energy, and much of the basis for planning in the natural gas industry.

But is it so? Now that reserves of gas have finally stabilized again after their long decline, the belief that gas may be plentiful, if not cheap, is beginning to acquire some support. The upturn in discoveries seems to be closely related to increased drilling effort: In 1973, 26,592 holes were drilled (the average depth was 5,150 feet). Of these, 9,902 produced oil, 6,385 produced gas, and 10,305 were dry. In 1980 the number of holes drilled had increased 128 percent to 60,845 (average depth 4,680 feet). The number of oil producers increased 174 percent to 27,026. The number of gas producers increased 147 percent to 15,730. The number of dry holes increased 75 percent to 18,089. These data indicate that drilling effort was concentrated on increasing production from relatively certain sites during the period of scarcity and high prices, which certainly seems like behavior that furthers the national interest while making successful drillers rich. Even with the bias toward development rather than exploration, reserves stabilized. The implication is that the continuation of the drilling effort into less conventional sites may pay off very handsomely.

A few authorities on mineral resources have begun to suggest that the quantity of natural gas accessible to deep drilling in the United States will dwarf other fuel resources. Foremost exponents of this view include Vincent McKelvey, former director of the United States Geological Survey, and Thomas Gold of Cornell. The explanation offered by the latter is that natural gas—methane—is not just a fossil fuel. Some of it is part of the original material of the earth, and much of this is still trapped within the earth waiting to be tapped by deep drilling.[12] It is too early to know whether such optimistic appraisals are correct, but some people and companies have made a lot of money by drilling into deep formations.

According to the natural gas distributors, the most likely possibility is that gas will continue to be available in modestly increasing amounts so that it will continue to account for roughly one-quarter of energy consumption in the United States. Because of the expected decrease in gas from conventional sources, the delivery of the gas will be achieved only at the higher prices necessary to encourage deep drilling, undersea drilling, and other unconventional sources. Such sources may once again include some imports of LNG via ship from overseas, as well as imports from Canada and Mexico, gas from Alaska, and gas produced from coal and perhaps from municipal or other waste. In addition, more attention will be paid to extracting gas from tight formations (rocks that require special fracturing to let the gas flow) or other currently marginal sources. Under this view the companies will be forced to keep scrambling for supplies, but the consumer will not notice any problems except for the rising price. Gas will continue to be available for industrial use.

If the extreme pessimists who dominated discussion and the Department of Energy in the 1970s are right, conditions will be difficult after the mid-1980s. Supplies of gas will not be adequate to meet demand except at exorbitant prices. The government will be faced with the choice of letting prices rise above the levels of alternate fuels to choke off demand or forbidding industrial use of natural gas. In any event, the result will be continued pressure on other energy sources such as coal, nuclear power, and imported oil. If this view is correct, the attempt to preserve the available supplies for use by residential customers makes some sense, but the failure to raise prices immediately to the level of marginal cost is harder to justify.

Finally, it is worth considering the possibility that natural gas will be available in abundance at a price no higher than $8 per thousand cubic feet. A natural gas price at the wellhead that stayed around $8 per thousand cubic feet would mean that consumer prices could still double. Nevertheless, it would cap the price of home heating oil just slightly above the 1983 level and make the construction of new nuclear power plants even less attractive than it is now. If the price were to be closer to $4 per thousand cubic feet, then oil would be overpriced at $30 per barrel. Even the coal mining industry would find its future clouded by the ready and cheap availability of the cleaner and more convenient fuel. Only in transportation—automobile and airplane—would oil have a significant advantage. That is such a small segment of the total consumption of petroleum that the power of OPEC to raise prices would be ended. The future of gas will clearly have a great deal of importance for the future of energy and the entire economy.

NOTES

1. The history and technology of the industry are discussed in British Petroleum Company, *Our Industry Petroleum*, chaps. 10 and 18 and *Oil and Gas Journal* (August 1977), pp. 279–315. For a detailed introduction to the industry see Arlon R. Tussing and Connie C. Barlow, *The Natural Gas Industry* (Cambridge, Mass.: Ballinger, 1984).
2. American Gas Association, *Gas Facts* (1966), p. 48.
3. Economies of scale are carefully analyzed by Leslie Cookenboo, Jr., *Crude Oil Pipe Lines and Competition in the Oil Industry* (Cambridge, Mass.: Harvard University Press, 1955). John A. Hansen, *U.S. Oil Pipeline Markets: Structure, Pricing and Public Policy*, Series on Regulation of Economic Activity, no. 6 (Cambridge, Mass.: MIT Press, 1983) provides a comprehensive analysis. The specific features of natural gas pipelines have not received as much academic attention. The *Oil and Gas Journal* publishes much current information in articles such as "Pipeline Economics," November 23, 1981, pp. 79–106.
4. For a brief analysis, see Henry D. Jacoby and Arthur W. Wright, "The Gordian Knot of Natural Gas Prices," *Energy Journal* 3, no. 4 (October 1982), pp. 1–9 or Robert S. Pindyck, "Prices and Shortages:

Policy Options for the Natural Gas Industry," *Options for U.S. Energy Policy* (San Francisco: Institute for Contemporary Studies, 1977), chap. 6. See also Paul W. MacAvoy and Robert S. Pindyck, *The Economics of the Natural Gas Shortage (1960–1980)* (North-Holland, 1975).

5. For some detail, see Bruce Egan, "Residential Energy Consumption, 1978 through 1981," *Monthly Energy Review* (September 1983), pp. iii–xii.
6. The entire issue of *Energy Journal* for October 1982 (3, no. 4) is devoted to disentangling the complexities of gas policy. This estimate is by Jacoby and Wright, p. 25.
7. *Wall Street Journal*, June 18, 1982, p. 49 and September 17, 1984, p. 21.
8. U.S. General Accounting Office, "Economics of the Great Plains Coal Gasification Project," RCED-83-210, August 24, 1983.
9. *Wall Street Journal*, February 11, 1982, p. 2.
10. "Lifeline" rates are often advocated as devices to help the poor. The idea is that some low level of service of a basic utility would be priced very cheaply. The details of such a plan are easier to work out for telephone service than for gas. The poor may after all rent drafty houses that are expensive to heat.
11. I. C. Bupp and Frank Schuller, "Natural Gas: How to Slice a Shrinking Pie," *Energy Future: Report of the Energy Project at the Harvard Business School*, eds. Robert Stobaugh and Daniel Yergin (New York: Ballantine Books, 1979).
12. Boston *Globe*, July 27, 1980, p. D-3; *Oil and Gas Journal*, February 5, 1979, p. 30; *New York Times*, November 29, 1981, sec. 3, p. 1.

PART THREE

The Energy Conversion Industries

CHAPTER ELEVEN

The Electric Utility Industry

This chapter departs from the pattern of the previous three because the electric utility industry converts energy rather than supplying it. Much of the output of the coal industry and some oil and gas are consumed by the electric utility industry, and so to add the output of the coal industry to that of the electric power industry would involve a substantial amount of double counting. Nevertheless, we should consider electric power now because nuclear power (the subject of the next chapter) is useful only for generating electricity and powering large ships. Furthermore, hydro power and geothermal power, although they have long histories of direct use, are today mainly important as sources of electric power. The ease with which different energy sources can be used to generate electricity and the great versatility of electricity for almost every use of energy also suggest the significance of this topic. Even if the importance of electric power generated at central stations is eventually reduced by such innovations as windmills or solar cells on every rooftop or widespread use of cogeneration, central station power will still set the standard against which other possibilities will be compared, as well as providing much of the electric power used by an urban industrial society.

Most of the data and discussion in this chapter refer to electricity produced by electric utilities. Such companies are in the business of selling electricity to numerous separate industrial, commercial, and residential customers. A significant amount of electricity is produced by other entities, however. It has been estimated, for example, that the 108 million motor vehicles registered in the United States in 1970 had an aggregate generator capacity exceeding that of federally owned electric utilities.[1] Nevertheless, for practical reasons it seems better to ignore the electricity that is produced as an incidental aspect of the use of various kinds of equipment or by standby generators that operate during power failures.

Large industrial plants, however, often generate significant amounts of electric power using equipment similar to that used by the utilities. Figure 11.1 indicates that industrial establishments generated more than half of all electricity in 1902. Although the absolute amount increased over the years, by 1970 industrial establishments generated less than 7 percent of the total.

The decline in relative importance justifies the neglect of industrial generation in the remaining data, but it should be borne in mind that the utilities do face challenges to their monopoly of electric generation. The idea of cogeneration, producing electricity and heat for other uses in the same facility, has been promoted vigorously in recent years. This is discussed later in the chapter. In addition, many of the alternative energy sources discussed in Chapter 13 permit households to generate their own electricity. Losing even a small share of their traditional markets would hurt the electric utilities in an era of excess capacity.

HISTORICAL PATTERNS

The first commercial production of electricity began in 1882, and by 1900 electrification of major cities was well under way in the United States. The obvious convenience of electric lights powered by central generating plants led to the rapid diffusion of the industry wherever population density was adequate. The central generators were sometimes powered by water and experiments were conducted with wind-powered generators, but coal-fired steam plants soon became the largest single source of electricity, as is indicated in Figure 11.2.

It is scarcely an exaggeration to say that electric appliances liberated the twentieth-century housewife from the drudgery of household chores. Only those who have experienced the ceaseless heavy work entailed in firing up a wood or coal cookstove, heating water on it, scrubbing clothes by hand, ironing with heavy irons heated on the stove, and trying to preserve food without refrigeration can appreciate the freedom that electricity offered. Wealthier households, of course, had servants to help with these tasks, so the improvement in living conditions has been greatest for the poorer households that formerly not only could not afford domestic help but also had to provide it to the rich.

Electrification of the household and of industry proceeded at a steady, rapid pace right into the 1960s. It became a rule of thumb that power consumption would double every decade (a compounded rate of growth of about 7 percent per year). Utilities were able to plan accurately for their own growth by taking account of trends in local population and major changes in local industries.

With this process of steady and rapid growth came an extra benefit for the consumers. During the first half of the century, the new power plants were generally much larger than older units and the large units produced power at a lower cost. Smaller plants were also becoming more efficient as the technology advanced, but the

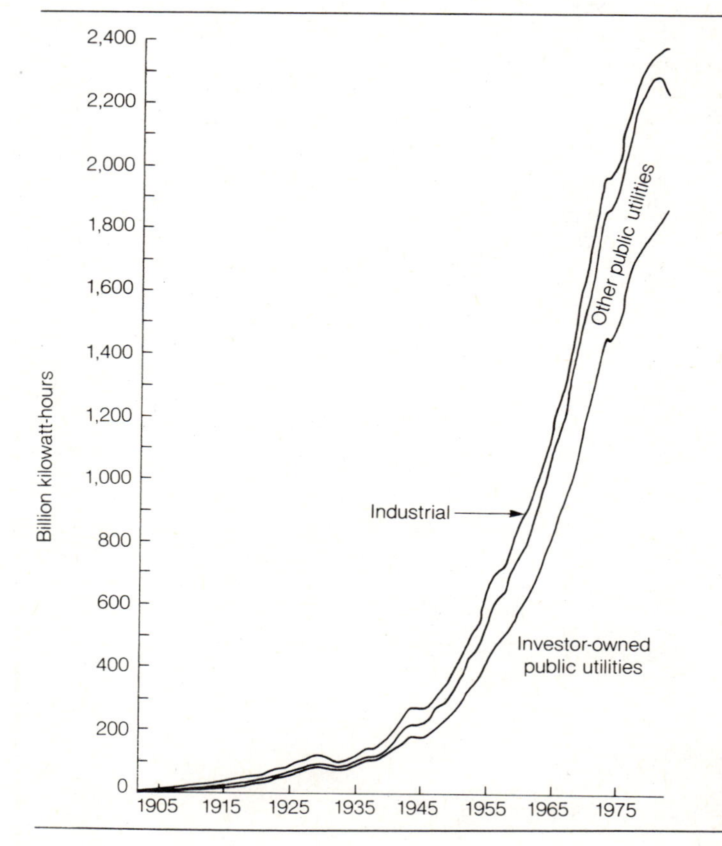

Figure 11.1 *U.S. Production of Electricity by Class of Ownership, 1902–1982*

Sources: U.S. Historical Statistics *(1975), p. 821, S 44–52;* U.S. Statistical Abstract *(various editions).*

association of size with efficiency was justified by the data, as well as being an article of faith of the industry. The relationships among size, vintage, and cost, however, have become more complicated in recent years.

Another key aspect of the industry is the very heavy fixed costs combined with the impracticality of storing output. The implication is that at all times except when the system is close to its maximum output, marginal costs will be substantially less than average costs. The generating capacity is there, the wires are in place, and the meters must be read, so the only cost is adding more fuel to produce more power. Although this is slightly oversimplified, it does catch the spirit that has shaped much of the industry.

The policy that recommends itself under such circumstances is to encourage the growth in consumption to provide everyone with the benefits from the economies of scale. It is particularly beneficial to encourage off-peak uses of power. Just as in the natural gas industry, if the electric power industry can even out the fluctuations in sales, it can increase the utilization rate of the expensive equipment that it needs to meet peak demand.

Since electric power cannot be stored cheaply and interrupting service is disruptive for most consumers, the utility must carry

Figure 11.2 *U.S. Production of Electricity by the Electric Utility Industry by Type of Energy Source, 1951–1981*

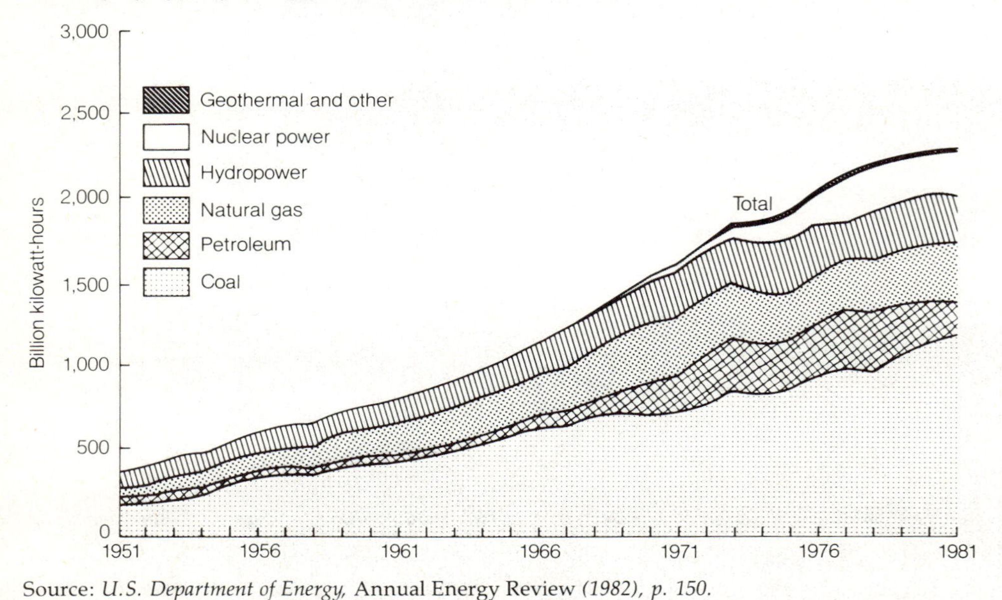

Source: *U.S. Department of Energy,* Annual Energy Review *(1982), p. 150.*

enough capacity to meet seasonal and daily peaks even after allowing for scheduled maintenance and breakdowns. Of course, tie-ins between different power systems can reduce the reserve capacity required by unscheduled breakdowns, but this does not solve the peak load problem. The utility must match the quantity of electricity generated to the quantity demanded at each moment.

As the industry developed, the problem of maintaining standby capacity for peak loads and breakdowns was eased by the continuing increases in efficiency of new units. The newest (largest) units could be dedicated to base load service; that is, operated essentially continuously at full capacity. Meanwhile, older units that still ran reliably but had higher fuel and maintenance costs could be kept ready for peak loads and emergencies. There is also a public relations advantage in operating an expensive new unit at full capacity to make the capital cost per unit of output lower in the eyes of the state public utilities commission.

These tidy patterns weakened during the 1960s and collapsed completely in the 1970s, although the industry has not fully adjusted to the changed circumstances. In particular, the long period of stability or decline in costs of building new capacity ended. This, coupled with the rise in interest rates to unprecedented levels, meant that new construction became a very expensive proposition. Power from new plants is now much more expensive than the average cost for the system.

Reinforcing the impact of the high capital cost of new facilities was the end of the steady trend toward greater efficiency in the use of the fuel. The consumption of fossil fuel per kilowatt-hour (kwh) of electricity had declined from 25,175 BTU in 1925 to 10,879 BTU in 1959. By 1981, however, it was still 10,353 (Figure 11.3). With these changes in the patterns on the supply side, the long decline in price to consumers also had to end, as is indicated in Figure 11.4. Consumers responded by changing consumption patterns. The steady 7 percent annual growth in use of electric power was replaced by smaller and less predictable changes for which the industry has had difficulty in planning. These changes are examined in more detail later in this chapter.

TECHNOLOGY

Nearly all electric power is generated by rotating a magnet inside a coil of copper wire. Although the basic approach can be demonstrated using crude materials and techniques, modern commercial generating units have evolved into complex, sophisticated, and efficient machines. The power input into a generator can be anything that will make the rotor spin, but in commercial practice most rotors are turned by a steam turbine that is attached to the other end of the same shaft. A large generating plant being built today would typically have either a coal-fired boiler or a nuclear reactor to generate the huge amounts of high-pressure steam to turn the turbine. Many of the boilers now in use burn natural gas or oil, but these became expensive fuels in the 1970s.

Various other possibilities have been well refined for special purposes. Preeminent among these is the water-powered turbine, which accounted for 11 percent of electric power in the United States in 1981 and 42 percent of that produced in the western third of the country. Small portable generators, of course, are often powered by internal combustion engines (gasoline or diesel). Units to provide peaking power also may be of special forms, such as gas turbines or diesel engines. These are useful because they can reach full power quickly and because the capital costs are low, so it is possible to tolerate the high fuel costs for short periods of operation.

Electricity can be generated using solar cells, wind-powered turbines, boilers fired by waste or wood, water turbines powered by tides or small streams, and turbines powered by geothermal steam. These approaches that are somewhat outside the mainstream are examined in Chapter 13. Although they do not now account for a very large fraction of total electric power in the United States, they may be important in specialized applications.

According to the laws of thermodynamics, the efficiency with which fuel is transformed into electricity is limited by the difference between the temperature of the steam entering the turbine and the temperature of the condensate as it flows out. Specifically, maximum efficiency $= 1 - T_2/T_1$, where T_2 and T_1 are the output and input temperatures, respectively, in degrees above absolute zero. If, for example, the temperature of the cooling water is 300°K and the turbine uses steam at 500°K, the maximum theoretical efficiency possibly attainable would be 40 percent. By raising the steam temperature to 800°K, the maximum possible efficiency is increased to 62.5 percent.[2]

The theory has several implications, including the following: (1) Thermodynamic efficiency can be increased by raising the temperature of the steam coming into the turbine. (2) Efficiency can also be increased by decreasing the temperature of the condensate as it flows out of the turbine. (3) Any feasible input and output temperatures will still leave a great deal of waste heat to be dissipated. Modern fossil-fueled plants dispose of about 2 BTU to the environment for every 1 BTU that is transmitted over the wires in the form of electricity. The losses are slightly higher for nuclear power plants because they are restricted by the government to lower temperatures for safety reasons.

Efficiency, in the sense used here, is a thermodynamic concept, not an economic one. The loss of BTUs in conversion is basically irrelevant to the economic decision of whether to buy coal directly or to buy electricity made from coal. The individual household will make the choice between heating with oil or heating with electricity on the basis of initial cost, operating cost, and convenience, so the losses in generation are only a small consideration. In many uses, such as lighting and powering small appliances, the household cannot undertake the same operation any other way.

In one interesting case, the heat that is lost in the conversion process becomes significant. If an industrial or commercial user requires large amounts of both electricity and steam for heating or some industrial process, the overall efficiency of energy use can be increased by producing the two jointly, a process known as *cogeneration*; that is, the electricity can be generated using a relatively small part of the total energy in the fuel (less than in a conventional electric power plant). Much of the remaining heat can then be used to accomplish some other useful function, rather than being dissipated directly into the atmosphere. The overall efficiency in a thermodynamic sense can be much greater than the efficiency of separate installations for power and steam.

Thermodynamic efficiency is never free, however, and cogeneration is no exception. Whether it will save dollars depends on a variety of circumstances. Scale is a primary consideration, because the operation must be quite large to use an efficient coal-fired boiler and steam turbine. Smaller generators are available, but they would ordinarily be powered by natural gas or diesel engines. With a diesel unit cogeneration takes the form of a heat exchanger to extract heat from the exhaust gases of the engine that turns the generator. The disadvantage of these systems is that any fuel other than coal is so expensive that the greater efficiency in energy consumption is more than offset. Other considerations in cogeneration include the regularity of the electrical and the steam requirements and the terms on which the power company will supply power for peak use or backup when the cogenerating unit is shut down.

Cogeneration is not different in principle from the old practice of electric utilities' selling steam to the large buildings in the center of many cities. The power companies have drifted away from the practice, and new generating plants are usually not located close enough to the central business districts to make a revival of central steam systems particularly advantageous. At the same time, consumers in congested areas do not find much public support for plans to install large coal-fired facilities or noisy diesels. Intense public opposition to noise and other externalities is part of the reason that an effort by Harvard University to construct a large cogenerating facility in a residential area turned out to be an expensive fiasco.[3]

Although the basic outlines of the conventional approach to generating electric power have remained constant since before 1900, the equipment has changed substantially. One striking result is the downtrend in coal consumption per kilowatt-hour displayed in

Figure 11.3. Many of the changes are difficult to pinpoint precisely, but the net impact has been an upward trend in steam temperature and pressure and an increase in the maximum size of generating units. Steam pressures in the vicinity of 500 pounds per square-inch gauge (psig) at the turn of the century had gone to 2,400 psig by 1950. Over the same period, temperature was increased from about 400°F to nearly 1,000°F. Similarly, the small plants (less than 50 mw, or megawatts) of the early period had yielded to much larger ones (a few even exceeded 200 mw) by 1950.[4]

Some of the important changes included the adoption of cast steel to replace cast iron boilers around 1912 and the shift to alloy steels in the late 1930s. The latter change was particularly important because it permitted one boiler to be made large enough to supply all the steam for one turbine, eliminating complex piping and controls. In addition to the advances in metallurgy for turbines and boilers, there were improvements in electrical insulation that permitted generators to run hotter. Improvements in cooling systems for generators permitted increases in scale. Hydrogen cooling replaced air in 1937. In the 1950s oil cooling was adopted for the stator (outer casing of the generator from which the electrical power is drawn). Later, water-cooled stators became common.

The 1950s saw the availability not only of nuclear-powered steam generators but also of supercritical fossil-fueled plants. Supercritical boilers have pressures in excess of 3,206 psig, at which water turns directly to steam without boiling. Although such plants are theoretically more efficient than lower-pressure plants, they impose much more stringent metallurgical requirements. This rapid

Figure 11.3 *U.S. Overall Heat Rate at Electric Utility Plants, 1925–1982*

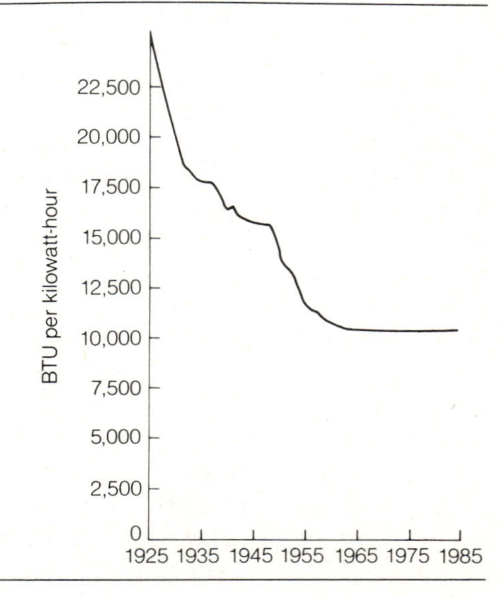

Sources: U.S. Historical Statistics *(1975), p. 826, S 107; U.S. Department of Energy,* Annual Energy Review *(1982), p. 238.*

increase in pressure was the continuation of a long history of such increases at a more moderate pace. The benefit of the higher working pressure was the increase in electrical output from units of similar physical size. Whereas capacities exceeding 200 mw were rare before 1950, by the 1960s the 1,000-mw plant had become the standard large plant, either fossil fueled or nuclear.

UNIT CAPACITY COST

The importance of the technological changes in generation is suggested by the record of *unit capacity costs* (UCC) in the generating sector of the industry. Unit capacity cost is defined as the total cost of a new capital facility divided by some measure of its capacity. In this case, the costs of putting a new electric power plant on line include planning, site acquisition and preparation, construction, and purchase of equipment including boilers, turbines, and generators. The capacity figure should be the maximum rate of power output that the plant could sustain for a long period of normal operation. This is usually measured in kilowatts, or thousands of watts, and as noted earlier a typical large new plant would have a capacity in the vicinity of 1,000 mw (1 million kw). Just to keep the orders of magnitude in mind, a household might use 500 kwh of electricity in a month. If the 1,000-mw plant actually operated at capacity every hour of the

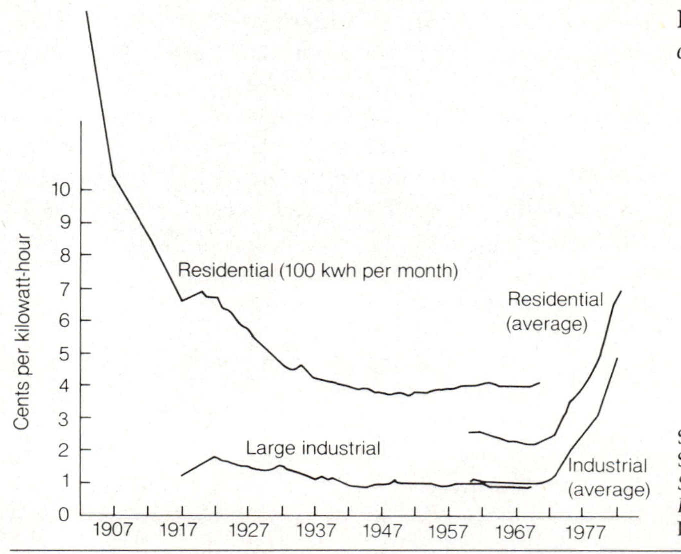

Figure 11.4 *U.S. Prices of Electricity, 1907–1982*

Sources: U.S. Historical Statistics *(1975), p. 827, S 113, S 118; U.S. Department of Energy,* Annual Energy Review *(1982), p. 165.*

month (twenty-four hours per day times thirty days in the month, or 720 hours), it would produce 720,000 mwh or 720 million kwh of electricity. This would be enough for nearly 1.5 million of those typical households.

Naturally, the real world is not so simple. Since the electricity cannot easily be stored, the plants have to accommodate peak loads, which means that they cannot run at full capacity at other times. Furthermore, the system must have some reserves for breakdowns, as well as extraordinary peaks during extreme weather conditions when heaters or air conditioners are running continuously. Finally, industrial and commercial users consume two-thirds of the power, so one plant would not be enough to serve a community of 1.5 million households and their associated activities. As a practical matter, the utility confined to such a small market would prefer ten small plants to one large one so it could continue to supply power through shutdowns for routine maintenance or unscheduled repairs.

To make this example more realistic, let us consider actual operating rates in the industry. In 1981 electric utilities produced 2,294,812 gwh, or gigawatt-hours (*giga* means billion) of electricity using generator capacity that equaled 613,546 mw at the start of the year.[5] If all of the generators had been run at full capacity, the year's electricity would have been produced in 3,700 hours. This is equivalent to saying that the industry operated at 42 percent of capacity, but it does not mean that the industry has excess capacity. The problem of meeting peak loads in the absence of storage facilities and the trade-off between capital cost and operating cost complicate the analysis.

Nevertheless, the concept of unit capacity cost is important and it provides the starting point for considering the impact of technological changes and economic conditions on the electric power industry. It is not the only important measure, of course, because the industry has costs other than capital costs and we cannot move from UCC to cost per kilowatt-hour without knowing the rate of capacity utilization, as well as some capital charge factor. For example, if capacity costs $150 per kilowatt and the capital charge is 10 percent, then the utility must spread the $15 annual charge over the kilowatt-hours produced. If the plant operated all 8,760 hours of the year at full capacity, the capital cost per kilowatt-hour would be $15/8,760, or less than 2 mills (1 mill is one-tenth of a cent) per kilowatt-hour. More realistically, the plant might produce only 60 percent of that amount of power, so the capital cost per kilowatt-hour would come closer to 3 mills.

Those assumptions are not very far from actual conditions in the 1960s, but if we alter them to correspond with the 1980s, the differences are striking. Nuclear plants have recently had UCCs exceeding $2,000 and interest rates are so high that a realistic capital charge is 20 percent, making the annual capital cost about $400. Furthermore, large modern plants (both nuclear and fossil fueled) have been so unreliable that 4,500 hours of use per year may be more realistic than the figures used in the past. The capital cost per kilowatt-hour therefore is $400/4,500 or about 9 cents per kilowatt-hour. The implications of this extraordinary change are only beginning to be confronted by utilities and regulatory commissions.

The UCC data are shown in Table 11.1. It must be stressed that the data are in current dollars. There has been no deflation by price indexes or any other form of adjustment for changes in prices. Although the Wholesale Price Index doubled between 1923 and 1968, the average UCC for all fossil fuel plants of all sizes declined from $117.3 in the period 1923–1929 to $110.5 in 1966–1968. The highest average UCC of this entire era was recorded in 1946–1949 ($189.5), and numbers in the vicinity of $150 were common in the 1950s and early 1960s. Over the long period, the technological advances in areas including design, available materials, and fabricating techniques were sufficient to offset the very large increase in wage rates and other inputs to the construction process. During the 1970s, however, UCC increased rapidly and preliminary reports indicate that the increase continued into the 1980s.

Economies of scale call for more specific treatment because of their importance in the standard view of the electric utility industry. The increases in scale themselves, however, were dependent on the advances in metals and other materials and cannot be considered as some factor apart from technological change. Not until the 1940–1945 data do such economies appear. If the very smallest plants are excluded, moreover, the relationship is much weaker.

The data shown in the table can be summarized as follows: First, the impact of a whole series of technological changes in keeping UCC from rising, and even reducing it in the 1960s, is extremely impressive. The role of scale is less clear-cut; only the very smallest units seem to have priced themselves entirely out of contention. In the late 1960s, however, it appeared that some gains were available only to units of 200 mw and above. The startling deterioration of conditions after 1969 requires closer examination.

Unit capacity costs rose rapidly during the 1970s as is indicated by such estimates as $1,500 UCC for a lignite-fired plant scheduled for operation around 1990.[6] This is fifteen times the level that seemed plausible in the late 1960s. Nuclear plant costs too have ballooned since the late 1960s, but these are not examined until the next chapter.

The easiest explanation for increases in cost of conventional plants is the general inflation of the 1970s, which seemed to strike the construction industry with particular virulence. This can explain a doubling or tripling, but not a ten- or fifteenfold increase, however. More significant, the comparison with the behavior of UCC during the other great era of inflation—the 1940s—indicates that although the initial upsurge should have been expected, the continuing increase is surprising. Why has technological change not yet been adequate to reverse the growth in UCC?

One factor that has been costly for the industry is the change in public attitudes toward pollution. The electric utility industry was forced to take expensive measures to control the emission of air

Table 11.1 *Unit Capacity Cost for Fossil-Fueled Electric Plants by Size, 1923–1982 (dollars per kilowatt)*

Year completed	Generating capacity (megawatts)					
	0–49	50–99	100–199	200–399	400–999	1,000–1,999
1923–29	117	121	122	123	–	–
1930–39	100	110	113	–	–	–
1940–45	104	93	91	–	–	–
1946–49	193	163	155	151	–	–
1950	160	135	129	–	–	–
1951–52	184	156	150	148	146	–
1953–54	218	158	143	136	131	–
1955–56	173	155	148	144	142	141
1957–58	225	162	146	138	134	–
1959–60	239	184	171	164	160	–
1961–62	–	267	202	170	151	144
1963–65	–	168	144	132	125	123
1966–68	–	–	140	119	108	103
			1973	190		
			1974	194		
			1975	228		
			1976	320		
			1977	275		
			1978	361		
			1979	537		
			1980	541		
			1981	602		
			1982	702		

Sources: *1923–1968, David A. Huettner,* Plant Size, Technological Change, and Investment Requirements *(New York: Praeger, 1974), p. 67, reprinted by permission; 1973–1982, U.S. Department of Energy, Energy Information Administration, "Historical Plant Cost and Annual Production Expenses for Selected Electric Plants, 1982" (August 1984), p. 96.*

Note: *Data for 1923–1968 are estimated for subcritical, conventional indoor, coal-fired plants by using regression techniques. Data for 1973–1982 are averages for all fossil-fueled steam plants of 300 mw or larger.*

pollutants and also to reduce the impact of its cooling water on aquatic environments. The latter has generally been achieved by transferring the heat to the air (rather than directly into a lake or river) by using expensive cooling towers. The air-cleaning equipment adds more than 10 percent to capital costs and increases operating costs by almost 1 cent per kilowatt-hour.[7]

The requirement that air pollution be reduced has posed continuing problems. The direct approach is to convert from coal to oil, gas, or nuclear power. But after the initial flurry of conversions to oil and gas, energy policy of the 1970s foreclosed those possibilities. With fears of a shortage of oil and gas commonplace, government policy forbade new installations in major power plants and even called for reconversion of formerly coal-burning plants from oil and gas back to coal. (This part of the law was never implemented.) The nuclear alternative became increasingly expensive during the 1970s, as well as encountering growing public resistance.

Thus coal-burning plants were forced to clean up. The reduction of particulates (soot and ash) is relatively straightforward, although not cheap. It involves proper combustion controls and draft and the use of electrostatic precipitators. Elimination of sulfur dioxide emissions has proved to be more troublesome. One choice is to eliminate most sulfur from the fuel, as is now routinely done with gas and oil. Whereas simple washing removes about half of the sulfur in coal, removal of the rest has so far proved to be difficult and expensive. It has not progressed beyond the experimental stages. At the other extreme, sulfur dioxide can be washed from stack gases. This raises UCC as well as operating cost. The washing processes have also suffered from a reputation for poor reliability, although this should improve as operating experience accumulates. Experiments are now underway with fluidized bed combustion, which burns the coal in close association with limestone to capture the sulfur dioxide.

Utilities that have not been willing to risk pioneering with scrubbers have had to seek out coals with naturally low sulfur content. These have often included coal that is remote or expensive to mine, increasing fuel costs. An alternative is to use a cheap, poor quality, local coal (such as lignite), which requires greater boiler capacity and larger investment in handling facilities because of the low heat content. This raises the UCC.

Average costs have also risen because of the elimination of the cheaper alternatives. During the period before 1960, coal-fired plants were built where they were expected to provide the cheapest power. Gas-fired plants were built in the Southwest and oil-fired plants were common in areas remote from coal and gas. Nuclear was an experimental alternative that looked increasingly attractive, especially where the conventional fuels were expensive. As each of these other possibilities was ruled out, reliance on coal increased even where it was expensive.

Regardless of the changes that have occurred in recent years, the analysis that was developed during an era of declining costs is important to an understanding of how the industry has evolved. The standard old view is the decreasing cost firm shown in Figure 10.8 (page 178). This had a good deal of validity, considering the greater thermal efficiency of new plants as well as the lower UCC. Once a plant was built, of course, it made sense to encourage use, since the marginal cost was low (mainly fuel) as long as sufficient capacity was available.

Over the years the industry promoted the use of electricity vigorously. This was socially desirable as long as average cost was declining, because whenever any customer increased consumption it reduced the cost of serving all others. Not only did the utilities advertise and otherwise try to persuade people to consume more electricity, they also adopted declining block rate structures. Such a rate structure imposes a high price per unit on the first few units purchased, then progressively lower prices per unit on successive additional blocks purchased.

This rate structure made a great deal of sense when average cost was decreasing. The higher rates for the initial units paid for the fixed costs of distribution and billing for each user, and the lower rates for higher use brought the marginal cost to the consumer closer to the marginal cost of electricity to society.

More recently, however, declining block rates have become increasingly controversial. The earlier arguments centered on the inequity of requiring small users ("the poor") to pay more per kilowatt-hour, whereas later the conservationist argument of the irresponsibility of encouraging use of energy became dominant. The utilities argue that in fact many costs remain fixed and that efficient, modern generating equipment used for base loads has a low marginal cost that should be reflected in low prices for large users.

In an attempt to satisfy the valid argument about fixed costs but still to move away from encouraging consumption, the public utilities commissions of some states have adopted two-part tariffs. These provide a fixed customer charge plus an unvarying rate per kilowatt-hour. In any month therefore a household faces a constant marginal cost, but its average cost will decrease with added consumption because the fixed charge is spread over a larger quantity.

The rate per kilowatt-hour will change from month to month as fuel costs vary. This is strongly influenced not only by fluctuations in the coal and oil markets but also by the amount of time that any nuclear plants owned by the utility actually operate. The nuclear plants have very high capital costs, which must be paid every month anyway, but very low fuel charges. If the nuclear plant is shut down for repairs or refueling, the consumer pays both the high capital charge for the construction of the nuclear plant and the high fuel charge for the replacement power.

To allow for the extra cost of building rarely used capacity adequate for peak loads, many utilities have adopted peak season and off-peak season rates. For many utilities the maximum power demands now come late in the afternoon of extremely hot days. Before air conditioning was widely used, winter demands were generally greater. The exact seasonal pattern varies with climate and the amount of air conditioning and electric heating, as well as the characteristics of the commercial and industrial load.

The last vestige of the declining block rate structure in the sample schedule shown in Table 11.2 is the low base rate for very heavy winter consumption (in excess of 1,000 kwh), which means in practice the use of electric heating. Since the generating capacity has to be built to meet the summer peak, it is perfectly appropriate not to charge winter users for building it. We might question, however, the exact distribution of peak summer charges and off-peak winter benefits among small and large users.

Peak and off-peak rates do not have to be defined by the seasons. It is possible to have time-of-day metering with special low rates for certain hours such as midnight to 6:00 A.M.[8] Some utilities in the United States have done this, often in the form of a separate electric meter that operates only at certain hours, to which the hot water heater or other appliance is connected. In some other countries the concept has been extended to electric space heating by using a very well-insulated heater lined with special heat-retaining bricks that are heated by an electric element during off-peak hours of the night. As heat is desired during the day, the cabinet is opened and air is warmed by being drawn past the hot bricks. Until the utility companies in the United States offer a bigger advantage for off-peak use, however, it would be surprising to find much use being

Table 11.2 *Residential Electricity Rate Schedule, Winter*

Monthly usage (kilowatt-hours)	Total cost (dollars)	Average cost (cents per kilowatt-hour)	Marginal cost (cents per kilowatt-hour)
50	5.65	11.3	
			5.2
100	8.24	8.2	
			5.2
250	16.03	6.4	
			5.2
500	28.99	5.8	
			5.2
1,000	54.94	5.5	
			3.7
1,500	73.58	5.0	

made of such heat storage devices in the United States. Peak rates high enough to encourage consumers to transfer their use of power to off-peak periods are most efficient when peak loads are dangerously close to capacity and new capacity is very expensive. Although the latter condition certainly exists now, it appears that the industry will have ample excess capacity during the 1980s.

The other concept that has surfaced time and again in the rate-setting process is lifeline rates. The general notion, which is based on equity rather than efficiency, is that the first block of electric power (and perhaps other utilities as well) should be made available at very low cost so that the poor would be able to consume some minimum necessary amount. The equity of such proposals is not as obvious as it seems in view of the poor correlation between consumption of any one form of energy and income. Moreover, poor people may not have independent electric meters for each household. In the days of decreasing costs, the efficiency argument against lifeline rates was very strong: Setting price equal to marginal cost is efficient, marginal cost is below average cost, so price should be low with the difference made up by high charges for connection or initial blocks of power.

The situation now has changed for electricity, just as it has for natural gas. Because of the astonishing increases in UCC, the cost of new electric-generating capacity is far above the average cost for power produced in existing facilities. In an unregulated, reasonably competitive industry, we would expect price to rise until it equals marginal cost. If the public utilities commission actually sets price equal to marginal cost, the firm will make substantial profits. Still, the price is efficient, for it will force the consumer to ask whether an additional kilowatt-hour of power is really worth that price to him, and that is the question he should ask since the price is the cost to society of producing it. There is no virtue in having a person consume a kilowatt-hour that gives him 7 cents' worth of pleasure if it actually costs society 15 cents to produce.

The excess profits could be taxed away, although traditionally excess profits taxes have been difficult to design or implement in a fair and efficient manner. One alternative is to distribute some of the profits to consumers by setting the rate for the first few kilowatt-hours very low. In any event, the use of a fixed monthly base charge has very little to recommend it under contemporary conditions.

Billing for industrial users of electricity often involves a different form of two-part tariff. In addition to an energy charge for the number of kilowatt-hours of electricity used, the industrial bill generally has a demand charge for the maximum rate of energy consumption during the month. This makes a good deal of sense if we think of it as a charge for building and maintaining a unit of capacity dedicated to the particular user, but it is an imperfect device unless it is also adjusted for the time of day when the peak load occurs. An industrial peak that occurs in the middle of the night is much less expensive than one that occurs in the late afternoon.

Firms have reacted predictably to the demand charge by adopting load management techniques and even peak-shaving equipment to reduce the maximum demand during the month. Scheduling the use of certain pieces of equipment is a common practice, as is the use of small private generators at times of peak loads. Whether these adaptations are socially useful is questionable. They would be if all industrial and other peak loads were simultaneous, but if the load management techniques of individual firms use scarce resources to flatten peaks that could have been met anyway by the utilities, they have benefited only the firm, not society.

PEAK LOADS AND STORAGE

The peak load problem is much easier to manage in industries that can accumulate inventories. This raises the question of whether electricity could be stored to meet peak demands. The answer so far is that it is not presently feasible on a large scale, although the problem has attracted some serious research effort and it is not impossible that a useful solution could emerge. Electric utilities might someday store energy for peak periods by compressing air into underground caverns or spinning huge flywheels. Generators would still have to meet the peak loads, however.

The various types of storage batteries now available can be very convenient, but are rarely used in situations where we calculate cents per kilowatt-hour. Two prominent exceptions to this generalization involve electrically powered vehicles and wind-driven generating equipment. Both of these have carved out special niches in which they are routinely used. The area of application of each could also be greatly expanded by improved storage techniques. Wind-driven generating equipment is discussed in Chapter 13.

Vehicles powered by storage batteries have been used chiefly in mining and industrial applications where internal combustion engines would be hazardous and trailing electrical cables or trolley wires are awkward or expensive. Since the big increase in petroleum prices of the 1970s, efforts have been made to widen their area of applicability. The most promising application seems to be local delivery, where distances traveled per day are small, the vehicle returns to its own garage for recharging every night, and gasoline or diesel engines must endure much idling while covering the route. Clearly, improved electrical storage techniques could extend the limits of applicability of electric vehicles, perhaps to include household use for errands and commuting.

The closest thing to storage of electricity now practiced by utilities is pumped storage of water. The basic idea is to build a reservoir on top of a hill. During the off-peak hours water is pumped up to the reservoir. During peak hours the water runs back through a turbine generating electricity. By 1979 pumped storage turbines had a capacity of 12.9 million kw compared with total generating capacity of the electric utility industry of 616 million kw.[9] The conversion process always imposes some losses, so as a practical matter the amount of energy available from pumped storage is only about two-thirds of the energy put in. Opponents of particular pumped storage facilities sometimes suggest that it is foolish to invest large amounts of capital (and use up scarce or scenic land) to end up with less energy than you started with.[10]

The same argument could be made about other phases of the energy industries, however. Electric utilities invest substantial amounts in a distribution network to carry energy from the generator to individual consumers. If this results in a loss of 10 percent of the energy produced, is it therefore irrational? Ordinarily it is not, if the consumers will pay enough for the energy they receive to repay all the costs of the utility. Transformations of time and place are among the most basic ways to increase economic value, so the correct test of plans for pumped storage is economic: (1) Is it cheaper to build pumped storage capacity or new generating capacity? (2) Is there a cheaper storage possibility than the one under consideration? (3) Is it cheaper to shave peak loads than to accommodate them? Although the questions are simple enough, if the pumped storage facility requires chopping the top off a mountain, someone will have to put a price on the change in the scenery. *That* is difficult.

Ordinary hydroelectric installations have storage built into the system in the form of the reservoir. It is simple and relatively inexpensive therefore to incorporate peaking capacity in a hydro system by installing excess turbine and generator capacity. If the turbines were operated at capacity all of the time, there would not be enough water. The output is concentrated during the daily peaks. The use of hydro power to meet daily peaks is easier than its use to meet seasonal peaks because of fluctuations in water flows over the year, limited reservoir capacity, and the use of the same reservoir for flood control and recreation as well as electric power.

GROWTH IN DEMAND

The electric power industry met with extraordinary success during the first seventy years of this century. The diffusion of usage throughout the population is shown in Figure 11.5. With the rapid price increases of the 1970s (Figure 11.4), the old relationships began to break down. Nevertheless, looking at the broad trends gives us some perspective on changes in the use of energy.

Figure 11.6 shows total sales and sales to residential, commercial, and industrial users in the United States. The data are plotted on a semilog chart to show the steady rate of growth until the final period. It is clear that the 1970s saw an interruption in the steady and predictable growth of the preceding four decades. Electric power plants, however, take an excruciatingly long time to put into production. A coal-fired plant requires about a decade now, whereas the period from start to finish for a nuclear plant is at least fourteen years and may even be infinite with the increasing opposition of vocal segments of the public. Given that long planning horizon, the utilities must try to guess what will happen to consumption of electricity during the next decade and more. This has become increasingly difficult because the steady growth trend of the past has ended.

Forecasts are always perilous, but the basis can be improved by bringing in other relevant information. It is possible to look at consumption per capita or consumption per unit of GNP, but these show roughly the same pattern of smooth and predictable growth in the earlier period, combined with a recent breakdown in the relationship. This is hardly surprising, since both population and GNP are also generally smoothly growing series.

In this case, as in so many others, it is useful to consider price when we try to analyze consumption. The era of great growth in household consumption of electrical energy corresponded with the era in which electricity became cheaper in real terms. Conversely, the stagnation in consumption growth occurred during the era of price increases.

The analysis is complicated by other factors. The first is the behavior of other energy prices. In certain cases (notably home heating and some industrial uses), electricity can substitute for fossil fuels. The 1970s, of course, were also the era of rapid increases in fossil fuel prices. Figure 11.7, showing electricity prices and average

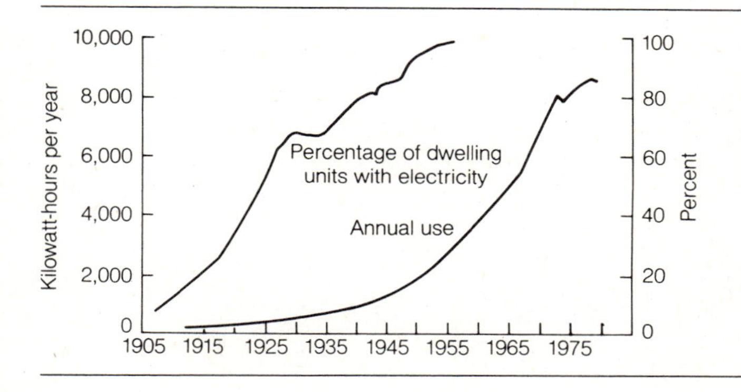

Figure 11.5 *Percentage of Dwelling Units with Electrical Service and Use per Customer in the United States, 1907–1979*

Sources: U.S. Historical Statistics *(1975), p. 827, S 108–109;* U.S. Statistical Abstract *(various editions).*

household consumption in selected parts of the United States, indicates the strength of the relationship. In the cities with cheap electricity, consumption has far exceeded that in expensive power areas.

General recognition of the ease with which electricity can substitute for direct use of fossil fuels in all household and most industrial uses lay behind some of the early optimism about nuclear power. If nuclear fission had fulfilled its early promise of providing cheap and inexhaustible power, then any remaining energy problems would be easy to solve. The same sorts of forecasts that were once heard for fission power are still being made for fusion and solar. But until such a state of bliss is achieved, we must confront the reality of several scarce and expensive energy sources serving as partial substitutes.

The fact that most electricity is produced from fossil fuels or from nuclear reactors (which have become more expensive) does not mean that electricity will continue to grow in price as rapidly as fossil fuels. In fact, the percentage increase in electrical prices has been less than that in other fuels. Moreover, since fuel costs amount to only about one-third of total costs, if only fuel costs were rising the net impact on total costs would be attenuated. As noted earlier,

Figure 11.6 *U.S. Use of Electric Energy, 1912–1981*

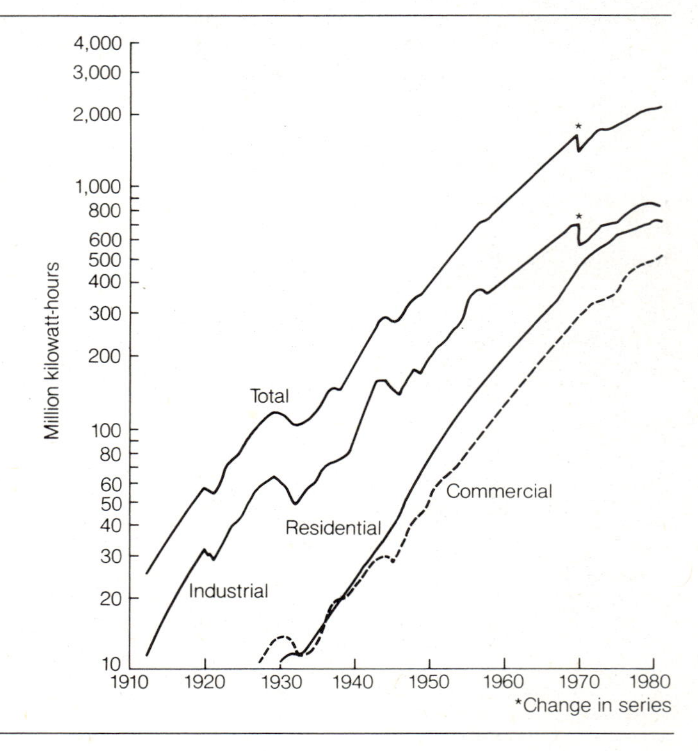

Sources: U.S. Historical Statistics *(1975), p. 828; U.S. Department of Energy,* Annual Report *(1981), p. 155.*

however, UCC has risen far more dramatically than average fossil fuel prices. Only the fact that electricity prices reflect average costs, which are dominated by cheap old capacity, has kept electricity from becoming far more expensive.

The owner of an electrically heated home who saw electricity rise from 3 cents per kilowatt-hour to 7 cents per kilowatt-hour during the 1970s had no incentive to switch to oil, which was increasing from 16 cents a gallon to $1 a gallon during the same period. The incentive to insulate and turn down the thermostat was certainly increasing, however. With fossil fuel prices apparently stabilizing for the 1980s and the most extreme impacts of the increase in UCC yet to be reflected in electric rates, it looks as though electricity will be growing more expensive relative to fossil fuels during the second half of the decade. If this analysis is correct, then electric utilities will not need to build much additional capacity during the rest of the 1980s. The utilities, of course, must spend these years planning for the new capacity required in the 1990s and beyond, which is a more difficult problem.

The problem is made more tractable by the fact that consumption responds so strongly to the price of electricity. If utilities were free to set their own prices, there would be no reason to fear a shortage of capacity. By raising prices to the level of marginal cost,

Figure 11.7 *Price and Consumption of Electricity in Various Cities, 1981*

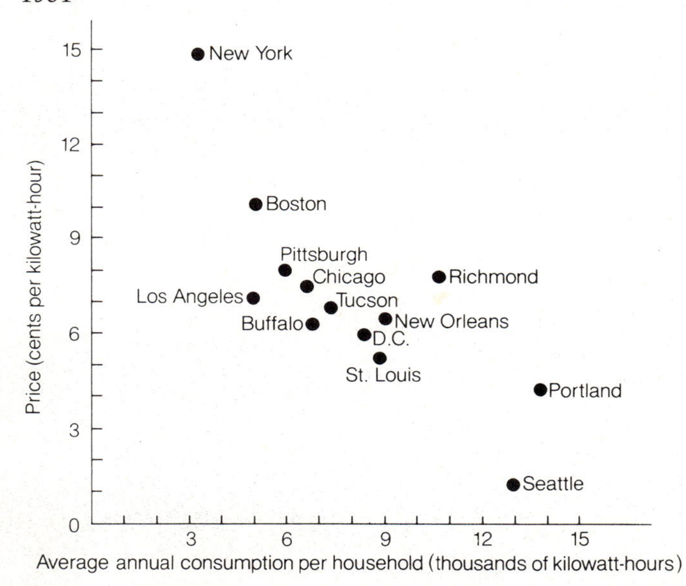

Source: *U.S. Department of Energy,* Electric Power Annual *(1982), p. 143.*

enough consumption could be eliminated so that very little new capacity would be needed during the next several years.

This view of the experience of the 1970s is sketched in Figure 11.8. The assumption is that the underlying processes determining the demand curve did not change at all. The demand curve has continued to shift out at the same rate as it did during all those years when consumption doubled every decade. The difference between 1970 and 1980 is that price rose enough to offset most of the increase in quantity consumed that would have occurred at a constant price. The question for the next decade is whether the price will continue to rise enough to halt the growth in consumption or whether it will stabilize near the 1980 price permitting the quantity consumed to resume its steady growth.

Of course, we might argue that the world has changed enough in ways other than price so that in fact the demand curve is not steadily marching to the right. But when a trend has continued for several decades, the burden of proof should be on anyone who thinks that a change has occurred.

PROSPECTS FOR THE ELECTRIC UTILITIES

The electric utility industry entered the 1980s in a financially stressed condition.[11] High interest rates are always difficult for an industry that borrows so heavily. The rapid escalation of construction costs was particularly burdensome. If a utility plans to double

Figure 11.8 *Demand for Electricity*

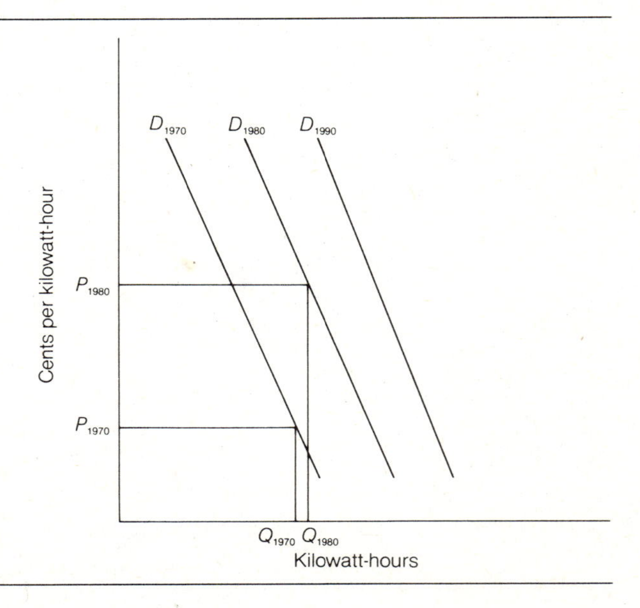

capacity during a decade when UCC increases tenfold, then it must plan for a twentyfold increase in fixed investment. Nuclear facilities encountered special problems from the public, from cost overruns, and from lack of reliability. Furthermore, as the expensive new facilities came on stream, the rate of capacity utilization for the industry declined.

Although forecasts of the growth rate for electric power have become increasingly difficult, the decision about the kind of facility to add when additional capacity is thought to be needed has become increasingly easy. The law does not permit burning of natural gas or oil in new central station power plants. They are also too expensive. Some additional hydroelectric sites became profitable during the 1970s because of the rapid increase in the cost of alternatives and renewed interest in the development of efficient small turbines. Some commercial geothermal plants are now on stream, and utilities that are located near potential sites would certainly have to think seriously about such a possibility as well. There have been many installations of wood-burning small industrial boilers, but the locations in which electric power plants could rely on wood (or other biomass) are limited by the cost of transporting such bulky fuel. Vermont has one such facility.

For most utilities in most locations, the choice was coal or nuclear. At the beginning of the 1970s, the monetary costs of the two seemed roughly comparable, although this topic requires further consideration in the next chapter. At this point we must raise the question of whether there are social costs involved that do not enter the private cost-benefit calculation. Clearly, both coal and nuclear power plants can be built in a dirty and unsafe way. The important question is the safety and cleanliness of a plant built today according to standard practice. Such plants will be influenced by government regulations designed to protect people and other features of the environment. The plants may not actually meet standards, however, because both people and equipment are fallible.

It is only a slight exaggeration to say that those who have studied nuclear power have concluded that coal is the better choice, whereas those who have studied coal have opted for nuclear. Federal government policy before the Three Mile Island (TMI) incident was to encourage both coal and nuclear. The atmosphere since TMI has been one of resounding ambivalence for reasons that are discussed in the next chapter.

Many of the environmental impacts of coal have already been mentioned, but these concerns are briefly sketched here as they relate to electric power. We have in Chapter 8 discussed the principal problems with coal mining, including the death and illness of miners, difficulty of reclaiming strip-mined land, water pollution from acid mine drainage and siltation, as well as a variety of local hazards and inconveniences.

Coal-fired power plants require a huge quantity of coal each year (2 million tons per year for a 1,000 mw plant that has a plant factor of 60 percent). Moving that mass of material creates a considerable nuisance and even some traffic hazard.

More to the point, when such a large quantity of anything is processed, fractional percentages of hazardous materials can add up to substantial amounts of poison that may be released into the atmosphere. Coal contains at least small amounts of every element, as well as an array of toxic organic compounds. The exact list of substances emitted will depend on the coal and the conditions under which it is burned. Attention has been focused on the sulfur content of coal and the sulfates emitted in the stack gas, which are thought to be a prominent contributor to acidity in rainwater. Coal also contains mercury, lead, arsenic, and beryllium.[12] Some coals, if burned without special controls, release more radioactive elements into the air than would any nuclear power plant under normal conditions.

Some of the emissions can be controlled, of course. The earlier technique was to spread the worst effects away from the immediate neighborhood by building a high enough stack to disperse fly ash and other pollutants widely. More recently, regulations have prescribed precipitators and now scrubbers, unless low-sulfur coal is used.

For the ultimate catastrophe, however, there are no simple solutions. The ultimate catastrophe is a significant change in the earth's climate. One hypothesis is that carbon dioxide resulting from the combustion of coal (or other fossil fuels) will retard the radiation of infrared rays into space, leading to a gradual rise in the temperature of the earth's atmosphere. Even a slight change could be disastrous if ice caps melted and agricultural regions became deserts. Despite the lack of firm evidence, the scenario is frightening enough to recommend appropriate vigilance. Recent indications that termites emit more carbon dioxide than the combustion of fossil fuels suggest that vigilance need not yield to panic right away.[13]

NOTES

1. *Historical Statistics of the United States* (Washington, D.C.: U.S. Government Printing Office, Bicentennial ed., 1975), pp. 812–813.
2. These temperatures are expressed in degrees Kelvin; that is, degrees of Celsius size but starting from absolute zero (−273°C or −460°F). In the Kelvin scale water freezes at 273°K, room temperature (68°F) is 293°K, and water boils at 373°K. The example of 800°K (or 1,000°F) corresponds to good operating practice of recent years.

3. The problems of the Harvard plant are described in the *Boston Globe*, September 6, 1981, p. 41; for a general analysis of cogeneration see Sam H. Schurr et al., *Energy in America's Future: The Choices before Us* (Baltimore: Johns Hopkins University Press/Resources for the Future, 1979), pp. 159–176; the economic analysis is developed by Paul L. Joskow and Donald R. Jones, "The Simple Economics of Industrial Cogeneration," *Energy Journal* 4, no. 1 (January 1983) pp. 1–22.

4. For further information and references, see David Huettner, *Plant Size, Technological Change, and Investment Requirements* (New York: Praeger, 1974).

5. These data are from U.S. Department of Energy, *Electric Power Annual* (November 1982), pp. 14, 35.

6. *Wall Street Journal*, December 3, 1981, p. 6.

7. Robert H. Shannon, *Handbook of Coal-Based Electric Power Generation* (Park Ridge, N.J.: Noyes, 1982), pp. 303, 306.

8. David Huettner, Jack Kasulis, and Neil Dikeman, "Costs and Benefits of Residential Time-of-Use Metering," *Energy Journal* 3, no. 3 (July 1982), pp. 95–112.

9. *U.S. Statistical Abstract*, 1981, p. 594.

10. Charles J. Meyers and A. Dan Tarlock, *Selected Legal and Economic Aspects of Environmental Protection* (Mineola, N.Y.: Foundation Press, 1971), pp. 286–311.

11. For an analysis of some of the implications, see Peter Navarro, "Public Utility Commission Regulation: Performance, Determinants, and Energy Policy Impacts," *Energy Journal* 3, no. 2 (April 1982), pp. 119–139.

12. U.S. General Accounting Office, "U.S. Coal Development—Promises, Uncertainties," EMD-77-43 (September 22, 1977), p. 6.17.

13. *Smithsonian* (April 1982), p. 90.

CHAPTER TWELVE

Nuclear Power

Nuclear power serves as a perfect reminder that *plentiful* does not mean *cheap*. The early dreams of electricity "too cheap to meter" have given way to serious questioning of the profitability of building nuclear power plants at all. Even people who have no emotional opposition to peaceful uses of nuclear power and are satisfied by the existing provisions for safety are dismayed by the high costs. Naturally, the cost issue is also raised by people who want to stop nuclear power for other reasons.

Although the question of cost is the easiest one to describe in numbers, there are other important questions as well. All other sources of energy kill people singly or in small groups at a slow but steady rate: Rock above a coal miner's head suddenly gives way and crushes him; a miner breaks regulations by putting a cigarette in his mouth and as he spins his lighter the entire mine explodes and thirty people are killed; a faulty valve at an oil well releases a cloud of hydrogen sulfide gas that kills two people; an ocean-drilling rig buckles and capsizes in a furious storm and 150 people are drowned. In comparison with the steady accumulation of death and injury in other energy industries, nuclear plants appear to be extraordinarily safe. But many people—even those who are not obsessed by the mushroom clouds that arose over Hiroshima and Nagasaki—keep asking the question: What if the worst possible accident at a nuclear plant occurs? Vainly, the proponents of nuclear power speak of the expected number of casualties. Is some other factor involved? This is a question to which we must return.

The other great issue is the relationship between nuclear power and nuclear weapons. Power reactors of the ordinary sort do not turn themselves into atomic bombs. The fear, however, is that the radioactive materials fueling reactors and contained in the spent fuel will be available for fashioning into explosives. The risk of diversion of fuel is also one to which we must return.

TECHNOLOGY

After achieving success with nuclear weapons, the U.S. Atomic Energy Commission began immediately after the Second World War to seek out peaceful uses of nuclear power. Despite talk of nuclear excavation and a great research effort on the nuclear-powered airplane, it soon became evident that the only users of nuclear power would be large ships, electric power plants, and perhaps a handful of extremely large industrial consumers of energy, such as steel mills.[1]

The two major constraints are the bulk of the reactors and the hazards associated with them. Even a reactor that is designed to produce a small amount of power requires massive shielding, so the early dreams of powering automobiles and trains directly with nuclear reactors faded fast. Still, the dream of cheap power suggested that trains could be electrified and automobiles powered by storage batteries or other techniques for making use of electrical power. Research funding for the nuclear plane continued into the 1960s, although the implications of a crash by such a vehicle are not pleasant to contemplate.

It was another military use, the submarine, that showed that nuclear fission could be controlled to produce useful energy. By getting rid of internal combustion engines and the limitations of storage batteries, submarines could stay submerged for weeks at a time and travel for months without refueling. These powerful military advantages motivated the research and development funding for nuclear reactors.

The earliest electric utility nuclear plants used the reactors designed for submarines. It was not until 1970 that the new generation of specially designed electric power plants became available in the United States. It is one of the many ironies of the nuclear power program that those early plants performed more reliably and cheaply than the ones that were specially designed to capture economies of scale and satisfy the particular requirements of the utilities.

All of the nuclear power plants in operation today use conventional steam turbines and generators to produce power. They differ from fossil-fueled plants in using a nuclear reactor to produce the heat to make the steam, rather than using a boiler fired by coal or some other fuel. Nuclear reactors can be designed in a number of different ways, but they have certain common features. The basic principle is that when a neutron encounters the nucleus of uranium-235 or certain other fuels, the fuel atom is split (fissioned) into different elements (fission products) plus neutrons and heat. The heat must be transferred out of the reactor to keep the reactor from overheating and to produce steam to power the turbine. The newly released neutrons will keep a chain reaction going if they encounter other fissionable atoms. A control material (such as boron steel or cadmium) that absorbs neutrons can slow or stop the chain reaction. A moderator (such as graphite, water, or heavy water) will slow the neutrons so that they are absorbed more easily by the fuel.

The reactors commonly used to produce electricity in the United States use uranium as a fuel. It is fabricated in the form of uranium dioxide pellets, each about 1 inch tall and less than ½ inch wide. The pellets are stacked up in thin metal tubes into rods that are one pellet in diameter and about 12 feet tall. About 200 fuel rods are grouped into a fuel assembly, which could also include control rods and instrument rods. The reactor might have close to 200 fuel assemblies. A large power reactor will have about 100 tons of fuel of which about one-third is replaced each year during the twenty or thirty years that the reactor runs.

The reactor is started up by raising control rods using magnetic clamps. Heat output is increased by raising more control rods and raising them higher. In an emergency, if the magnetic field is broken, the control rods will drop and by absorbing neutrons stop further fission. Even after all control rods are fully down, however, the reactor continues to produce a great deal of heat because of the radioactive decay of the fission products. It is thus vitally important that the cooling system continue to transfer heat away from the reactor core. If the fuel rods become hot enough to melt, radioactive material is released into the coolant and may escape into the atmosphere.

Reactors used in the United States use ordinary water as both coolant and moderator. The two main types in use are the pressurized water reactors (PWR) and the boiling water reactors (BWR). The BWR uses a single steam loop; that is, the water that circulates around the fuel rods in the reactor core goes out through a steam line to power the turbine, then is cooled in the condenser and pumped back into the reactor. This system is less complex, but there is more risk of spreading radioactivity to the turbines and other parts of the system.

The PWR has a separate primary cooling loop serving the reactor core and a steam loop powering the turbine. The heat is transferred from primary to steam loops in a heat exchanger called a steam generator. This obviously requires additional equipment. Furthermore, the primary loop operates at a higher temperature (330°C compared with 300°C) and is kept at higher pressure (about 2,000 pounds per square inch compared with 1,000 pounds per square inch) to keep the water from boiling. The higher temperature raises the theoretical maximum thermal efficiency and the separation of the two loops is additional protection against leaks of radioactive material, but the extra equipment and higher pressures increase the equipment cost and the higher temperatures increase corrosion problems. In general, corrosion is a particularly severe problem in nuclear reactors because materials must withstand the intensely radioactive environment, as well as high temperatures and pressures. The fuel rods must also be made of a material that permits neutrons to pass through readily, which limits the range of choices.

Although the BWR and PWR dominate the industry in the United States, other designs have been important in other countries. Pressurized heavy water reactors (PHWRs) use heavy water instead of ordinary water. The Canadian version, the only commercial PHWR, is called CANDU. It is efficient enough to use ordinary uranium, rather than the enriched uranium resulting from an expensive process of increasing the proportion of the fissionable U-235 mixed with the more common U-238. The CANDU reactor can also be refueled while operating, which can be expected to increase the number of hours per year that it can operate.

Gas-cooled reactors (GCRs) using carbon dioxide as a coolant were developed by Britain and France. The early versions are no longer being built, and France has shifted to PWRs. Britain has built a new version, the advanced gas-cooled reactor (AGR). The United States built two high-temperature gas-cooled reactors (HTGRs) using helium as a coolant before scrapping that approach.

The remainder of this chapter concentrates on the experience in the United States with BWRs and PWRs. The basic principles of generation of electricity using nuclear power are simple, as this bare outline of the technology suggests. Nevertheless, the actual process of generating the power, including the entire fuel cycle, involves a great deal of complex engineering using exotic materials. All power plants must contend with high temperatures and pressures. The nuclear power plant also has the serious problem of controlling radioactivity. It must also be equipped to operate safely during the long period before a reactor can be brought to a cold shutdown. Any process involving lags is more difficult to control precisely than is one that responds instantaneously to the controller. This inherent control problem is made more difficult by the radioactivity, which means that changes have to be accomplished by remote controls and operators must rely on remote readings from various sensing devices.

Whenever anything goes wrong in a nuclear power plant, repair is excruciatingly slow because of the need to minimize releases of radioactive materials and limit exposures of particular individuals. Often a job that would require a small crew working for a day or two in a conventional plant must be cut up into fifteen-minute segments so that no one is exposed very much. No job can be done very efficiently under such conditions.

The description of the technology at the plant site is only a part of the total picture. The fuel is the other part.[2] Mining uranium ore is much like other forms of hard rock mining. Miners are subject to such hazards as roof falls and haulage accidents. If a mine is poorly ventilated, the radon level can build up enough to create a lung cancer hazard. At the mine site the ore is processed into "yellowcake," which is uranium oxide (U_3O_8). This stage can present problems with contamination by radioactive dust. The tailings, or waste, from the concentration stage should be disposed of safely, because radon will be produced as a decay product of the remaining uranium.

The yellowcake resulting from the concentration step at the mill has uranium mainly in the form of U-238, which does not fission. There is a small percentage of U-235, which does fission, but this is not a rich enough fuel to use in a BWR or PWR. The standard way of enriching uranium to increase the concentration of U-235 is the gaseous diffusion process. The uranium oxide of the yellowcake is converted into the gas uranium hexafluoride. The gas is passed through ceramic filters. Since U-235 is lighter, it passes through the filters slightly more readily than U-238. By building a big enough series of such filters and processing and reprocessing the uranium hexafluoride, the proportion of U-235 is raised from 0.7 percent to 3 percent. France is now using a centrifuge as a cheaper alternative to gaseous diffusion, but the United States was slower to initiate construction of such a plant. The enriched uranium hexafluoride is converted into uranium dioxide (UO_2), from which fuel pellets are manufactured. These are assembled into the fuel rods.

As the reactor operates, the percentage of U-235 in the fuel declines. The spent fuel rods are still highly radioactive, however, because they contain a variety of decay products of the reaction that are themselves radioactive. Today these old fuel rods are stored in tanks of water at the nuclear power plants awaiting government action.

The choices are to find a safe way to dispose of the highly radioactive waste or to reprocess spent fuel into new reactor fuel. At the time when it seemed as though the nuclear power industry would grow rapidly and that uranium reserves were sparse, reprocessing attracted a lot of attention. Indeed, the government subsidized research on the breeder reactor, which was intended to produce more fuel than it consumed. At the present, natural uranium is abundant and cheaper than reprocessing or breeding fuel, so these receive little attention here.

GOVERNMENT ACTIVITY

The Atomic Energy Commission was heavily involved in promoting, subsidizing, and regulating nuclear power in its early days. Growing recognition that the functions may be inconsistent contributed to the separation of activities into the Nuclear Regulatory Commission (NRC) and the Energy Research and Development Administration in 1975. The major subsidies are listed in the following paragraphs, but the encouragement of the subsidies has been at least partly offset by the regulation to which the industry has been subject. All large power projects are regulated to control costs to the consumer and the environmental impacts of such large projects. The nuclear industry has the additional burden of the regulations motivated by the hope of keeping this complex technology safe. Explicitly considering the trade-offs among cost, aesthetics, and safety might have produced safer and cheaper plants than those that resulted from the actual mix of promotion, subsidy, and regulation piled on regulation.

Fuel

In the early days of the nuclear power program, the Atomic Energy Commission (AEC) retained ownership of the fuel, leasing it to utilities. The AEC also promised to do something with the waste fuel, even after the leasing program ended and utilities purchased their own fuel. The net effect of early AEC involvement was to guarantee a lower cost to utilities than they could otherwise have counted on and to reduce capital requirements, since fuel could be paid for as it was used.

Research and Development

Several other subsidies were also used to encourage adoption of nuclear power. The largest of these was the entire R&D program that led to the development of power reactors. Much of this would have been done for military purposes anyway. Some opponents of nuclear power have suggested that the costs that the taxpayers have already borne somehow prove that nuclear power is not economically justified. Since the early subsidies cannot be recovered by discontinuing a nuclear power program, however, the decision on a power plant should be made on the basis of current costs, forgetting the past. We might want to argue that we would be better off if the R&D had been directed toward solar or other sources of energy. That is an interesting historical argument, but it is irrelevant for current decisions.

Liability

The Price-Anderson Act (1957) limiting the liability of plant operators for nuclear accidents is one of the subsidies to the industry. Utilities were reluctant to own nuclear power plants unless they could buy liability insurance against the unknown risks. The insurance companies also were reluctant to take on the risks of the untried process. Uncertainty about the total liability was ended when the Price-Anderson Act limited total liability to $560 million. The federal government assumed part of that, utilities and insurance companies sharing the rest. Some such legislation may have been necessary to launch the industry, but practically all economists and other analysts have considered the maintenance of such a subsidy unjustified. The question of who bears the liability for the first $560 million is minor. The crux of the issue is the limitation on total liability, which means not that the consequences of an accident will be limited but that they will be borne by innocent bystanders, rather than by the company that causes the problem.

In other industries, of course, individuals who have suffered losses as a result of the behavior of some company cannot collect unlimited amounts. After the insurance carried by the firm is exhausted and the assets of the firm have been sold, the victim has no further recourse (unless the government will pass special legislation to spread the burden to all taxpayers). The Price-Anderson Act limits the financial consequences to the company for its own behavior. Ironically, even with that protection an accident serious enough to harm the public would probably bankrupt the company. The Three Mile Island incident, which involved only trivial releases of radioactive substances and did not damage property outside the plant site, nevertheless has wreaked financial havoc on Metropolitan Edison, the operator of the plant.

A more rational law would require utilities to carry some minimum amount of liability coverage and would leave the utility open to suits for any additional damages. In the general industrial setting, such requirements have had the added virtue of inspiring insurance companies to train experts on safety who provide an independent inspection of plants for the purpose of reducing losses by the insurance company. In view of the poor quality of training in parts of the nuclear power industry and the negligence of the NRC, the assistance of an interested insurance company could be useful.[3]

Waste Disposal

A subsidy may be implicit in the government's promise to do something about nuclear waste, but the point is uncertain because the government seems unable to come to grips with the problem at all. The waste continues to accumulate at the power plants. Already many plants have filled their routine waste-holding facilities and have resorted to building more storage capacity to continue operating. The Department of Energy now charges the utilities 1 mill per kilowatt-hour of electricity generated as a waste disposal assessment. It appears that this will be adequate if it is adjusted for inflation.[4]

Wastes from commercial nuclear reactors constitute such a minuscule volume that it is difficult to see why some solution would not be technically satisfactory. The standard estimate is that 100 acre-feet (less than 4,500,000 cubic feet) will accumulate from all electric power reactors in the United States by the year 2000.[5] With the slowing of the rate of nuclear plant construction, such an estimate is undoubtedly far too high. Of course, it is not the volume of reactor wastes, or even the toxicity, but rather the persistence of the radioactivity that gives some pause. Although the fission products will decay to harmless levels within 1,000 years, plutonium has a half-life of 24,000 years and so must be sequestered "forever."

We should bear in mind that the problem of disposing of radioactive waste will not go away even if all nuclear power plants are shut down. Much of the waste now awaiting disposal arose from the nuclear weapons program or from nuclear reactors powering ships and submarines, so we will have to learn to deal with the problem somehow.

Spent nuclear fuel could be reprocessed to extract uranium-235 and plutonium for use as new fuel for power reactors. This makes the waste problem more manageable because it leaves only elements that will be harmless within a thousand years. The objection to reprocessing is that it moves the United States closer to a plutonium-based economy, which could present hazards because plutonium is the easiest material to use in constructing a nuclear bomb. The risks of diversion of plutonium by terrorist groups or agents for a foreign government would require extremely tight controls over all stages of transportation and reprocessing.

This same concern lies behind the opposition to the breeder reactor, for the breeder is designed to produce a great deal of plutonium, which could then be separated from the waste and used as fuel. Since other countries are building breeders and reprocessing facilities anyway, it is not obvious that the refusal of the United States to do so will contribute to the safety of the world.

Whether the waste is reprocessed or not, we are left with something that must be put out of reach. Some have conjured up the image of our barbarous descendants flinging nuclear waste, the dung of our electric draft horses, at each other in their tribal battles. Should we worry about that? If humanity loses its civilization so completely after the start we have given it, perhaps our descendants *should* wallow in our nuclear wastes in the hope that it produces some better mutations.

Few would treat such churlishness seriously, however. Civilizations do decline into barbarism, and perhaps we do have an obligation to lock up wastes to keep even barbarians from defiling their environment. Attention has been focused primarily on locations for permanent burial, such as old salt mines or other stable rock strata. Ocean burial is also a possibility in areas of rapid siltation. Every such permanent solution always arouses opposition, however, so retrievable storage in some underground warehouse might be more acceptable. If a better location is found or some serious objection develops to the existing location, the wastes could be moved.[6]

In any event, the government has an obligation to make some decision, carry it through against the inevitable political furor, and then set appropriate prices so the industry pays the full cost of purchasing fuel and disposing of waste. The situation where government refuses to carry out its policy-making role on waste, but offers subsidies to encourage the activity that generates it, guarantees inefficiency.

The problem of disposing of old plants has caused almost as great a frenzy. The containment structures and equipment will accumulate some radioactivity that will make them unsafe to occupy when they have outlived their usefulness. Discussion of the problem usually assumes that plants will be dismantled and all radioactive materials disposed of or that they will be blocked up with concrete and guarded permanently. The present value of perpetual guarding is low relative to other costs, but it can be reduced even further if the new plant is built adjoining the obsolete one or even incorporating many of the structures and components of the old plant. In such a case, the old plant would require no additional guards. It is, of course, important that decommissioning costs be estimated at the outset of the plant's life so they can be borne by consumers of the power. Otherwise, today's consumers are being subsidized by whoever will pay the decommissioning costs in the future. The small reactor at Shippingport, Pennsylvania, that generated the first commercial nuclear power in 1957 was expected to cost $79 million to dismantle when the work began in 1984. It had less than 10 percent of the capacity of a modern reactor and was located where the parts could be shipped by barge to Hanford, Washington, for burial.[7]

THE OPERATING RECORD

Nuclear power plants are expensive to build and cheap to operate compared with coal-fired plants. Whether the electricity they produce is cheaper than electricity from coal is strongly influenced by the number of hours the plant operates. Because the variable costs are so low, nuclear plants are generally intended as base load plants; that is, the utility plans to produce as much electricity as possible with a nuclear plant once it has built it. If we multiply the design capacity of a plant (say, 1,000 mw) by the number of hours in a year (8,760), the product (8.76 billion kwh) is the amount of electrical energy the plant would produce in a year if it were operated at full power all the time. This is a plant factor of 100 percent. This same statistic is sometimes called capacity factor, load factor, or operating rate.

No plant operates at 100 percent plant factor year after year. The BWRs and PWRs must be shut down periodically for refueling. Other maintenance operations and replacement of various pumps, valves, and instruments with known limitations to reliable life will be planned for the scheduled maintenance periods. Then there are the shutdowns for unscheduled repairs. These are troublesome, even if they are regular enough to estimate in a statistical sense, because they do not come at convenient times. The advantage of scheduled maintenance after all is that it can be done at times of slack demand when the maintenance crews are not engaged in other projects.

Nuclear plants are also subject to precautionary orders by the Nuclear Regulatory Commission not to exceed some fraction of full power when the NRC is concerned about a possible safety problem. A utility heavily dependent on nuclear power might have to run nuclear plants at less than capacity during slack periods when the demand for power is low. In a technical sense, a nuclear power plant is well suited to such load-following operation, because power output can be varied quickly and easily. Unfortunately, however, the economic success of a nuclear power plant requires a high plant factor.

The calculations by which nuclear plants were justified assumed 70 percent plant factors. These have not been achieved. As Table 12.1 indicates, 60 percent comes closer to actual experience, but the range of results has been extraordinarily large. The implications both of the lower average plant factors and of the variation are serious; a utility that expected a 70 percent plant factor and obtained only 50 percent would find that capital cost per kilowatt-hour of output had increased 40 percent. When this is combined with the huge increase in UCC, much of which was not anticipated when plants were begun, the increase in capital cost per kilowatt-hour has been breathtaking.

Table 12.1 is restricted to the U.S. experience, but Surrey and Thomas performed a detailed analysis of data for 152 commercial nuclear plants in seventeen countries completed by 1978.[8] Only the PHWRs were able to achieve plant factors exceeding 70 percent with any regularity. Of the reactors commonly used in the United States, the PWRs were more reliable than the BWRs, with one-third of the reactor-years below 60 percent plant factor in the former case and more than one-half of the reactor-years below 60 percent plant factor in the latter. The gas-cooled reactors relied on by the British avoided the extremes of poor performance.

Surrey and Thomas tested numerous hypotheses and common speculations about factors in the reliability of nuclear plants. The strongest relationships were those relating to very large plants and to differences by manufacturer. During the period they studied, turbine generators exceeding 800 mw did not work reliably in either conventional or nuclear plants. It is not obvious why scaling up to the larger machines posed such difficulties or why the utilities persisted in ordering them.

The major differences in plant factors of power plants built by different manufacturers are indicated by the data in Table 12.2 compiled by Surrey and Thomas. Of the manufacturers in the

Table 12.1 *U.S. Nuclear Power Plant Operations, 1973–1983*

	Reactors licensed (number)	Electricity generated (million kilowatt-hours)	Nuclear share of all electricity (percent)	Capacity (million kilowatt-hours)	Capacity factor (percent)
1973	40	83,479	4.5	13.850	63.2
1974	53	113,976	6.1	29.921	43.5
1975	56	172,505	9.0	35.671	55.2
1976	62	191,104	9.4	40.642	53.5
1977	67	250,883	11.8	45.554	62.9
1978	71	276,403	12.5	49.385	63.9
1979	71	255,155	11.4	50.604	57.6
1980	71	251,116	11.0	51.941	55.1
1981	74	272,674	11.9	55.051	56.6
1982	77	282,773	12.6	59.552	57.2
1983	80	293,677	12.6	62.809	54.8

Source: Monthly Energy Review *(September 1984), p. 87.*

United States, Westinghouse achieved a plant factor of 66.9 percent, whereas Babcock & Wilcox managed only 53.7 percent. Since these data apply to the 1975–1978 period, before Three Mile Island, recent data for Babcock & Wilcox would undoubtedly be worse. Both of these firms produce PWRs, so the data are not confused by the generally poorer performance of BWRs. The range of performance worldwide is even more dramatic, the PWR plants built by Siemens/KWU averaging 78.7 percent and the BWR plants of AEG/KWU averaging 43.4 percent.

This strong association of reliability with particular manufacturers raises the interesting question of why utilities have been so slow to react to the differences. Part of the problem may be the long lead times, since meaningful commercial experience dates back only to the mid-1960s, whereas the plants that were on stream in 1975 resulted from decisions made a decade or so earlier.

Most of the failures that have forced the temporary shutdown of reactors have been in conventional parts of nuclear plants. The reactor core and the nuclear auxiliary and emergency systems have functioned reasonably well. Failures of such components as steam generators, turbine generators, and valves bespeak inept engineering and slovenly management, rather than peculiar problems of managing a complex technology. A close examination of the Three Mile Island incident confirms that impression: The problems involved human failure to leave valves in the correct position, mechanical failure in one valve, and general confusion in operation during the incident, combined with an inability of the operators to sit back and allow the machine to solve its own problems.

Table 12.2 *Nuclear Power Plant Capacity Factors, 1975–1978 (by manufacturer)*

Pressurized water reactors		Boiling water reactors	
Manufacturer	Capacity factor	Manufacturer	Capacity factor
Westinghouse	66.9	General Electric	55.7
Babcock & Wilcox	53.7	ASEA-Atom	63.8
Combustion Engineering	64.4	AEG/KWU	43.4
Siemens/KWU	78.7		
Mitsubishi	59.9		
All firms	65.8	All firms	56.1

Source: *John Surrey and Steve Thomas, "Worldwide Nuclear Plant Performance: Lessons for Technology Policy,"* Futures *12, no. 1 (February 1980), p. 12. Reprinted by permission of Butterworths, Guildford, U.K.*

The fact that poor management by equipment manufacturers and utilities is responsible for low plant factors does not mean that the problems are less serious. While government vacillates over energy policy and demonstrators chant their slogans, costs continue to exert their pressure on the daily decisions of utilities, regulatory commissions, and especially the consumers of electricity.

COSTS OF GENERATING ELECTRICITY

An electric utility planning for future capacity will in most cases be forced to choose between coal and nuclear power. It will also have to guess how much electricity it will be able to sell at the high price necessitated by expensive new capacity. Of course, utilities should look more widely to see whether particular circumstances permit cheaper use of wind, sun, water, waste, geothermal, or other sources. For the next two decades, however, these do not seem likely to displace coal and nuclear as the main sources of new electric capacity.

Table 12.3 gives a comparison of costs of coal and nuclear power. These costs were expressed in 1975 (RFF) and 1979 (Komanoff) dollars but were forecasts for plants to be completed in 1985 and 1988, respectively. The total costs for the two types of power were close, as might be expected for a time when utilities were actively building both. The RFF data suggest a slight advantage for nuclear in the national average cost data. Specific plants, however, face their own conditions, so it would be expected that coal plants would be cheaper in some locations and nuclear plants in others. Komanoff,

Table 12.3 *Cost Estimates for Electric Power from Coal and Nuclear Plants (cents per kilowatt-hour)*

	Resources for the Future		Komanoff	
Costs	Coal	Nuclear	Coal	Nuclear
Capital and fixed	0.80	1.03	1.34	3.07
Fuel	0.87	0.48	1.96	1.09
Operating and maintenance	0.21	0.15	0.62	0.62
Total	1.88	1.66	3.92	4.78

Sources: *Sam H. Schurr et al.,* Energy in America's Future: The Choices before Us *(Baltimore, Md.: Johns Hopkins University Press/Resources for the Future, 1979), p. 288; and Charles Komanoff,* Power Plant Cost Escalation: Nuclear and Coal Capital Costs, Regulation, and Economics *(New York: Van Nostrand Reinhold, 1981), p. 281.*

in contrast, found that coal was the cheaper source of power even in the Northeast where coal is relatively expensive.

The distribution of costs is also significant. Nuclear plants have heavier capital costs and lower fuel costs than coal plants. Operating and maintenance costs are similar. The data from the mid-1970s suggested that coal and nuclear were very close in cost, meaning that the choice was hard to make but not too important. By 1980 the more rapid growth in nuclear costs gave the edge to coal.

Such an analysis is fraught with difficulties that a simple table of numbers glosses over. One astute analyst concluded a detailed review of the available data with the observation that they could be used to support any conclusion.[9] The three major reasons why the picture is so cloudy are the following: (1) The relative environmental costs of coal and nuclear have not been accurately identified. (2) The fuel and waste disposal costs cannot be fully known until government makes some policy decisions. (3) The appraisal of capital cost per kilowatt-hour requires a host of tenuous assumptions. Item (1) is considered later and (2) is ignored. Item (3) is by far the most significant, so the major issues are summarized here.

Unit Capacity Cost

In attempting to estimate capital cost per kilowatt-hour, the major question marks are UCC, the timing of investment, the appropriate capital charge, and the capacity factor. The uncertainties in each of these are immense, and our accumulating experience has shifted all of the estimates in an unfavorable direction. The trend in unit capacity cost was mentioned in the previous chapter. In short, it has leaped upward. The UCC compiled by Huettner for the 1960–1970 period ranged from $155 to $457, although the lower costs for the early years resulted in part from an outright subsidy by the Atomic Energy Commission in the name of research and development.[10] Reactor manufacturers in the early years also offered low prices in the expectation that they were thereby establishing themselves in an industry that would become very profitable. Most of these early units were small, so there were reasonable grounds to expect that large reactors specially designed for power plant use and produced in some quantity would have a lower UCC, as indeed had been the experience with fossil fuel plants as the industry matured.

By the time of the RFF study, summarized by Schurr et al., the record of cost escalation gave cause for worry. Nevertheless, the authors used an average UCC for the United States of $531. These are 1975 dollars for a plant available in 1985 and are "overnight" construction costs; that is, they do not allow for interest on the funds tied up in the project between the moment it is initiated and the moment it is operating. These are significant matters because of the rapid inflation as the 1970s drew to a close, but if all prices and wages had moved at the same rate it would have been irrelevant for the choice between coal and nuclear. Prices do not move at the same rate throughout the economy, however. It is generally assumed that construction costs were increasing at a particularly rapid rate during the 1970s. Moreover, coal and nuclear fuel prices have undergone substantial short-run changes that have made forecasting perilous.

Since nuclear plants are so expensive to build, the treatment of capital costs is very important in determining profitability. It is precisely the accounting for capital costs that is always most distorted by inflation. The rapid increase in construction costs added particular difficulties in this case, as did the delays in construction. As construction periods stretch out and interest rates rise, the impact of interest during the construction period becomes significant. For example, if the interest rate is 10 percent, an expenditure of $100 million to initiate a plant will have accumulated to $200 million of principal and interest after seven years and will have doubled again to $400 million by the time the plant is completed if completion takes the fourteen years that now seems to be average. Over the plant's operating life, it must repay with profits the $400 million—not $100 million—if the project is to be considered an economic success. If interest rates were lower or construction periods could be shortened, such extraordinary burdens could be reduced.

The implication of this is that UCC cannot be determined without knowing the interest rate. This problem became particularly acute toward the end of the 1970s as nominal interest rates soared and construction times lengthened. In its political form the issue is argued before the state public utilities commissions when they determine the time at which construction work in progress can enter the rate base. If plants could really be built overnight, or even in one year, the concern would be minor. With construction requiring many years, utilities would like to add construction costs to the rate base as soon as the money is spent so that the company can start earning the allowable rate of return right away. The technically correct procedure is to accumulate the construction costs and the interest, compounding the interest until the plant is finished. It will then be a much more expensive plant entering the rate base.

In effect, the latter procedure treats the utility as two companies: an operating company and a plant construction company. The operating company can be thought to purchase the completed plant from the construction company the day it is ready to go on line. At that moment the plant enters the rate base. What price should the utility pay for the plant? If this were a market transaction, the correct price would not be the cost of the facility but rather the market value. This does not help much because there is no regular market in such assets, but it focuses attention on the two crucial questions: (1) Was the plant built at the lowest possible cost? (2) Should the plant have been nuclear or coal? Since nuclear plants must meet more exacting standards than conventional plants, it is sometimes argued that the utilities are not capable of coping with the extra care and complexity. The Zimmer plant (near Cincinnati) and the Midland, Michigan, plant give examples of allegations of faulty construction and in the latter case tremendous cost overruns. Under such conditions, should the entire cost of the plant be added to the rate base?

If the utility and the construction company were negotiating a price for a completed plant, we could analyze the value of the nuclear plant relative to a coal-fired one. Unfortunately, the premium that the firm should pay for a nuclear plant to reduce the fuel and operating costs of a coal plant depends on the interest rate. How much would you pay to reduce operating costs $60 million per year for twenty years? At an interest rate of 10 percent you should be willing to pay up to $520 million, but at an interest rate of 20 percent the saving is not worth more than $270 million.

It is possible therefore to compare the cost of a coal or nuclear plant at a given interest rate. The construction branch of the utility could see whether it was worth building a nuclear plant to sell at that target price based on coal. The construction company, however, would have to allow for the accumulation of interest on all capital charges from the beginning of the project until it can sell it to the operating branch to include in the rate base. In equilibrium, the value of the nuclear plant at completion must be enough to compensate the construction branch for building it, without exceeding the premium over a coal plant that the operating company should be willing to pay.

The Interest Rate

Since both of these values depend on the interest rate, there is no way to avoid confronting that knotty issue. It has become commonplace to distinguish between the *nominal* interest rate quoted in financial markets and the *real* interest rate, which is approximated by subtracting the inflation rate from the nominal rate. In a year when corporate bonds yield 15 percent and prices rise 10 percent, the real rate of interest would be the difference of 5 percent. This approach makes a good deal of sense to individual investors who want to know how much income they can spend without impairing their capital. Investors who start the year with $100 and earn and spend the $15 of interest income will find themselves poorer at the end of the year because the price increase of 10 percent has decreased the purchasing power of their capital to $90. They can use the real interest calculation as an easy test to see how much they can spend without becoming poorer each year.

The case of public utility accounting is not exactly analogous. The only justification for separating the observed corporate borrowing rate into "real" and "inflation" components is an assumption that the prices of inputs, assets, and products are all rising at the rate of inflation. This assumption does not apply to any specific industry. Traditionally, UCC decreased from decade to decade. It would have been highly misleading to assume that capacity built in 1950 would have been worth more in 1960 because of general inflation. Similarly, over long periods the price of electricity was falling, rather than rising with general inflation. Meanwhile, the interest that the utility had contracted to pay its bondholders was as real as any other obligation that the utility had to meet.

The only justification for the use of a lower real interest rate is an explicit forecast of the behavior of the prices that the particular firm will be able to obtain in the future by selling its product or of the cost of replacing its assets. By this criterion the real rate of interest during the 1960s was very low and often negative. Prices of replacement capacity were about to rise so rapidly that it would have been profitable for utilities to go deeply into debt to build surplus capacity at the very low prices then prevailing. Such forecasts are difficult to make or justify. In the 1970s, moreover, the situation changed dramatically. Utilities borrowed large amounts at high nominal rates to build plants that were worth less when completed than the investment that had been sunk into them. It is true that UCC was rising, as were prices of electricity and most other items, but the fact remains that the expected value of the stream of quasi-rents from the incremental capacity was less than the cost of construction. The real rate of interest is a misleading and even dangerous concept in decision making by the firm.

Utilization of Capacity

The reality has been high nominal rates combined with high UCC. This has put the utilities under great pressure. Moreover, the more rapid increase of UCC for nuclear, combined with its relatively high capital cost in the best of times, put nuclear power at a special disadvantage during the late 1970s. At that time, however, the decisions had already been made based on the low costs and reasonable operating record of earlier years. The worst problems, of course, are encountered by utilities that find themselves with so much capacity that even finishing plants that are under way seems pointless.[11] Public utilities commissions are reluctant to put these in the rate base: After all, unregulated firms cannot ordinarily charge consumers for excess capacity. In fact, excess capacity generally depresses prices. Nevertheless, the worst failure of a public utility in the United States is often assumed to be the failure to deliver whatever quantity the consumer will buy at the regulated price. If the public and the utilities commissions believe this, they must be prepared for the impact on cost of having all errors in the direction of excess capacity. The debate over who pays for unfinished plants (many of which may never be completed, whereas others represent a substantial investment that was made a decade too early) rages in several states.

Even without the burden of unfinished plants, which are a symptom of the temporary problem of adjusting to a lower rate of growth, the burden of plant costs has grown rapidly. The UCC data are shown in Table 12.4. The leap from the $400–500 range to recent estimates in excess of $3,000 reflects the cumulation of all of the problems mentioned earlier: (1) the general inflation in construction costs, (2) high interest rates that have caused early spending on a long project to compound at a faster rate, (3) the long delays in the entire process that have resulted from more cumbersome licensing procedures and more determined opposition by the enemies of nuclear power, (4) changes in design during construction as a result of heightened concern about safety or environmental effects, and perhaps (5) insufficient managerial capacity in some utilities to carry out both the complex construction project and the increasingly difficult public relations campaign necessary to ensure its acceptance by an increasingly skeptical public.

Table 12.4 *Unit Capacity Cost for Nuclear Power Plants (dollars per kilowatt)*

Year of first commercial operation	Unit capacity cost	Plant (actual average unless specified)
1960	270	Rowe
1962	542	Indian Point
1971	147	
1972	319	
1973	261	
1974	358	
1975	462	
1976	508	
1977	690	
1978	639	
1979	653	
1980	615	
1981	1,097	
1982	757	
1985	3,200	Perry

Sources: *David A. Huettner,* Plant Size, Technological Change, and Investment Requirements *(New York: Praeger, 1974), p. 89; U.S. Department of Energy, Energy Information Administration, "1983 Survey of Nuclear Power Plant Construction Costs," 0439[83] (December 1983), p. 8. The estimate for the Perry, Ohio, plant was given by a spokesman for the Cleveland Electric Illuminating Company in October 1984.*

The high UCC combined with the high interest rates implies a correspondingly high annual capital charge. If in the early 1970s we had assumed a cost for a 1,000 mw plant of $500 million and an annual capital charge (interest plus amortization) of 10 percent, it would not have seemed unreasonable. Now it would be more appropriate to assume a $3 billion cost for a plant of that size and a 15 percent capital charge in view of the interest rate that an electric utility typically will pay today. Thus the annual cost per kilowatt of capacity has gone from $50 to $450.

A key question is how many kilowatt-hours of energy will be generated in a year with the kilowatt of capacity that now costs $450 to own for the year. Many of the early studies of nuclear power assumed that plants would operate 70–80 percent of the hours in a year, on average. (Shutting down for refueling at intervals slightly greater than twelve months will influence the annual rates.) At a 70 percent plant factor, a plant would generate 6,132 kwh of electricity for every kilowatt of capacity. If the capacity cost $50 per year, the capital charge would be about 8 mills per kilowatt-hour. In view of the operating experience, a safer estimate of plant factor would be about 55 percent. This means that the plant would generate 4,818 kwh of electricity during the year for each kilowatt of capacity.

If it also costs $450 per year for that kilowatt of capacity, the capital cost alone for power produced from a new nuclear plant is about 9.3 cents. In other words, the combination of higher UCC, higher interest rates, and lower reliability has raised the capital cost of nuclear energy tenfold over the predictions that seemed plausible based on the experience of the 1960s. Since the residential price of electricity in 1981 was in the vicinity of 6 cents per kilowatt-hour, nuclear power seems to have priced itself out of the market, at least for the moment. This is reflected in the cancellations of plants in various stages of planning and construction. Even before Three Mile Island, utilities had ceased to order nuclear plants.

The Cost of Nuclear Fuel

Although capital costs now dominate the total cost of nuclear power, the other costs of nuclear energy are worth brief mention. In particular, since nuclear fuel is derived from the mineral uranium, we should consider the adequacy of reserves and the responsiveness of reserves to price increases.

Uranium is a metal rather than a fossil fuel, and in many ways the analysis developed for the common metals like copper and iron is useful in analyzing the economics of fuel reprocessing and breeder reactors. For uranium as for most other metals there is a trade-off between the expenditure of scarce labor, capital, and mineral reserves in mining the crude ore and the expenditure of labor and capital in more precise separation of useful material from the crude ore. There is also a trade-off between mining more crude ore and reusing and recycling the existing stock of material.

Various rocks contain some uranium, generally in the form of U_3O_8. A typical mine today might process 8 or 9 pounds of ore for every pound of U_3O_8 (yellowcake) shipped. If the price were to rise substantially, however, it would be worthwhile to mine leaner ores and to process the ores more carefully so that a smaller proportion of the uranium would be lost with the tailings (waste rock).

Production and marketing of yellowcake is reasonably competitive. About fifty firms produce yellowcake, and although large firms are well represented among the producers, the twenty-six largest energy producers accounted for only 35 percent of uranium in 1982.[12] Furthermore, the international market is now well developed, with exports and imports occurring routinely.

The electric utilities buy yellowcake in the spot market and by contract for future delivery. Prices have fluctuated greatly; transactions have been reported at less than $15 per pound and more than $120 per pound within a few years' time. New transactions recently, however, have been in the vicinity of $20 per pound. Even after the market recovers from its depression and adapts to the lower rate of nuclear power production now forecast, there seems to be no reason for prices to exceed $40 per pound. Resources that can be exploited at that price are being discovered in many countries.

Before the utility can use the uranium as fuel for an ordinary power reactor, it must have it enriched. Uranium as it is found in the earth is mainly the stable isotope U-238, with about 0.7 percent in the form of the unstable U-235. The proportion of U-235 is raised to about 3 percent for use as fuel (and higher for weapons). The U.S. government built the enrichment plants to produce material for weapons, but the utilities can have enrichment done by paying for the SWUs (separative work units) performed.

The utility has some choice between buying a lot of enrichment services and discarding relatively little U-235 in the tailings or discarding a higher proportion of the uranium but using a smaller amount of enrichment services. As an approximation, however, 600 tons of natural uranium, containing 4.2 tons of U-235, will be separated into 100 tons of enriched uranium, containing 3 tons of U-235, and 500 tons of depleted uranium, containing the remaining 1.2 tons of U-235.

Enrichment services in 1980 cost about $400,000 per ton of product, which means that the cost of nuclear fuel to the utility is almost equally divided between mining and milling, on the one hand, and the manufacturing operations of enrichment and fabrication, on the other. The gaseous diffusion process used for enrichment in the United States consumes a great deal of electricity. The newer centrifuge process, used in France and some other countries, consumes less electricity. A still newer process using lasers has attracted some research effort. Because of the availability of the newer technologies, it seems doubtful that the cost of enrichment services or of completed nuclear fuel will rise very much.

Even allowing for the cost of disposing of spent fuel, mining, preparation, and use of fuel approximates 1 cent per kilowatt-hour. As long as that cost seems unlikely to rise sharply, it is difficult to justify taking extreme measures to conserve fuel further. In particular, the breeder reactor and reprocessing of spent fuel could be justified only if natural uranium were scarce, since their main economic characteristic is the use of large amounts of capital and other inputs to conserve uranium.

NONMONETARY COSTS OF NUCLEAR POWER

In an earlier era the method of generating power would have been left entirely to the discretion of the utility company, as long as supplies were adequate and reliable and the price seemed reasonable. With the increasing concern about environmental and social costs of various kinds, the time when firms could make their own decisions based on monetary costs has ended. Ideally, of course, the government would identify the external costs of all activities and levy those on the firms in the form of a tax. If emitting a pound of sulfur from the smokestack does 40 cents' worth of damage to society, then the firm would be charged 40 cents for every pound of sulfur it emits, but it would be free otherwise to make its own decisions.

In our less than ideal world, things are much more complicated. In most cases, it is not possible to specify the amount of physical damage from an economic activity (how many people will contract cancer as a result of a particular power plant) or to put a dollar price on a particular impact (what is the cost of a case of emphysema). Some of the questions about physical impacts can be answered by further research, but most of the problems in attaching price tags to impacts are conceptual and so defy easy solution.

Since no solution is possible to most of these problems of measurement and analysis, society has adopted a quite different approach. Each energy source is subject to specific restrictions and standards, adopted in total disregard of the net impact on the health and safety of the public of the combinations of regulations on different industries.

Looking first at the actual operating records of the industries, it is easier to estimate the risks associated with normal operation of nuclear than of coal power plants. After looking at normal operations, it will be necessary to say something about the possibility of major disasters.

All human activities are stained with the blood of those who died or were injured as a result of them. Absolute data on number of deaths and illnesses therefore mean nothing. We do not give up the freedom and convenience of the automobile to avoid the 50,000 deaths a year that it causes; but it is possible to improve our chances of survival (don't drive when drunk, drugged, or drowsy) even while recognizing that some risk remains (we may encounter someone who *is* drunk, drugged, and drowsy). There are always safer ways to do anything, but sometimes the cost is too high. The 55 mph speed limit has saved some lives—but by increasing the time spent on every trip by a few minutes it aggregated to a cost of 100 years of additional driving time per life saved. Only someone passionately devoted to driving at 55 mph would consider that a satisfactory benefit/cost ratio.

The standard comparison with nuclear power is the coal-fired plant that will be built if the nuclear plant is rejected. Within a broader framework, however, all energy sources are substitutes. The economy could rely more heavily on natural gas, which is very clean and safe, if lingering doubts about availability can be resolved. At the other extreme, wood is still plentiful in many areas and will probably remain so considering the rate at which its regular users burn themselves up.

In comparing the risks for different energy sources, controversy surrounds even the question of what activities to include: accidents, routine operation, the fuel cycle, construction of the plant, and manufacturing the materials from which it is to be built. Herbert Inhaber tried to draw together the most inclusive set of numbers for the largest group of possible sources of energy.[13] After revising his results in response to the criticisms, he presented the data included in Table 12.5.

The most striking feature of the table is the wide range of uncertainty about the effects of power production by each of these sources. This is certainly not surprising for wind and solar power, since there is not yet enough experience on which to base firm estimates. It is noteworthy, however, that the hazards of such clean sources of energy are far from negligible. People do fall off roofs and towers and undoubtedly will succumb to flying fragments of broken windmill blades. Natural gas is the cleanest and safest energy source, despite the number of people who have been killed in gas explosions and in the hazardous drilling operations.

The main point is the comparison between coal and nuclear power. The range for coal is 3.3–107 deaths to produce 6.75 billion kwh of electricity, the amount generated in a 1,000 mw plant that has a capacity factor of 75 percent for a year. The range for nuclear power is 0.17–1 death to produce the same amount of electricity. Even if we take the high end of the range for nuclear and the low end for coal, nuclear is far less hazardous. The only source estimated to be less hazardous than nuclear is natural gas.

Since this result is contrary to popular belief, it is worth looking more closely at the basis for such estimates. The estimates of Resources for the Future are also presented in Table 12.5.[14] RFF has managed to raise nuclear and reduce coal hazards enough to have the two ranges overlap, but any reasonable examination of the data still leaves nuclear power looking safer than coal. Among the big question marks (the widest range of estimates) is the impact of air pollution from coal plants. This is a complicated question because the newest coal plants are supposed to meet stringent new standards that would eliminate most of the problems. It is not yet certain, however, how well the new equipment will actually work. More fundamentally, even the best of the studies of the impact of air pollution on human health have not been able to produce precise results for a variety of reasons: (1) The deleterious effects show up after many years, while people move and conditions change in the meantime. (2) The particular list of pollutants in the air varies from time to time and place to place. (3) Very few measurements of air pollution and health effects have been preserved. (4) The statistical relationships are masked by numerous other factors. Public policy has treated air pollution as a very serious problem, but many observers feel that the impact on health is unproved. One opponent of nuclear power argues that the deaths from air pollution should not be treated very seriously since most of the victims were sick anyway.[15]

Table 12.5 *Deaths from Generating 1 Standard Plant-Year of Electricity in Various Ways*

	Fatality estimates	
Energy source	Inhaber	RFF
Coal	3.3–107	0.8–14
Nuclear	0.17–1	0.2–3
Oil	1.3–93	
Natural gas	0.07–0.27	
Wind	8–15	
Solar space heating	5.3–6	

Sources: *Inhaber's revised data are adapted from the column by Suzanne Weaver, "The Passionate Risk Debate,"* Wall Street Journal, *April 24, 1979, p. 20. The RFF estimates were developed by William Ramsay and presented in this form in Schurr et al.,* Energy in America's Future, *p. 367.*

Note: *A standard plant-year means the amount of electricity that would be generated in one year by a 1,000-mw plant with a capacity factor of 75 percent.*

The range of estimates for occupational fatalities of coal-generated power is also high. This is mainly due to fatalities in coal mining. The two reasons for the uncertainty despite the years of data on mining accidents are, first, that we can always hope the new laws will bring about a significant reduction in accidents per ton and, more important, that the new laws have required measures designed to decrease the heavy toll from black lung disease. Many years will be necessary to prove that such measures work. In any event, we can hope that coal mining will become safer. As long as the trend toward surface mining continues, it is also a reasonable expectation.

Nevertheless, even if the toll from black lung and air pollution can be eliminated, just the general safety problems of coal exceed the most likely level of all nuclear fatalities. These general safety problems consist mainly of accidents between automobiles and the trains hauling coal. They are scarcely likely to be eliminated by new knowledge or regulations, although a large enough investment in underpasses could reduce them.

The nuclear fatalities are a more varied lot. On the high end of the estimate, the fatalities are dominated by reactor accidents. This topic is deferred for the moment. If we are concerned only with normal operations, by far the largest problem is the occupational fatalities. On closer examination of this category, it turns out to be dominated by conventional problems such as construction accidents and rock falls in mines. However, there is the additional problem of radon in the mines. Radon is a natural decay product of uranium that is constantly filtering up through the ground. It accumulates in wells and poorly ventilated basements, but particularly in poorly ventilated mines. Somewhere between 0.04 and 0.2 fatalities per plant year among miners and millworkers have been attributed to lung cancer from radon and uranium dust. At the high end of the range of estimates, the risk to a uranium miner from radon is comparable to the risk to an individual from smoking cigarettes.

Routine radiation exposures are a negligible risk. When plants are operating normally, even those people who live right outside the gates receive additional exposures that are a small fraction of those to which people are exposed anyway because of normal background radiation. Annual exposures are comparable to those from cosmic rays during a transcontinental airplane flight. The problems with assessing the damages from minute increases in radiation exposure are discussed later.

A more serious conceptual question is the treatment of the consequences to the public of the radon gas and its decay products. The radon that is liberated from the ground when uranium is mined and milled will drift around in the atmosphere and can be inhaled anytime in the future. Furthermore, the tailings that remain after the uranium oxide is separated from the ore will also continue to emit radon for thousands of years. In the past the tailings were often treated very casually—sometimes even used as aggregate for concrete to build homes and schools. If the tailings are returned to old mines or otherwise stabilized and covered with earth, the problem is of very minor consequence. But we are still left with the question of how to evaluate a fatality that is expected to occur several hundred years from now as a result of actions we take today.

RFF applies a discount rate to fatalities that will occur in the future, just as it would to monetary costs that will be incurred in the future. This seems reasonable: Which would you prefer, a technique that costs 999 lives this year or one that costs one life per century for the next 1,000 centuries? If the latter seems preferable to you, as it does to me, then it suggests that people really do apply discount rates to future fatalities. If that procedure is not acceptable, then the costs of nuclear power would have to be raised considerably, but it would also be necessary to raise the costs of coal mining to allow not only for radon that passes into the atmosphere from any mining activity but also for the many organic compounds—some of which are carcinogens—that are released into air and water as a result of mining, processing, and burning coal.

CATASTROPHIC ACCIDENTS

This leaves the question of reactor accidents, which has dominated popular concern since the incident at Three Mile Island. The standard technique for comparing the risks of various activities is to estimate the expected values of various events. If there is 1 chance in 2 that 5 people will be killed in a particular industrial process this year, then we can say that the expected value of the number of deaths each year is 2.5. The technique of multiplying a probability of occurrence by the consequences of the occurrence can always be used to give the expected value. Suppose, however, that the probability of occurrence is 1 in a billion years and the consequence of the particular accident would be the death of 1 million people. The expected number of fatalities would be 0.001 per year (1 in 1,000 years). This is clearly much lower than the 2.5 per year from the first process. Should we therefore select the second? It is not at all obvious that the average person thinks of probabilities and compares the expected numbers of fatalities. The popular imagination (and those who pander to it) will probably seize upon the million deaths and revel in the gory detail.

This is the sort of feeling with which the nuclear power industry should have grappled. The nuclear power industry on a year-by-year basis during routine operations does far less damage to people and other parts of the environment than do other possible sources of new power. But what about the nonroutine? What about the big accident? The nuclear industry should have conducted itself with such efficiency and precision that the public would never even have asked the question. It did not, however, and now must face the consequences of ever-increasing regulation by bureaucrats so scared of being blamed for a disaster that they cannot depart from procedure in the interest of reason.

What kind of disaster could occur in a nuclear power plant? (This discussion is restricted to the light water reactors that generate power in the United States.) The uranium in the plant is not enriched enough to explode like a nuclear weapon. The most serious problem results from the huge rates of heat production in a power reactor. A 1,000 mw reactor produces enough heat every minute to raise 8,000 tons of water 100°F. When trouble occurs, the magnetic clamps holding up the control rods are supposed to release. Once the control rods drop, the primary reaction stops. The products of the fission of U-235, however, continue to fission; this continues to produce heat. Although the heat produced is only a small fraction of full power output, a small fraction of a big number is still a lot of heat. One second after the reactor is shut down, heat production is well below 10 percent of the full power rate. It has dropped to less than 2 percent by the end of the first hour. But even after one month it is still at about 0.1 percent of full power, which is still a lot of heat.[16]

If the regular cooling system that transfers these immense amounts of heat for power production is working to any significant degree, the heat can be removed from the core easily. But if the cooling system is crippled by massive leaks or blockages, the core could overheat. In the event of a loss of coolant accident (LOCA), an emergency core cooling system is supposed to take over. If the emergency systems are adequate, then no problems to the public will result.

If the cooling capacity is not adequate, the reactor core will overheat enough to be damaged, as happened at Three Mile Island. This will be costly in terms of physical damage to the reactor. Because of the costs involved, utilities (and their insurance companies) will want to know how often they can expect such problems.

Of more concern to the public is the threat of catastrophic loss of life. What is the probability that a reactor will release large quantities of radioactive gas or particles? How many people would be exposed? What would be the level of exposure? How much land would be affected?

The starting point for the acrimonious debate on nuclear safety is the *Reactor Safety Study: An Assessment of Accident Risks in U.S. Commercial Nuclear Power Plants* (known as the Rasmussen Report or WASH 1400).[17] It is difficult to assign probabilities to exceedingly rare events in a way that will satisfy everyone. It is particularly difficult in this case because we have so little experience with the behavior of reactors when things go wrong. The information from Three Mile Island will furnish some insights, but no one will pretend that it was a controlled experiment.

The Three Mile Island experience did confirm that a lot can go wrong without the release of damaging amounts of radiation. It also suggests that the basic reactor design is safe enough to give the public some protection from faulty components, slack management, and gross mishandling of an emergency. Whether this is comforting depends on the amount of mechanical and human failure we expected to find anyway.

If melting of the core also leads to some sequence of events that breaches the reactor shell and the containment structure, the public is in danger. Ideally, we would like to know (1) the probability of substantial meltdown, (2) the probability that containment will be breached if meltdown occurs, (3) the likelihood of various climatic conditions leading to exposures to particular areas, and (4) how many people are in the area. It becomes apparent that the probability of catastrophic consequences can be reduced by locating the reactors downwind of densely populated areas.

Although the worst possible consequences of an accident will vary with location and weather, and the probability of occurrence of a serious accident will depend on the quality of the construction and the management of the plant, most attention has been focused on an average number of fatalities per reactor-year and a description of the worst possible accident. The worst-case release of radioactivity in the Rasmussen Report was expected 5 times in 1 million reactor-years. The probability of the worst weather pattern coinciding with the worst release was 1 in 1,000, giving a probability of 5 in 1 billion reactor-years for the worst exposure. This would be expected to cause 48,000 cases of cancer eventually. Thus the fatalities would amount to 0.0002 per reactor-year from the worst-case accident. To arrive at total fatalities, of course, we would also have to consider the results of all the smaller accidents.

In the years since the Rasmussen Report, every time people thought of a problem that had not been analyzed explicitly in the report or became concerned about some possible bias, they have multiplied the Rasmussen fatality figures by 10 or some equally reasonable factor. This explains the high end of the range shown in Table 12.5.

More recently, the Sandia National Laboratories conducted a study for the Nuclear Regulatory Commission that estimated the probability to be 1 in 100,000 reactor-years of a meltdown of the core combined with failure of all safety systems and breach of the containment systems. If this occurred at the worst location (the Salem, New Jersey, nuclear plant) and weather conditions were such that the cloud of radioactive debris moved directly over the most populated area and then suddenly rained down, as many as 102,000 early deaths could occur. For any particular reactor, the probability of worst weather combined with worst reactor accident is something in the order of 1 in 10 million each year of operation. Even if every worst-case reactor accident were to cause 100,000 deaths, the expected loss of life would be 1 per plant-year of operation (compared with a range of 0.8–14 for coal), but the newspaper headline was a stark "Nuke Accident Could Kill 100,000."[18]

EFFECTS OF RADIATION

All of the discussions of the nonmonetary costs of nuclear power rest implicitly or explicitly on estimates of the amount of damage to human beings from the expected radiation exposures. It is worth asking how sure we are of the relationship between the exposures of people and such effects as cancer, thyroid nodules, or genetic defects. Before being carried away by any remaining uncertainties in the data or relationships, it is important to stress that we have much better knowledge of the effects of radiation than we have of most other factors in the human environment. It is easy to talk about radiation exposure that accompanies nuclear power because radiation is the major external consequence, so the entire safety program has been designed to control and monitor radiation exposures. A coal-burning plant, in contrast, emits varying mixes and amounts of a lot of things, some of which are known to cause illness (including cancer), while others are only suspect. The fact that most of the deleterious substances released during coal mining, washing, and combustion are natural (that is, they were in the vegetation that formed the coal) should not blind us to the fact that they are harmful. That does not mean we should shut down the coal-fired plants, but only that if the coal-fired plants could be monitored as closely as the nuclear plants, they would undoubtedly be found to be emitting substances as likely to produce cancer and genetic defects as the carefully measured radiation from nuclear plants—even allowing for an occasional catastrophe.

Different forms of radiation have different impacts on people. Gamma rays are the most penetrating. Alpha radiation, emitted by plutonium, can be stopped by skin or paper. If a particle emitting alpha radiation is eaten or breathed in, however, it may lodge in the body and emit a steady stream of low-level radiation until the surrounding cells become cancerous. This appears to be the mechanism that makes plutonium so extremely carcinogenic.[19]

To summarize the varying impacts of different types of radiation on the body, the rem (roentgen equivalent man) is used as a standard measure. It weights the different types of radiation by their biological effect on people. One rem is any dose of radiation that is thought to have the same effect on a human being as 1 roentgen of gamma radiation. The rem is a measure of the radiation received by a person (or group of people), not of the intensity of the radioactive source.

The effects of large doses of radiation have been documented. A level of 500 rems will bring about the death of about half of the people exposed. At 250 rems most people suffer severe radiation sickness and many die. An exposure to 100 rems is enough to cause radiation sickness in most people. When the Rasmussen Report calculated that the worst accident would result in 3,000 early deaths from radiation sickness plus about 45,000 later deaths from cancer, it was the former figure that was most firmly based on observation.

At the other extreme—the very low exposures—the effects of small doses are difficult or impossible to discern statistically. The three principal reasons for this are the long delay between exposure and detection of a cancer, the apparently random nature of the relationship, and the masking effects of the radiation exposures of daily life and of the other sources of cancer. A typical American is exposed to roughly 140 millirems (0.140 rems) per year in addition to medical X-rays, which average another 60 millirems but are not evenly distributed over the population. This exposure accumulates from many sources—practically everything is radioactive to a slight degree. Living at sea level, a person is subject to 44 millirems from cosmic rays. By moving to Denver, a person increases that exposure to 97. Living in a wooden house subjects one to 35 millirems. By moving into a stone house, that figure is raised to 50. Plane flights add some exposure to cosmic rays. Although the occasional traveler can ignore this, maybe a pregnant pilot should not. The public exposure standard at the reactor site is 5 millirems per year and actual exposures are much less, so no statistical health effects could be detected.

In the intermediate ranges of exposure, some statistical relationships can be found. Of 67 Pacific Islanders heavily exposed to fallout from tests of hydrogen bombs in 1954, 23 had by 1982 developed thyroid nodules, of which four were malignant. Further downwind, 9 of 165 developed nodules, of which three were malignant. Such data have led to a rule of thumb that an additional 100 rems of exposure cumulated over a lifetime will double the naturally occurring cancer rate. Similarly, on the basis of studies of mice, the same doubling rule has been assumed for genetic mutations. A recent study of babies born in Japan after Hiroshima and Nagasaki, however, suggests that 468 rems are required to double the rate of mutation.[20] Even at these high levels of exposure therefore, a considerable measure of uncertainty persists.

The empirical problems are illustrated by the controversy over the question of whether workers who overhaul nuclear submarines at the Portsmouth Naval Shipyard have excess cancer rates. It is not even clear that the rates are excessive, but if they are, it would be hard to attribute the excess to radiation. Many of the same people were also heavily exposed to asbestos—a known carcinogen—and surely they have worked with other toxic substances. Besides, many of them live in communities where the water supply has unacceptably high levels of the carcinogens that result from the reaction of chlorine with organic material.

As a result of the impossibility of obtaining empirical evidence of the effects of low-level radiation, the standard analysis proceeds with the physicists' rule—When in doubt, assume that all relations are linear. (Economists, by contrast, always look for increasing marginal costs.) Thus if it is estimated that a population exposed to 100 additional rems will suffer a doubled cancer rate, then it is assumed that exposing the population to only 1 additional rem will bring about a 1 percent increase in the cancer rate. Conversely, increasing the exposure of half of the population by 200 rems is assumed to produce the same number of cancers as increasing the exposure of the entire population by 100 rems. It is also assumed that at low levels doses are cumulative, so 0.5 rem per month for a year is equivalent to 6 rems at once, although it is obvious that such a relationship cannot apply to very large exposures.

The linearity assumption is probably the easiest to defend of the infinite number of relationships that might be assumed. It could be, however, that there is some level below which radiation is totally harmless because the body can repair damage as it occurs. Conversely, damage to the body might not decline as rapidly as dosage, although the biological mechanisms that would lead to that result are harder to imagine.

The exact form of the relationship is really irrelevant to the question of nuclear safety. In routine operation a nuclear plant is so much safer than a coal plant that it would take an implausibly bad record of catastrophic failures to alter the assessment. Perhaps there is something especially sinister about dying as a result of an unseen particle that plants itself in the lung and years later produces a cancer. That may be more difficult for people to accept than being crushed or trapped by a rock fall in a mine, colliding with a train hauling coal, or gradually losing lung capacity to emphysema. These are questions about which an economic judgment cannot be rendered. The procedures for assessing the monetary cost of human fatalities are controversial enough without attempting to adjust the results for the quality of the death involved.

One category that suggests an economic distinction, however, is the difference between occupational and public deaths or illnesses. An opponent of nuclear power has argued that the risks to coal miners should not be weighed very heavily because they would be doing other dangerous things if they were unemployed and, besides, they are compensated well for the risk.[21] As noted in the discussion of coal mining, the economic theory of compensating wage differentials has not been fully accepted by society.

If we really believed in compensating wage differentials, then we would not worry at all about occupational hazards in the various energy industries, except to publicize the risks associated with every occupation. Similarly, we would not worry about the thousands who will be swept away if a major hydroelectric dam breaks; after all, they took the risk of living in a hazardous area. Is there any more reason to be concerned with the people who move into the neighborhood of a nuclear plant? In the unlikely event of a disaster, they will suffer. In the meantime they have a clean, quiet neighbor that pays high taxes and requires few community services. The test of whether any existing property owners deserve compensation when a nuclear plant is built should be the market test—does the nuclear plant raise or lower property values? If we disregarded risks of occupation and proximity on the grounds that they were already reflected in market prices, then only the truly random events, such as the grade-crossing accident or the stray atom of radon, would be worth considering. Not only are these negligible in magnitude but also they show nuclear power to be far safer.

NUCLEAR PROLIFERATION AND TERRORIST BLACKMAIL

The United States has refrained from proceeding with the early plans for a complete nuclear fuel cycle including breeder reactors and full reprocessing of nuclear wastes. Although the economics of the breeder reactor have been questioned, the most telling political argument against the whole program has been based on the quantity of plutonium that would be handled routinely. Having so much plutonium around might increase the number of nations that build nuclear weapons or even permit small terrorist groups to construct nuclear weapons.[22]

To deal first with the nuclear proliferation issue, the exact mechanism by which the United States can keep other nations from arming for battle by refusing to build civilian reprocessing plants is not clear. Other countries are moving toward breeder reactors and reprocessing facilities, so the technology will be available in the world. Our example does not do much good when, for example, France will send weapons-grade enriched uranium to Iraq for what is labeled a power reactor. Similarly, the United States has an announced principle of refusing to ship nuclear material to countries that are working on nuclear weapons, but it broke this rule in the case of India. We must recognize that nuclear weapons will soon be available to irresponsible, fanatic, or psychotic leaders. This is a dismal prospect. It is totally independent of the nuclear power program in the United States and is only remotely related to our own present and future decisions about nuclear weapons. There is no reason to allow questions of international nuclear proliferation to influence civilian uses of nuclear power within the United States.

This is not to say that international considerations have no place in decisions about nuclear power. The decision of the French to stress nuclear power heavily, including the breeder reactor, is undoubtedly due to a desire to retain independence in foreign affairs by becoming less dependent on oil from the Middle East. Very few countries have the luxury of vast domestic energy resources (such as the coal in the United States) that permit a substantial degree of energy independence without turning to nuclear power. In assessing the risks of nuclear power, perhaps one of the most realistic comparisons is the risk of an oil war over the Middle East if the industrialized nations do not develop cheap, safe, and reliable alternative power.

Diversion of nuclear material to domestic terrorist groups is also a serious worry and one that might be influenced by decisions of the United States government. Plutonium would not be the material chosen by a rational, profit-maximizing criminal of reasonable intelligence. Nevertheless, events of recent history suggest that enough crazy, cunning revolutionaries are at large to pose a real threat. Some group could steal enough plutonium to kill many people or even to make a crude bomb. Even the high risk of self-destruction would not discourage some of the fanatics from such missions. However, fabricating a bomb is a difficult undertaking requiring years of work even for a nation such as India or South Africa.

What is questioned is the ability of government to control the handling of plutonium tightly enough to prevent theft. Some people have even suggested that the federal plutonium police would turn the entire country into a police state and destroy civil liberties.[23] It seems more plausible that they just would not be able to detect a conspiracy to steal small amounts of plutonium over a long period of time from diverse installations. No matter how tightly the flow of plutonium (and the lives of people who process it) are controlled, it is hard to imagine much of a general threat to civil liberties, but it is easy to imagine ineptness, poor record keeping, sleepy and stupid guards, and generally slovenly administration.

SAVING NUCLEAR POWER

Nuclear power provides a good example of government action. The basic research was competently done, although it may have been too expensive. But the decisions and regulations and maintenance of standards for routine operations have been slipshod and shortsighted. In particular, the technically simple but politically charged decision on waste disposal has never been made, which creates a very real threat that the industry will choke on its own wastes or, more realistically, that the decision on waste disposal will be made under conditions of crisis rather than being carefully thought out.

After the Atomic Energy Commission envisioned the grand scheme of civilian nuclear power, it proceeded to substitute rhetoric for analysis. The details of plant design and safety systems were not settled. The policy decision about appropriate siting of plants to avoid the worst catastrophes was ignored. The economic analysis was dismissed with some wishful thinking about learning curves. The whole program was oversold to the public and crowded out other research on energy that might have offered a higher payoff. Governments are expected to substitute rhetoric and public relations for analysis since the only reality in the politician's world is votes. Votes depend on the voters' perceptions of the world, rather than the world itself.

It still remains disturbing that the government showed such total disregard for externalities—especially those not expected to be evident for many years. This characterized the nuclear weapons program even more than the nuclear power program, with the laxity in waste management and lack of concern for fallout. The civilian program shares with the military the early disregard for radon exposures in mining and, especially blatant, the failure to require simple controls on the disposal of tailings.

Despite the mismanagement of the details, the grand scheme of nuclear power was correct. It did turn out to be safer and cleaner and (with the breeder) as close to inexhaustible as a contemporary energy source needs to be. The error in the economic analysis was corrected when OPEC gained control over oil pricing, the Federal Power Commission ended the search for new reserves of gas, and the coal industry was hobbled by the new environmental rules. The utilities that were early adopters of nuclear power made the correct decision even if most of the underlying assumptions were incorrect.

Should nuclear power be saved? France has decided that it should be and has proceeded with the entire system including conventional plants, fuel recycling, and breeder reactors. In part, that decision was forced by the relative scarcity of other energy sources in France. The ability of the government to push ahead with the program also reflects a difference in political tradition that makes it easy for the national government to overcome opposition of local areas and groups. The United States obviously is not going to change its federal structure of government to save nuclear power.

Nevertheless, nuclear power would save lives in comparison with coal. Although there is no other great urgency in pressing its immediate expansion, it may be important to keep the nuclear option open for the day when the coal requirements of an expanding electric utility industry begin to impose excessive monetary or environmental costs. In view of the fact that no new reactor orders have been placed since 1978 and every reactor ordered since 1974 has been canceled, the civilian reactor industry in the United States will soon disappear unless some new orders start flowing in. Would it be any loss to let the industry die and then have to rebuild it again in two or three decades?

Would the nation lose much if the existing four reactor manufacturers leave the business? Their products are very expensive and not very reliable. The basic design is adequate, but questions of metallurgy, instrumentation, control, and ease of repair have not been resolved satisfactorily. If the United States should choose to return to nuclear power some years from now, it might be preferable to start with a clean slate so that various possibilities could be considered. These would include an updated version of the circa-1980 commercial PWR and BWR, but also might include the best standardized and tested version being used in naval vessels. More to the point, a fresh look would include a survey of foreign experience. The Canadian heavy-water reactor has done very well, and the French are accumulating experience rapidly with their active program. Perhaps this is the occasion to retire from the perils of pioneering, use our coal, consider other choices, then return to nuclear power when its prospects become clearer.[24]

This does not require any kind of formal moratorium. The main arguments against starting more nuclear plants are their high capital costs and their low availability. These are factors that the utilities are now very much aware of. The de facto moratorium therefore has already begun, and the only issue is whether government will take positive action to reverse it in the interest of saving the industry. It is hard to see much justification for that.

NOTES

1. Sam H. Schurr and Jacob Marschak, *Economic Aspects of Atomic Power* (Princeton, N.J.: Princeton University Press/Cowles Commission, 1950); Palmer Cosslett Putnam, *Energy in the Future* (Princeton, N.J.: Van Nostrand, 1953); and I. C. Bupp, "The Nuclear Stalemate," in *Energy Future*, ed. Robert Stobaugh and Daniel Yergin (New York: Ballantine Books, 1979). The story of the nuclear airplane is told by John Tierney, "Take the A-Plane," *Science 82* (January/February 1982), pp. 46–55.
2. For a good discussion of uranium resources, see Eric S. Cheney, "The Hunt for Giant Uranium Deposits," *American Scientist* 69, no. 1 (January–February 1981), pp. 37–48. Market conditions are

discussed in United States Department of Energy, Energy Information Administration, *1982 Survey of United States Uranium Marketing Activity*, DOE/EIA-0403 (September 1983).

3. President's Commission on the Accident at Three Mile Island, *The Need for Change: The Legacy of TMI* (Washington, D.C.: U.S. Government Printing Office, October 1979) details the problems of the industry.
4. For a summary of the status of nuclear waste disposal, see *Science*, January 20, 1984, pp. 258–260. No solution to the problem of highly radioactive waste is expected before 1998. Costs are discussed by the U.S. Congressional Budget Office, "Nuclear Waste Disposal: Achieving Adequate Financing," Special Study (August 1984).
5. Sam H. Schurr et al., *Energy in America's Future: The Choices before Us* (Baltimore, Md.: Johns Hopkins University Press/Resources for the Future, 1979), p. 356.
6. R. Philip Hammond, "Nuclear Wastes and Public Acceptance," *American Scientist* 67, no. 2 (March–April 1979), pp. 146–150.
7. Matthew L. Wald, "Retired Reactor to Be Dismantled," *New York Times*, September 27, 1984, p. 13.
8. John Surrey and Steve Thomas, "Worldwide Nuclear Plant Performance: Lessons for Technology Policy," Occasional Paper no. 10 (Sussex, England: Science Policy Research Unit, University of Sussex, January 1980). A short version is available in *Futures* 12, no. 1 (February 1980), pp. 3–17, and a follow-up study is in the same journal, 14, no. 6 (December 1982), pp. 517–532.
9. Bupp, "The Nuclear Stalemate," p. 146.
10. David Huettner, *Plant Size, Technological Change, and Investment Requirements* (New York: Praeger, 1974), pp. 87–89.
11. The most extreme case is the Washington Public Power Supply System, which started to build five plants, postponed or cancelled four, and defaulted on its bonds. For a popular summary of the industry's problems, see "Pulling the Nuclear Plug," *Time*, February 13, 1984, pp. 34–45. A more detailed survey of problems and prospects is available in *Spectrum*, a journal of the Institute of Electrical and Electronic Engineers (April 1984), pp. 25–55. Managerial failures are stressed in the cover story of *Forbes*, February 11, 1985, pp. 82–100.
12. U.S. Department of Energy, Energy Information Administration, "Performance Profiles of Major Energy Producers, 1981" (Washington, D.C.: U.S. Government Printing Office, 1983), p. 71; "1982 Survey of United States Uranium Marketing Activity (Washington, D.C.: U.S. Government Printing Office, 1983), p. 15.
13. Herbert Inhaber, "Risk with Energy from Conventional and Nonconventional Sources," *Science*, February 23, 1979, pp. 718–723. The response by John P. Holdren, Kirk R. Smith, and Gregory Morris charged Inhaber with distortion, error, and inconsistency; see *Science*, May 11, 1979, pp. 564–567. Inhaber's revised estimates

were quoted by Suzanne Weaver, "The Passionate Risk Debate," *Wall Street Journal*, April 24, 1979, p. 20.

14. Schurr et al., *Energy in America's Future*, p. 367. These estimates were based on the work of William Ramsay, *Unpaid Costs of Electrical Energy; Health and Environmental Impacts from Coal and Nuclear Power* (Washington, D.C.: Resources for the Future, 1979).

15. John P. Holdren, "Energy Hazards: What to Measure, What to Compare," *Technology Review* (April 1982), p. 36.

16. The data are from U.S. Atomic Energy Commission, *Reactor Safety Study: An Assessment of Accident Risks in U.S. Commercial Nuclear Power Plants*, WASH-1400, NUREG-75/014 (Washington, D.C.: U.S. Nuclear Regulatory Commission, October 1975); this document is often called the Rasmussen Report or WASH 1400. The data are summarized by Norman C. Rasmussen and David J. Rose, "Nuclear Power: Safety and Environmental Issues," in *Options for U.S. Energy Policy* (San Francisco: Institute for Contemporary Studies, 1977).

17. Ibid.

18. *Cleveland Plain Dealer*, November 1, 1982, p. 1.

19. Edward A. Martell, "Tobacco Radioactivity and Cancer in Smokers," *American Scientist* 63, no. 4 (July–August 1975), pp. 404–412; and D. T. Crawford and R. W. Leggett, "Assessing the Risk of Exposure to Radioactivity," *American Scientist* 68, no. 5 (September–October 1980), pp. 524–536.

20. William J. Schull, Masanori Otake, and James V. Neel, "Genetic Effects of the Atomic Bombs: A Reappraisal," *Science*, September 11, 1981, pp. 1220–1227.

21. Holdren, "Energy Hazards," pp. 74–75.

22. For a discussion of the issues, see Albert Carnesale, "Nuclear Power and Nuclear Proliferation," in *Options for U.S. Energy Policy*.

23. J. Gustave Speth, Arthur R. Tamplin, and Thomas B. Cochran, "Plutonium Recycle: The Fateful Step," *Bulletin of the Atomic Scientists* 30, no. 9 (November 1974), pp. 19–20.

24. For the view that nuclear power *is* worth saving, see Richard K. Lester, "Is the Nuclear Industry Worth Saving?" *Technology Review* (October 1982), pp. 39–47. Worldwide experience with nuclear power is surveyed by Nigel Evans and Chris Hope, *Nuclear Power: Futures, Costs and Benefits* (Cambridge: Cambridge University Press, 1984).

CHAPTER THIRTEEN

Other Sources of Energy

SOME OLD AND NEW CHOICES

During the years of cheap oil and natural gas, the more expensive or less convenient energy sources that had once been commonplace were pushed aside. Coal was being displaced from traditional uses but managed to hang on in the electric utility industry because of its cheapness in some locations. Even there, however, it was under attack because of pollution and handling costs. Water power had almost disappeared except at a few exceptionally good sites used to produce electricity; the old practice of powering mills directly from streams and tides had been eliminated by the sheer convenience and low cost of electricity. Nuclear power, although it had been developed for powering ships and suggested for such industries as steel and concrete, in fact was able to penetrate only the electric utility market. So it is not much of an exaggeration to say that oil and gas were on the way to becoming the sole energy sources after World War II, except that the production of electric power could also in selected circumstances be accomplished with hydro, coal, or nuclear power.

Old techniques do not die a sudden death. Some designers still located buildings to make the most of sun and wind. Wood, although banished to purely decorative combustion in the typical fireplace, still supplied useful heat in numerous rural households and even a few sawmills that had escaped the attention of the clean air vigilantes. Coal still retained some specialized industrial uses, the largest being the manufacture of coke for fueling the blast furnace, and was burned in many industrial boilers. Even a few homes used coal for heating. In all of these uses, however, new installations almost always used less or no coal and more oil, gas, and electricity than their predecessors. It was even possible to find a handful of water-powered manufacturing plants or farms powered by small water turbines, but these were just curious remnants of an earlier time. Similarly, numerous wind-powered pumps continued to fill watering troughs on the Great Plains, but new installations were rare.

Cheap gas and oil also dampened the enthusiasm for new sources of energy. Some experiments with solar power were under way because of the vision of a few people who foresaw scarcity of fossil fuels. Most government support of research and development was devoted to nuclear fission, which promised to provide unlimited electric power at reasonable cost. The limited stock of fossil fuels could be used to produce liquid fuels for transportation. Everything else could be electrified. Nuclear fusion would serve as a backstop in case nuclear fission hit unexpected snags.

Even in an economy dominated by cheap energy, however, there are some unusual situations where approaches that are outside the mainstream can survive. The National Aeronautics and Space Administration (NASA) funded research on photovoltaic cells because it wanted a device that could reliably produce a small amount of power for a long time without adding much weight to space vehicles. Solar cells met these criteria, even though the cost of a kilowatt of capacity is high in comparison with earthbound power plants. Also outside the mainstream, Iceland learned to develop its plentiful geothermal resources because of the absence of domestic alternatives. Similarly, a few people located far from the power lines installed modern wind-powered electrical plants.

These obsolete techniques and the new ideas that were economically justified in specialized applications were not of large enough magnitude to influence the aggregate statistics of energy use in the United States during the 1960s. They are important, however, because they suggest the variety of possible leads for innovation after the increases in oil and gas prices of the 1970s. The initial impact of those increases was to slow some evolutionary changes and give new economic life to formerly obsolete techniques.

The key to understanding the bewildering array of other energy sources discussed in this chapter is their limitation to particular situations. As the prices of oil, gas, and electricity have risen, the total cost of using other sources of energy in special situations has become favorable. For example, occasional remote applications (emergency telephones in the desert) can now make use of solar cells, rather than relying on batteries or generators. In rural areas, where gas is not available and oil is expensive, wood has become a cheaper fuel for home heating and in a few instances even for industrial uses. We should not expect any one of these other sources of energy to provide a large fraction of total energy consumption very soon. Instead, we can expect a few experiments followed either by rejection or by application in uniquely favorable situations. Some will never amount to anything more than that, whereas others will find their area of usefulness expanding as further experience refines the innovation and reduces the uncertainty in estimates of cost and quality of output. It would be a mistake to belittle the importance of some energy source that accounts for only a tiny proportion of total energy use. After all, 1 percent here and 2 percent there can add up to enough to make a significant difference in the markets. Even more important, every approach to producing or conserving energy that meets the market test in some realm, regardless of how restricted, represents a point from which rapid growth can occur if oil and gas prices once again jump up or if technological changes loosen the constraints on the applicability of that energy source.

DIRECT SOLAR TECHNIQUES

It is clear that the amount of energy reaching the earth from the sun (insolation) far exceeds the amount of energy that people are ever likely to use. One guess is that by the year 2000, the United States will be consuming about as much energy as is arriving from the sun on 0.4 percent of the U.S. land area.[1] There is some ambiguity in the numbers because the energy data for the United States do not include the energy embodied in agricultural produce and lumber, which capture solar energy, or the sunlight that warms my study through its large south window. Nevertheless, the magnitudes suggest that solar energy is plentiful.

Plentiful does not mean cheap. Capturing that free energy can require huge amounts of capital because the energy is so widely diffused. This suggests immediately that we should seek out ways in which natural processes have already concentrated the direct solar energy into more easily used forms. People have traditionally done this, most notably by mining the stored-up remains of solar energy of years past as preserved in fossil fuels. The fact that the stock of fossil fuels must be finite, however, suggests the wisdom of seeing whether we can survive on the current income of energy from the sun.

Green plants concentrate solar energy. It is perfectly straightforward to use trees or other crops as collectors of solar energy, then harvest the crops and burn them or process them into more desirable fuels. These approaches will be considered under the topic of "biomass," which sounds more impressive than "cordwood."

Physical phenomena also can concentrate solar energy. The wind and falling water both result from the action of the sun, and they are both traditional sources of energy. The power of ocean currents also derives from the sun, and we may eventually be able to harness that. Before considering such possibilities, let us examine the direct collection of energy from sunlight in some collector fashioned by people.

Heat Collectors

The simplest solar collectors are the windows in greenhouses and cold frames used to grow plants, as well as the glassed-in "Florida rooms" commonly built on houses in the 1920s. In the era of cheap fossil fuels after World War II, such architectural touches passed out of fashion. With the recent revival of interest in solar heating, "passive solar" features again command a premium. Even passive systems have been improved by technology. The basic principle of most systems is that window glass permits much of the ultraviolet energy from the sun to pass through, but then traps the air that is warmed by the infrared energy emitted by the objects in the room. Thus more energy goes in the window than comes out, at least while the sun is shining. Various innovations in glazing have recently been marketed that are supposed to improve the ratio of ultraviolet entering to infrared leaving.

Active solar devices for heating water or heating and cooling houses are also available, and both the potential gains and the costs are significantly higher. Although nearly every new house can incorporate some passive solar features at little cost, active solar heating plants and storage facilities can easily add $10,000 or more to the cost of a new home and, more important, may impose severe constraints on the appearance, size, or layout. Since it is no great problem to insulate a new house of typical size heavily enough to keep heating bills under $500 per year, the economic justification for spending $10,000 on a solar heating installation is absent at prevailing interest rates. The true believer in solar energy or in energy conservation should be in favor of saving massive amounts of national income to drive down the interest rate to the level at which "free" sources of energy are worth exploiting.

This same point can be made for all of the efforts to conserve fossil fuels or to capture the free energy from the sun, its terrestrial effects, tides, or the heat of the earth. These are all activities that require very heavy capital inputs to reduce operating costs. Such activities are best undertaken when interest rates are low (capital is relatively plentiful) and fossil fuel prices are about to rise. Those who invested in energy conservation and nuclear power plants as well as solar power in the 1960s did well. Construction costs were about to rise rapidly, interest rates were low, and fuel prices were about to begin their rapid increase. By the 1980s, however, the situation had changed. High interest rates and construction costs combined with steady or declining fossil fuel prices made major investments to save fossil fuels less attractive.

Much of the remaining interest in active solar devices results from the income tax, which is biased in favor of cost-saving investments in general and energy-saving investments in particular. If a household borrows $10,000 at 16 percent to install a solar heating system, $1,600 of annual interest payments are deductible from gross income in calculating taxable income. If the household has to pay 50 percent of the marginal dollar in income taxes, the interest on its loan for the solar collector cuts its tax bill by $800. Yet the annual savings in fuel costs as a result of owning the collector are not taxable. (If they had borrowed the money to buy a business, the profits of the business would have been taxable.) This is an example of the general point that "a penny saved is a penny divided by $(1-t)$ earned," where t is the marginal tax rate on earned income.

In addition to this general subsidy, the federal government has used the income tax law to provide special subsidies for conservation and renewable energy. The tax law in 1981 furnished a tax credit of 15 percent of the first $2,000 in conservation spending and 40 percent of the first $10,000 of spending on renewable (solar, wind, and geothermal) energy. Thus if the rich homeowner spent $10,000 on a solar heating system, the other taxpayers would contribute $4,000 of that on April 15 of the following year. The entire savings from the system would belong to the homeowner and would not be taxed. To summarize, a household could install solar heating equipment worth $10,000 at a cost to it of only $6,000. If it expects a return of 16 percent on ordinary, taxable investments, it would be content with half of that on the solar system because savings are not subject to tax. It needs to save only $480 a year to make this a profitable investment. Meanwhile, of course, an investment of $10,000 of the scarce capital of society that returned only $480 per year (4.8 percent) would be rejected immediately in other sectors of the economy.

We could construct a defense of such heavy subsidies on four different grounds. The first is the conservationist one; that is, fossil fuels are limited, so the sooner we reduce the dependence of residential heating on fossil fuels, the longer they will be available to power transportation or other uses where substitution is difficult.

This argument denies the validity of the economic theory of the mine, which concludes that the market assigns appropriate prices and leads to an optimum pattern of use over time.

A second noneconomic argument is that conversion to solar will move us toward energy independence. If the United States imports less oil, then both the United States and its allies will be able to follow more nearly independent foreign policies, especially toward the Middle East, and will be less damaged by interruptions in supply that may occur even if we maintain good relations with governments of producing states. This argument also denies that markets are acting appropriately; if the risks of having supply interrupted are so great, someone could make money by stockpiling oil and selling it at a high price during the shortage.

In addition to the two noneconomic arguments, there are two interrelated economic ones. The first is that encouraging the development of a market will spur innovation. The second is that expanding the market will permit economies of scale in manufacturing and thus bring down the cost of production and the price. In short, these are the standard arguments for giving special protection to infant industries that the nation has chosen to foster. The increase in oil and gas prices alone would have been enough to encourage experimentation with solar energy, which indeed was proceeding rapidly even before the special tax subsidies.

Table 13.1 gives some indication of the growth of the solar collector industry. It seems doubtful that economies of scale in

Table 13.1 *Solar Collectors: U.S. Manufacturing Activity, 1974–1983*

	Number of firms	Collector area (thousands of square feet)		
		Low temperature	Other	Total
1974	45	1,137	137	1,274
1975	131	3,026	717	3,743
1976	186	3,876	1,925	5,801
1977	321	4,743	5,569	10,312
1978	340	5,872	4,988	10,860
1979	349	8,395	5,857	14,251
1980	276	12,233	7,165	19,398
1981	282	8,677	11,456	20,133
1982	274	7,476	11,145	18,621
1983	224	4,853	11,975	16,828

Source: *U.S. Department of Energy, "Solar Collector Manufacturing Activity, 1983," DOE/EIA-0174[83] (June 1984).*

manufacturing are important in this industry. The reason for this is the basic simplicity of the collector units themselves combined with the necessity for individually engineering each application. The prospects for maturation in the industry are better in the installation end, where the existence of experienced contractors and architects may do much to make the installations feasible for the general public. Unless the cost comes down, however, it is neither likely nor desirable that they become commonplace.

As is true of other innovations, the economic advantage varies widely with different circumstances. Solar systems for domestic hot water were popular in Florida and other sunny southern locations until they were displaced by cheap natural gas. A few operating units, as well as the tradition and experience, survived and provided some base for renewed growth when conditions changed. Hot water is obviously a particularly favorable use of solar power because the amount consumed does not vary much over the year. Storage is necessary for nights and rainy weather, but the amounts are manageable or, more frequently, some supplemental heat can be used.

Heating the home requires large collector areas and large amounts of storage or supplemental heat. Solar energy is least plentiful in winter when it is most needed for heating. This consideration suggests that solar power is better suited to cooling houses than to heating them. After all, it is precisely on the hot sunny days of summer that the supply of solar energy and the demand for air conditioning are both greatest. If solar energy is used to cool a building, however, a heat-powered machine is necessary. Such equipment has long been available for refrigeration and air conditioning, but experimentation with it raises the technical level of the activity beyond that of simple backyard tinkering. The great advantage of routine solar collectors is that they have typically consisted of a shallow black box with the top glazed and exposed to the sun. The rest of the equipment consists primarily of pipes or ducts to move a liquid or air through the box to transfer heat from the box to the living space. Such devices can be pieced together with scrap, salvage, and other cheap materials by those who are so inclined. It may be that the commercial industry cannot be built on such a crude and inefficient basis. To avoid having to blanket the earth with collecting apparatus, it may be necessary to adopt more sophisticated approaches, of which the heat pump is only one possibility.

Concentrating Collectors

Although the flat plate collectors just described are adequate for domestic heating, many industrial processes require that the diffuse energy of sunlight be collected from a large area and focused to produce higher temperatures in a smaller area. Nonconcentrating collectors usually heat the air or water to the range of 100°F to 200°F. A parabolic dish reflector can heat a small spot to 500°F to 2,000°F. A parabolic trough can heat a pipe to the range of 150°F to 750°F. Thus the concentrating collectors can generate steam but require the added complexity of focusing mechanisms.

The technical feasibility of such devices has been demonstrated and some industrial demonstrations are now under way.[2] It is the cost that is open to question because of the need for a large amount of material and construction labor to complete the project. Any solar electric project that uses a conventional steam turbine will have most of the capital costs of a fossil-fueled plant plus the added capital costs of the collectors. The question is whether we can collect 1 million BTU from the sun for less than the $1.65 or so that 1 million BTU of coal would cost. (The comparison is not quite that bad—the coal plant would require storage, handling, and pollution control facilities.)

The coal plant is available at any hour of the year, whereas solar is either limited to sunlit hours or requires heat storage facilities. This point is irrelevant for the first few plants connected to a large grid, however. Better methods for storing electricity could improve the efficiency of the entire electric utility industry. Nevertheless, storage seems to be a really crucial issue only in connection with remote applications and electrification of transportation.

Photovoltaic Collectors

Sunlight can be used directly to make electricity by photovoltaic cells. Solar cells are routinely used in satellites to produce electricity. They have been used in a few other isolated applications. There is no doubt that they work. The question is one of cost. For use in outer space, cost is almost no object. For generating power to compete commercially, however, cost becomes all-important. The cost that is significant here is cents per kilowatt-hour actually generated. Attention is usually focused on dollars per kilowatt of capacity. It appears that the latter figure is now below $10,000.[3] The gap between photovoltaic and conventional sources has narrowed substantially from the days when photovoltaic was in the vicinity of $100,000 per kilowatt and fossil fuel plants appeared to be headed toward $100 per kilowatt. Today a fossil-fueled plant can easily approach $2,000, and a nuclear plant may exceed $3,000 per kilowatt.[4] If photovoltaic cells have negligible operating costs, then a single major manufacturing breakthrough, on which several firms are actively working, could bring them within sight of being competitive with nuclear power. Naturally, we would expect to see a gradual widening in the area of applicability, beginning in remote areas with high conventional costs and then spreading toward the mainstream. It is too early to say whether they will ever provide a direct challenge to conventional methods of generation, as distinct from their current use as substitutes for batteries in calculators and watches.

The solar cells now in use are made from carefully grown silicon crystals. The underlying principles and the manufacturing techniques are those of the transistor. Current research is proceeding in two directions. One is to increase the efficiency of the cells in converting sunlight to electricity, reducing the area that must be covered to obtain a kilowatt of capacity. The other is to devise ways to rely on amorphous (noncrystalline) materials that can be produced so much more cheaply per square foot that their UCC would be lower despite their lower efficiency (about 8 percent compared with about 15 percent) in converting sunlight to electricity.

What could be done with solar cells if they could be manufactured cheaply? The characteristic of solar cells that has fixed the imagination of the romantics is the absence of significant economies of scale. It seems now as though every rooftop could be a power plant for the occupants of the building. This incidentally would solve two large economic problems: how to find enough land to locate the vast arrays of cells and how to build the supporting structure cheaply enough.

Some have discussed the decentralization of electric power production as though it would be a veritable social revolution. No longer need we be enslaved by the giant capitalist (or state capitalist) monopolistic power company. Cut the wires. We are free! Even if rooftop photovoltaic cells are successful, of course, the effects will hardly be so dramatic. If a household actually chose to buy its own solar cells, it would be faced with an investment of thousands of dollars. Right away, those who rent homes and the poor in general would be ruled out of the solar power business. The wealthier household, once having installed the cells, could proceed with the entire troublesome installation of sufficient storage capacity (conventional batteries?) to provide power at night and meet peak loads, but it would find it cheaper not to cut the wires connecting it with the utility company generators. (Builders of backyard wind generators or water turbines find themselves in the same situation.)

The exact conditions under which utilities will provide backup power and buy the homeowner's surplus are topics of discussion now among the few firms and individuals that have their own plants, the companies, and the state public utilities commissions. The retail buyer would like to "run the meter backwards," which implies that he can sell power to the utility at the price he has to pay when he buys, or the standard retail price. If such a customer ends up the month having sold back as much power as he has purchased, his bill would be zero, except for any minimum connection charge that the utility might levy (perhaps $5 per month). If the homeowner generates surplus power, he can hardly expect the power company to pay retail rates for it, because the company must earn some return and cover the costs of buying power from household A and selling it to household B. Homeowners will see it as unfair if the company offers to sell power at 7 cents per kilowatt-hour and to buy it at 2 cents per kilowatt-hour, but some differential is to be expected.

The Public Utility Regulatory Policies Act (PURPA) requires the state utilities to determine an "avoided cost" at which "qualified facilities" can sell surplus power to the utilities. Implementation remains controversial. Avoided cost is conceptually the correct number to use since it is the cost of the cheapest available power.

The electric utilities will be much more eager to make such arrangements if they are short of capacity or if their marginal capacity is very expensive to operate. In the early 1980s, however, the utilities were caught with large amounts of excess generating capacity as a result of the sudden end to the once steady, rapid growth of consumption. As this excess capacity disappears and as the cost of installing new capacity far exceeds the cost of existing facilities, utilities will become more receptive to working out the necessary arrangements with private producers. In the view of the utility, of course, solar power is much more attractive than wind power because it is produced during the daytime hours of peak use and especially on sunny summer afternoons when air conditioners push most systems to their annual peaks. Nevertheless, we can foresee a long tedious interval of relatively slow adoption of solar cells even after all technical and economic obstacles to their use have been resolved.

Satellite Collectors

At the other extreme from photovoltaic cells on every roof is the proposal to build huge satellites to collect solar energy. There are two advantages of collecting the energy in space: (1) The satellite can be in an orbit such that it receives sunlight nearly all the time. (2) By being located above the interfering atmosphere, the light it receives has far more energy per square foot of surface. The two main disadvantages are the immense cost of launching and assembling the device in space and the difficulties of transporting the energy to earth.

The success of the space shuttle provides reasonably good estimates of the problems and costs of launch and assembly, so it is more important to think about getting the energy to earth. The original proposal called for converting the electrical energy to microwaves and beaming the microwaves to a receiving station on earth. Astronomers have worried about the brightness of such an object, but the major concern is the consequences for people and other features of the environment of such a large amount of microwave radiation. Although the receiving station could be located in a remote area, we have to consider the possibility that waves might be scattered by clouds or other phenomena and the consequences of a defect in the controls that sent the microwave radiation toward a populated area. Until the health and environmental concerns have been dealt with, there seems to be no point in a closer calculation of economic feasibility. This does, however, seem to be a project that must be carried out on a huge scale if it is to be done at all. It is not likely to be pursued seriously until the outlook for conventional sources of energy becomes much worse than it is today.

BIOMASS

Fuel Crops

The direct uses of solar energy are only a small segment of the total possibilities for the use of renewable resources. Among the most obvious indirect uses is the harvesting of biomass, that is, burning wood or otherwise using any vegetable or animal product. Anyone who has wandered through the woods of New England recognizes that a vast quantity of wood rots where it falls to the ground. It would seem better to harvest it than to leave it for the natural processes of decay.

When it comes to the realities of moving biomass from point of generation to point of use, the costs of labor, equipment, and liquid fuel can quickly become prohibitive. A cord of wood has roughly the same heat content as a ton of coal, but weighs almost twice as much and occupies 128 cubic feet—about five times the volume of a ton of coal. Furthermore, the wood comes in pieces that are not uniform in size or shape and hence are awkward to handle. Now that more attention is being paid to biomass, some of the handling problems will be overcome. In particular, trees can be grown to uniform size, clean cut with massive equipment, and reduced to uniform chips for easy handling. Chips can be blown into boilers in an airstream just as coal is. Once the focus has been shifted from the decorative logs used in fireplaces, the entire activity can be reexamined beginning with the question of which crop produces the maximum BTU per acre per year.

If we turn to conventional crops such as corn or sugar cane or to newly bred super varieties, then the production of fuel competes for land with the production of food. The advantage of old-fashioned trees grown and harvested in traditional ways is that the trees are an environmentally sound (and aesthetically pleasing) use of steep, rocky, or fragile land. Often forestry is the only activity that can be conducted on the terrain without environmental damage. As we move toward standard cropping techniques, especially mechanized clear cutting, then the biomass plantation will begin to compete with food crops for the limited amount of prime agricultural land, that is, the biomass farmer will need a huge uniform area with no obstacles to the passage of equipment.

This is not to deny that biomass has any future role in energy supply. Obviously, it does. Already in the wooded states such as Maine, New Hampshire, and Vermont, wood has become the primary fuel in one-quarter to one-half of the households, without any impact on cordwood prices or availability. As the more efficient modern equipment for harvesting, transporting, preparing, and burning wood becomes standard, the trend toward increased use will persist, but only where land has no good alternative uses.

Turning to biomass in the more general sense, it has been noted that nearly half of the new industrial boilers sold in 1980 were designed to burn wood or waste. This impressive statistic exaggerates the importance of wood because (1) most such installations were small, (2) the equipment to burn oil and gas was already in place, and (3) much of the equipment that is capable of burning wood is also capable of burning coal. Nevertheless, it suggests a significant change during the decade of the 1970s. In an earlier time even the plants that generated wood wastes (sawmills) were shifting from their traditional waste-powered boilers to modern oil to comply with air emission requirements. Now even the largest paper mills are using bark and other waste as boiler fuel.

This suggests a more general message for the future of biomass as an energy source in the United States: It is only in those situations where the biomass is otherwise a useless or troublesome waste that it is likely to become a competitive fuel. In such situations the competition for prime agricultural land is not an issue. Nor is it necessary to consider the expense and inconvenience of accumulating the large quantities of bulky fuel of low heat content. Instead, the virtue of being able to dispose of a troublesome waste and at the same time obtain some energy makes the operation worthwhile.

Municipal Waste

The clearest example of this advantageous use of problem waste is municipal rubbish disposal. Much of this waste is now hauled off and dumped in landfills, but environmentally acceptable dumping sites are becoming scarce and expensive. Collecting household waste requires so much time, energy, and equipment that it would not be a useful source of fuel in its own right.

Since the rubbish is collected anyway, however, the collected waste can be used as an energy source. If you wanted to build a boiler to produce steam, municipal rubbish would not be the preferred fuel because of the need to prepare the nonuniform mass, the need for a large boiler area due to the low ratio of heat content to volume, the problems of cleaning the stack gases satisfactorily, and the high volume of ash compared with coal. Where landfill sites are scarce and expensive, however, incineration of municipal waste has long been practiced. If we view the steam that can be generated as a by-product of the incineration process to dispose of waste, then the procedure can make economic sense.

The two main approaches to using municipal waste as a fuel are the mass burning approach, a derivative of incineration for disposal, and the recycling approach. The former accepts truckloads of raw unsorted municipal waste in whatever form the city collects from the households. It then mixes the waste somewhat to make a more consistent boiler feed. The refuse is then dumped onto a specially designed grate and burned with careful control of air flow. The hot gases are run through a heat exchanger to generate steam that can be used to generate electricity. The exhaust gases are cleaned by electrostatic precipitation and discharged. The ash collected at various points is cooled, the ferrous metals pulled out magnetically and large scrap screened out, and then the remaining material can be used for landfill or as a substitute for crushed stone.

Many potentially useful materials are lost by such mass treatment, so it is tempting to think of ways to subject the refuse to more intensive processing. The general approach is to put the entire mass of incoming refuse through a hammer mill to reduce it to small particles, which are then sorted. Not only the ferrous metals but also small pieces of nonferrous metals and glass can be salvaged. It is even possible that other materials like rubber and plastic would be worth separating for special treatment. The trouble with such an approach is that it requires a great deal of handling and processing of a large mass of material. Just as in other forms of reuse and recycling, once capital and human labor are considered to have some cost, it rarely pays to carry out elaborate processing of nonuniform inputs.

It is not surprising, then, that the commercially successful efforts to extract useful products from municipal refuse use the minimum of processing and extract nothing except heat and scrap metal. In effect, we can think of the refuse as a low-grade fuel—roughly equivalent to low-quality peat—with a BTU content of 4,500 per pound and 20 percent moisture. The ash content of 5 percent by volume and 11 percent by weight is fairly high, but some of this can be sold as scrap metal. The fact that makes such a low grade of fuel interesting—in spite of the corrosion problems, the need for a large grate area because of the low heat value, and the need to control air emissions—is the willingness of suppliers to pay the consumer of the fuel. Dumping charges at the Saugus (Massachusetts) plant exceeded $17 per ton in 1982, and it seems likely that the alternative for municipalities in congested areas—a landfill within cheap hauling distance that complies with air and water quality standards—will become increasingly difficult and expensive to find.

Liquid Fuels from Biomass

The general topic of liquid fuels from biomass illustrates the range of problems and possibilities of biomass. The flurry of interest that greeted "gasohol" during the 1970s yielded to more sober calculations suggesting that distilling alcohol from grain actually consumed more energy than it yielded. That in itself is not a crushing objection, because liquid fuels are convenient enough to be worth some premium. Furthermore, as farmers react to the higher level of energy prices, agriculture will become less energy intensive (as will industry). If agriculture uses less energy for the same yield, it must use more of something else, and the most likely possibility is land; that is, crop yields per acre will go down. If that occurs, the United States will not have enough surplus grain to produce a significant amount of alcohol. A limited amount of experimentation could still be justified as a way of gaining information that may be valuable in emergencies and also as a way to use sporadic agricultural surpluses, but it seems pointless to try to build an industry in the United States to convert grain to motor fuels.

Still, there may be places where sufficient waste is routinely generated to justify building plants to use it, especially if the alternative is to pay a lot to alleviate a pollution problem. Plant wastes can be fermented and alcohol distilled out. The manure that accumulates in overwhelming quantities in large feedlots can be fed into digesters to yield methane. All such techniques should be appraised chiefly as devices to dispose of troublesome wastes.

Biomass may have more potential as a source of liquid fuels in some foreign countries. Brazil in particular attempted to promote the use of domestically produced alcohol as a fuel for motor vehicles, although the results have been disappointing in inadequately modified conventional gasoline vehicles.

In general, we would expect the possibilities for substituting biomass for petroleum to be best in economies with low wage rates because of the necessity for a great expenditure of labor to collect, transport, and convert biomass. This very fact is creating a bleak future in poor countries where the rise in the price of kerosene has inspired a shift back to cooking with wood. The result in many areas has been deforestation and consequent lengthening of the time villagers must spend walking to the woods, as well as the devastation of floods and erosion. This experience should serve as a warning against some of the more exuberant forecasts of the role of biomass.

The diffuse energy from the sun evaporates water, some of which falls as rain on land of high elevation. Hence the power of falling water can be classified as solar. The old-fashioned waterwheels provided much of the energy for the industrial revolution, but with the advent of cheap fossil fuels and efficient steam engines most of the minor water power sites were abandoned. On the major rivers it was, of course, a different story. Turbines of extremely high efficiency were developed for applications involving a high "head" (drop) of water.

With the increase in fossil fuel prices of the 1970s, attention turned once more to the water power sites that had not seemed worth developing in the recent past. Turbines have been specially designed for low-head applications, so the efficiency with which the small sites can be exploited has improved. Nevertheless, hydroelectric power has two serious problems. One is the environmental impact of damming yet another river when so few still flow freely. The second is the economic disadvantage of having to design each installation for its specific characteristics, rather than relying on standard models, which adds greatly to the cost of very small installations. Although such considerations inhibit the adoption of hydropower, they do not end it.

An interesting offshoot of hydroelectric power is tidal power, although this is more lunar than solar. The old towns of the New England coast are built around tidal pools named Old Mill Pond or the like. Indeed, the rotting pilings that form the backbone of the milldam and the foundation of the mill may be visible during a very low tide. These were very straightforward tidal power projects. As the tide went out, the dam confined the water in the millpond. A tremendous volume of water was permitted to leave only by a narrow opening in which the waterwheel was situated. The head could not be very high in most locations since the maximum that can be achieved is the difference between the extreme high and the extreme low of the tide. In practice, the mill would be operated for several hours at a smaller head, perhaps averaging 4 feet, but varying greatly with location and somewhat by time of year. As the millpond emptied and the tide receded to its low and started back in, a moment would be reached when the ocean water would begin to flow back into the pond, reversing the wheel.

These mill sites were all abandoned when economic development doomed the small local mills. They are now worth a closer look, because not only have the costs of the alternatives risen but also they offer some hope that standard designs can be used in many locations. The reason is that where the head is negligible but a huge volume of water moves at high velocity, specially designed "run-of-river" turbines may be installed directly in the stream. The objective is not to capture most of the energy but just to capture some energy cheaply enough to compete with fossil fuels. It may be possible to capture some of the energy from ocean currents in the same way. The criterion for efficiency is much closer to that used for wind generators than to that used for conventional high-head projects.

A few locations, such as Passamaquoddy Bay on the border between Maine and Canada, are suited to more elaborate tidal power projects. These are the areas where the drop between high and low tides exceeds 20 feet and where the natural terrain offers some possibility of damming off large pools of water. The French have developed one such project at Rance. The United States and Canada have studied Quoddy Bay for half a century, but the rate of return it promises has always seemed too low to justify construction. The main problem is the huge amount of construction work that has to be done, which translates into a high UCC. With the rapid increases in UCC of nuclear and coal plants, Quoddy Bay will merit another look if interest rates fall to traditional levels. The increasing stringency of environmental standards, however, particularly applied to changes in coastal land use, has created another barrier to tapping tidal power.

The awesome power of waves has stirred the imaginations of dreamers for centuries. A few contraptions have succeeded in tapping that power for specialized uses. The only use that has ever become close to routine has been the powering of some navigational aids. The basic problems are the irregularity of wave action and the difficult conditions of the shore environment.

OCEAN THERMAL GRADIENTS

The preceding ideas have attracted ingenious tinkering over a long period of time, but ocean thermal energy conversion is a large-scale "big science" approach to extracting solar energy from the sea. The fundamental idea is to build some machine to take advantage of the differences in temperature between different masses of ocean water or between the air and the water. The Gulf Stream, for example, may be 75°F while the underlying waters are 50°F. Since these are enormous masses of water, the amount of energy involved is huge, but with temperature differentials so small the energy is expensive to upgrade into a usable form. One possibility is to use a turbine generator modeled on the conventional steam turbine but driven by a fluid such as propane or ammonia that boils at a low temperature.

Such turbines have been made to work, but there are at least three fundamental problems: First, the apparatus has to be huge to extract much energy from such a slight temperature differential. This immediately raises a difficult engineering problem in trying to keep UCC within an economically feasible range. The second problem is that the sea is a cruel environment for a huge, lightly built machine. Will it really survive both the ferocity of storms and the quieter destructiveness of corrosion and marine organisms? Third, what use is the power in open ocean? Various possible ways to store or transmit it have been suggested, but these do not alter the fundamental fact that energy is worth a good deal less when it is several miles offshore than when it is available at some home or factory. If we can really devise equipment that will tap tiny temperature differentials cheaply, then it would seem possible to use them on land where construction is simpler and markets for electricity are close by. The cooling water from existing nuclear and fossil-fueled electric plants, for example, is about as many degrees hotter than the surrounding air and water as the Gulf Stream is compared with the colder ocean currents. Very crude solar collectors produce even larger temperature differentials.

WIND

Windmills ground grain and drained polders in Europe. Later models watered the cattle on farms in the United States and even furnished electricity to isolated ranches. But like the small waterwheels, they could no longer hold their own when cheap fossil fuels became widely available. The decline of the American windmill was speeded by the government subsidy for rural electrification after 1935. A few attempts at using large windmills to generate electricity on a commercial scale to feed into the ordinary electric distribution lines were moderately successful. Denmark in particular operated windmills as part of the national electricity supply. In most countries, however, new installations were made only for households so remote that wiring was prohibitively expensive. Such home or farm-size installations have been improved to the point where they can compete against the improved gasoline or diesel generators available for domestic use but not with an easily accessible electric utility.

Experimentation with wind generators of commercial size in the United States dates from Palmer C. Putnam's installation on Grandpa's Knob, Vermont, in the early 1940s. That machine did supply significant amounts of electricity for the commercial grid. Unfortunately, it was located a slight distance away from the spot that had the highest winds. Winds vary substantially within short distances, especially in the mountains, and the power of the wind varies with the cube of its velocity, so slight errors in location are extremely expensive. Putnam's machine was damaged by heavy winds and was not deemed a cheap enough source of power to be rebuilt after materials again became available at the end of World

War II. It did, however, set the standard for later large-scale experiments, which have also generally been broken by high winds and too expensive to be worth retaining in use.

Despite the inauspicious experimentation, the wind generator appeared to be on the verge of commercialization in the early 1980s. Several large companies were designing wind turbines, and both the Pacific Gas and Electric Company and Southern California Edison had signed contracts to buy either the turbines or power produced by them. One such contract for sixty turbines of 500 kw capacity at a total cost of $200 million implies a UCC of almost $7,000 per kilowatt, and the expected output of 6,750,000 kwh per month suggests a plant factor of about one-third. Thus the windmills are expected to be about twice as expensive as a nuclear plant but even less reliable. The combination of federal and California tax advantages appears to account for most of the recent interest.[5] Another advantage is expected to be rapid completion – one to three years, or about a decade sooner than a nuclear plant. Future wind plants must do better than these pilot versions if they are to displace coal-fired plants in standard locations. According to these data, the capital charges alone would approach 25 cents per kilowatt-hour. Although this is a very high price – about four times the average residential price – it is close enough to being competitive that we can imagine the day when wind power could become a useful contributor to the electricity supply in favorable locations.

Large windmills present hazards and annoyances to the surrounding population. Construction and maintenance risks of high structures are serious. The blades attain extremely high velocities at the tip, which is a particularly serious hazard because of the tendency of the blades to crack. Even during normal operation, moreover, the machines interfere with normal TV reception and create annoying vibration. These problems are all limited to the immediate neighborhood, so they should not prevent development of large-scale wind power in appropriate places if it becomes economically attractive.

Various other suggestions for using solar energy directly or indirectly have been made, but at this stage in development it is not possible to identify the exact form in which the innovations will succeed, nor to forecast the extent to which solar energy will displace fossil fuels and nuclear power. In some cases, it appears that solar technology is close enough to being competitive that a single scientific breakthrough (in improving the efficiency or decreasing the cost of solar cells, for example) or some clever engineering and improved manufacturing techniques (in wind power or in combining solar collectors with a heat pump for heating and cooling) might be enough to make a significant difference. Until the product is on the market, it is difficult to proceed with an evaluation. We must keep in mind also that apparently promising innovations remain "almost ready" for the market for decades. Oil shale and coal liquifaction are cases discussed later, but perhaps nuclear fission should have been put in that category as well.

GEOTHERMAL

Geothermal energy is in a different category for a variety of reasons. In the first place, it is not renewable. The overall quantity of energy that can flow out of the earth must be limited, although on a human time scale that is not a matter of concern. Second, the particular sources of highly concentrated or otherwise favorable heat may be scarce and easily exhausted. Successful exploitation of geothermal energy can be found in the United States and other countries as well. With the increase in fossil fuel prices, the possibility of more intensive exploitation of geothermal energy has attracted attention.

The simplest geothermal technique is to find an area of hot springs and geysers, drill a well that spouts steam, and use that directly for heating buildings or driving a turbine. The number of locations where this is possible is obviously limited, but it seems like a good idea to take advantage of what is available. Certain problems intrude, however. First, some of the most promising sites are located in national parks and are not available for development. Second, the brine may be an environmental problem because it is often laden with corrosive minerals. Third, because the brine is corrosive, it may present severe metallurgical problems.

At present, geothermal energy is not worth exploiting for electricity unless water or steam of at least 180°C is available at a fairly shallow depth (less than 10,000 feet).[6] Lower-temperature water can be used directly for space heating and other applications if the market is close to the source. It is also possible that low-temperature turbines (for example, the freon turbine) will be used for geothermal power.

The most promising methods for increasing geothermal reserves are prospecting for additional sites and decreasing the cost of deep drilling. The incentive to search for geothermal sites has been lacking until now, so it would not be surprising if entire categories of promising locations had been ignored. The technology for drilling deep holes is a by-product of the oil industry. Holes in the 10,000–20,000-foot range have become reasonably common, but costs increase rapidly with depth. Oil wells exceeding 30,000 feet have been drilled, but to be sure of hitting a temperature of 200°C, we would have to drill to the base of the continental crust, which is fifteen to thirty miles. For the present at least, geothermal reserves are limited to a small number of favorable sites, but the resource base is substantially larger and ways may be found to extend reserves.

The United States has several hundred megawatts of geothermal electric capacity in operation, and more is planned or under construction. Since experience in this country goes back more than two decades and Italy was producing significant amounts of electricity using geothermal energy even before World War II, the technique is hardly experimental. As long as fossil fuel costs were low, however, the high capital costs of geothermal power inhibited development. In one project, Occidental Petroleum decided to build an 80 mw plant in the Geysers steam field in Northern California for $175 million, or more than $2,000 per kilowatt.[7] That is less expensive than a new nuclear plant. Furthermore, the construction time is only two or three years and convincing the public that it is safe is much easier. The Northern California Power Agency in 1982 began construction of a 55 mw plant in the same field.[8] Clearly, geothermal energy has carved out a commercially successful niche in the supply of electrical energy. Whether that contribution can become much larger will depend on the price of fossil fuels, success in exploration, and improvements in technology. Fortunately, the oil industry has an immediate interest in improving drilling techniques. That research will also serve to extend the reserves of geothermal energy.

TAR SANDS

Oil shale and tar sands are similar to geothermal in having long been known and in yielding products that are identical to those already being consumed, but requiring some new processes to do so. Tar sands are also similar in having been pushed into the realm of the economically competitive processes by the increase in prices of conventional fuels. Oil shale, however, seems to remain always a few dollars away from being worth exploiting, regardless of what happens to the price of conventional crude oil.

Canadian reserves of tar sands are immense. The largest plant mines and processes 90 million tons of tar sands per year to supply 5 percent of Canada's oil. The oil contained in Canada's tar sands exceeds the oil reserves of Saudi Arabia, so there is no danger of immediate exhaustion. The cost of production seems to be relatively reasonable as well—something on the order of $15 per barrel was reported once the plant was operating smoothly. If Canada were interested in establishing a major exporting industry based on tar sands, it would apparently be possible, but this is not consistent with the recent Canadian policy of limiting its energy exports to the United States.

In 1953, when crude oil sold for $2.68 per barrel, one of the most knowledgeable energy specialists wrote: "It is difficult to see how the cost of fuels from petroleum could advance more than 30 percent without creating wide competitive markets for synthetic fuel from shale and coal."[9] In the intervening decades the cost of oil from shale has always seemed to be a moving target that the crude oil price approaches but does not quite reach.

The total quantity of oil shale in the United States is also huge, but none of this can be classified as a reserve until it can be exploited profitably. Small amounts of shale were used in the past as solid fuel. When burned it leaves a large amount of ash, so it is inferior to coal as a solid fuel. The only use that makes sense at this time is as a source of oil to power the transportation system of the country. Before 1950 pilot plants had demonstrated that crushed shale could be heated in a retort to drive off the oil. This waxy kerogen must be processed before it is suitable for refining in a conventional refinery into the standard petroleum products.

The three main problems with oil shale are cost, water, and environment. (The latter two could be considered as costs, but it is more representative of the usual approach in the United States to keep them separate.) At a time when crude oil sold for $3.50 per barrel, it seemed as though shale oil would be about $5.[10] By the time that crude oil sold for $13, shale oil seemed likely to cost $23–$37 per barrel. With the crude oil price rising above $30, estimates of capital costs for shale oil plants continued to rise much faster. As oil prices slipped during 1982 and 1983, most plans to build shale oil plants were abandoned. One discontinued project was expected to cost $5 billion for 50,000 barrels per day of capacity, which means that the oil price would have to exceed $30 per barrel just to cover capital charges.[11] Since this plant was the culmination of a decade of intensive development effort during which several pilot plants were built, it seems doubtful that the major reduction in costs necessary to make shale oil competitive with conventional crude will occur very soon. Nevertheless, Union Oil has a project operating in Colorado that should soon provide meaningful data on costs, as well as experience with the technology.

Even if the costs can be brought down to a reasonable level, shale oil projects face intense environmental opposition. The first problem is that the volume of the spent shale exceeds the volume of the original shale in the ground. This would not have been seen as a problem one hundred or even fifty years ago. The waste would have been used to fill a nearby canyon or piled to build an extra hill. Such a simple solution is obviously suspect in our more complex era.

A second problem is the one that afflicts every industry that processes millions of tons of material each year: In a mass that large there are at least a few pounds of every harmful substance we can think of. The fear is that these harmful substances will leach out of the porous waste and degrade the scarce water of the West.[12] The retorting process can also pollute the air unless expensive control methods are employed.

Since the richest shale deposits are in the West, it has been assumed that development would proceed there first. Yet the scarcity of water adds another obstacle. There seems to be little point in trying to estimate whether water supplies in general are adequate to support large quantities of oil shale processing in the West. It is better to proceed plant by plant to see what is available if the other objections (such as cost) are overcome. Nevertheless, there has been a real fear that lack of water would inhibit development either directly, because plants would not have enough water to carry out the extraction process, or indirectly, because the water necessary to revegetate the waste dumps would not be available.

This too is a problem of cost. If the market for scarce water in the West were organized properly, profitable shale oil plants could bid scarce water away from less profitable uses (irrigation, for example). Even in our imperfect world where the government does not permit such markets to function, the water problem could be solved (if the price were right) by either piping water from a distance (the Mississippi River) or transporting the shale to a processing site having ample water. The fundamental problem, however, is that the shale oil industry has not yet given any indication that its development is justified at this time.

COAL GASIFICATION AND LIQUIFACTION

Synthetic fuels from coal (synfuels in the current jargon) have lingered just over the economic horizon for decades. Gas was produced from coal to illuminate many cities in the United States and elsewhere during the nineteenth century, but the manufactured product was displaced by cheap natural gas once the pipelines could be built across the country. The old synthetic gas had a heat content of only about 500 BTU per cubic foot, compared with about 1,000 BTU per cubic foot for natural gas, but the recent experiments have aimed at producing high BTU gas identical with natural gas. Such a product could be fed into the pipeline without disrupting distribution systems or appliances.

Various newer gasification processes have been developed to produce gas that is useful for industrial purposes, even though it is not of pipeline quality. These processes were advertised during the period of natural gas shortage and some are available on a "turnkey" basis. (The builder of the plant assumes all of the burdens and risks of construction; when it is complete, the buyer of the plant pays for it and turns it on.) Such plants permit manufacturing processes that require large amounts of clean heat to rely on coal. The gas can be cleaned before it is burned. Since the gasifier is located at the point of use, the cost of transporting low-BTU gas (150–180 BTU per cubic foot) can be neglected. Such plants do not help directly in the provision of pipeline-quality natural gas, although by reducing industrial demand they make more gas available for residential and other consumers with limited choices.

Pipeline-quality synthetic gas is another matter, and as usual the problem is cost. High-BTU gas is expensive to produce. The commercial-scale plant in Beulah, North Dakota, was designed to produce gas costing $10 per thousand cubic feet beginning late in 1984.[13] That is more than double the cost of conventional gas. Meanwhile, the combination of individual conservation measures, industrial conversion to coal, and the success of the new exploration efforts have resulted in a surplus of natural gas. This may be a temporary phenomenon, of course. We can argue for building one gasification plant of commercial scale as a hedge, but this does not seem like the time to sink huge amounts of capital into plants to manufacture something we have enough of. If geopressured methane is as plentiful as the optimists suggest, moreover, it will be many generations before anyone wants to manufacture natural gas.

Liquid fuels from coal may have more immediate importance. From a long-range view, it is easier to adjust to the absence of natural gas than to the absence of liquid fuels. Furthermore, a proved and cheap method for producing liquid fuels from coal will help to keep oil plentiful and cheap. The fear that oil will be priced out of many markets has certainly been a major constraint on the pricing policy of Saudi Arabia. If the United States were prepared to build coal liquifaction plants, it would put a ceiling on the price that the cartel could charge as long as the Middle East remains able to supply large amounts of oil. In the event that oil shipments are disrupted, then coal liquifaction (or shale oil) capability would alleviate the consequences for the United States. It seems unlikely that oil from coal (or shale) can compete with crude oil selling for less than $35 per barrel, so a federal subsidy will be required if a demonstration plant is to be built.

From a technical standpoint, it is well known that liquid fuel can be made from coal. This was done in Germany during World War II when other sources of crude oil were inaccessible. It is also being done today in South Africa. The South African plant uses a standard Lurgi gasifier to make low-BTU gas, then cleans the gas and uses it as feedstock for an additional reactor where the liquid fuel is produced. The costs are not published. A liquifaction project discussed for the United States was estimated to cost $700 million and to produce 140 million gallons of gasoline a year. This suggests that capital costs would approach $1 per gallon and implies that the project has no value except as a demonstration and a bargaining tool. During 1982 and 1983 plans to construct such plants were canceled.

FUSION

Successful fusion reactors are plentiful, but the closest one to us is the sun. Physicists have long been intrigued by the possibility of producing controlled fusion reactions in a power plant to capture the huge amounts of energy that are theoretically available. Until the experiments are successful, it is difficult to say anything about the economics of the nuclear fusion process, except to sketch out some of the limitations and potentials. The nuclear reaction that has attracted the most attention is the fusing of deuterium and tritium to yield a neutron and helium. Since there is not much tritium available, it must be bred by firing neutrons at lithium.[14] The tritium can be produced from the lithium, but confining the deuterium and tritium at high enough temperatures to bring about the reaction is extremely difficult. Until it is clear that a particular method—such as extremely powerful magnetic fields or laser beams—is going to work well enough to produce significant amounts of power, it is not even possible to estimate the bill of materials for fusion power, let alone make an economic analysis. A meaningful economic evaluation of fusion power will not be possible until well into the twenty-first century.

The implication of the long and expensive research and development period (fifty to one hundred years or more) is that any conventional accounting technique would have rejected fusion power at the outset. If you have to wait a century for the payoff, you really need to expect an infinite payoff. Yet the fusion reactor, as far as can be forecast at this stage, will be comparable in many respects with the fast breeder reactors that are currently under development in several countries. It is probably better to charge off current fusion research to insatiable curiosity, and if it also turns out to yield a reasonably safe source of large amounts of energy that is not outrageously expensive, then that will be a welcome dividend.

That cautious hope is considerably hedged from the common dream of unlimited amounts of safe, clean, cheap energy. The hedging reflects not only a sobering rereading of some of the forecasts made for fission power but also a consideration of what appears to be a fundamental of nuclear engineering, namely, the materials problem. In the absence of some fundamental breakthrough, the confinement of nuclear reactions will continue to require a large amount of exotic materials, much of which is discarded as radioactive waste. This large consumption of materials imposes a more stringent limit on total nuclear power production than does the availability of fuel. It also keeps the cost of nuclear-generated electricity from ever becoming "too cheap to meter," as the famous forecast of fission power would have had it. Preliminary indications are that the hazards of radioactivity from fusion would be less than those from the liquid metal fast-breeder reactor.

THE OUTLOOK

An economically relevant time period is one generation or less, because any investments made today will be fully amortized or obsolete within that period and the relative prices, tastes, and technology of the more distant future cannot be stated with sufficient precision to serve any purpose. With the interest rates that seem likely to prevail well into the future, no return that is realized a generation from today has any economic value worth considering now.

It follows, then, that many of the possibilities mentioned in this chapter will have little or no role in an economic forecast. Certainly, fusion power will not be commercial within that time period, and with less certainty the same forecast can be made for some of the more grandiose solar projects (orbiting collectors, for example). When we turn to the small solar possibilities, including biomass, wind, and ocean, as well as solar heating and cooling and photovoltaic cells, the situation is quite different. These are already chipping away at the fringes of conventional uses of energy. The interesting question is not their existence but whether they will become something more than an insignificant footnote to a minor miscellaneous heading in tables on sources of energy.

This economic forecast depends in large measure on a political one. It is not so much a question of whether OPEC can continue to function as a cartel but whether the Middle East will remain stable enough to produce and export substantial amounts of oil. The power of OPEC to raise prices in the early 1970s resulted from a particular set of circumstances including the rapid growth in oil use. OPEC power was already eroding by the late 1970s when the elimination of Iran as a significant producer gave it a new lease on life. The early 1980s saw the standard economic forecasts of the demise of the cartel justified. Alternate sources of energy were responding to the high prices set by the cartel. Consumption was declining also in response to those high prices. Most significant, the economies of the cartel members were being reorganized sufficiently to make the oil revenue seem like a necessity rather than a luxury or nuisance. Even the nations most suspicious of change found the oil revenues essential to pay for past, present, and future wars.

If the Middle East remains stable enough to export oil, the real price of oil will remain stable through the end of the twentieth century. This will certainly put a crimp in plans to extend the range of applicability of the various energy sources reviewed in this chapter. It will also mean that new nuclear fission plants will not be initiated in the United States and coal will not enjoy the robust growth that has sometimes been forecast for the industry. If the Middle East becomes so preoccupied with war or revolution that the flow of oil dwindles, then all of the other possibilities will come into their own.

Although the political question overshadows other considerations, there are still some important general concerns about the mixed collection of other energy sources. The first of these, already mentioned but deserving emphasis, is the necessity for huge amounts of capital to capture free energy from sun, wind, ocean, and earth. There is nothing perversely coincidental about this; it is exactly what we should expect. From an economic viewpoint, if energy sources with no fuel cost also have low capital cost, they would already have been exploited. (An example is the best hydroelectric sites.) From a physical viewpoint, the free sources of energy such as sunlight are very diffuse, which implies that a large quantity of material will be necessary to collect it. The one exception to the rule that a lot of capital is required is the case of biomass where, at least in the traditional version of burning wood, labor usually takes the place of capital. Of course, the inefficiency of biomass as a collector means that large amounts of land are required as well.

The heavy capital requirements have several economic implications. The most obvious is that a time like 1984, which was characterized by high interest rates and very low rates of price increase, was an extremely unpromising time to try to convert from the fossil fuel techniques that have high operating costs but relatively low capital costs. Those who made the transition in the late 1960s and early 1970s had the advantage of interest rates that were not only low but also in many cases turned out to be negative because of the ensuing increases in prices. It would not be prudent to expect such circumstances to bail out every decision that appears to be economic folly when it is taken.

A second general theme is that increased efficiency with which existing energy sources are used is generally the first step, but once it is taken, the second step becomes questionable. An increase in the efficiency of a furnace from 50 percent to 70 percent results in a 28 percent decrease in fuel consumption. That may be more rational than installing solar collectors or other such experimental apparatus. Similarly, the first requirement for most solar installation is very heavy insulation and otherwise tight construction. If those things are done anyway, the solar installation will not save much. On an industrial scale it seems more promising to devise nondisruptive ways to conserve energy, such as improved insulation and controls, rather than reworking entire systems. Similarly, often ways can be found to make use of waste heat, sometimes with simple heat exchangers, sometimes with more fundamental changes involving cogeneration or combined cycle approaches, to make the best use of different qualities of energy. Once such relatively simple techniques have been adopted (as they increasingly have been), then the economic incentive for more revolutionary approaches may disappear. If an untested solar process promises to cut your fuel bill in half, you may be interested when the fuel bill is $100,000 but not want to bother if you have already found simple ways to cut the bill to $50,000.

All of this suggests a third major theme, namely, that future sources of energy will be much more diverse than those of the recent past. With fossil fuels so much more expensive, the trend toward basing all economic activity on oil and gas, which was so apparent in the two decades following World War II, has ended. The brief flowering of nuclear power and the renewal of the importance of the coal-mining industry were the most visible impacts. But that same pressure of high prices has also provoked the study of dozens of minor, local, and limited options for increasing the supply or decreasing the demand for energy. Many of these small special cases require careful engineering of the individual use. Small hydro projects and retrofitting existing industrial processes provide examples. The full impact of such schemes is not felt at once, but accumulates gradually as the limited number of people qualified to plan such projects proceeds from one to another. Also, it may take some time before modern materials, controls, and fabricating techniques are adapted to long-ignored local sources of energy or to methods that have only approached feasibility with the new technology now available.

With the gradual dispersion in sources of energy, much of the concern for energy storage is misplaced. Obviously, a system that is totally dependent on sunshine must devise some ways to survive through the nights. If many different sources are connected to the same electrical system, however, the storage problem simply disappears into the general problem of maintaining sufficient reserves to meet peak loads and various outages. Daily peaks in particular do not need to impose any problem in a system that incorporates both solar and hydroelectric power. Even in the case of home heating, the sacrifice of living space and considerable capital to achieve storage capacity for several sunless days does not seem rational. A well-insulated dwelling having some backup heating system should use the backup system. After all, the objective is to decrease the total cost, not to conserve fossil fuels at any price. Experimental units obviously have an excuse for going to extremes, but good design does not.

Finally, with all of the enthusiasm for new (or newly redeveloped) energy sources, it should not be forgotten that they should meet the standards imposed on the mainstream of fossil fuels and nuclear plants. These standards obviously include the economic one. In the long run, no amount of enthusiasm can substitute for a favorable rate of return, although enthusiasm can certainly help in overcoming the obstacles to achieving that return. More to the point, the new techniques must be held to the same standards of safety for employees, the public, and users and of freedom from serious environmental impacts that are imposed on the mainstream. It is easy to forget that a soft energy source like a windmill can kill someone as effectively as would the radiation from a serious nuclear malfunction. Similarly, the smoke from wood stoves is no less damaging than the smoke from an electric power plant, just because it comes from thousands of dispersed sources using biomass.

NOTES

1. Alan B. Kuper, "Solar Energy Prospects: A Report for the Multi-Level World Model Project" (Cleveland, Ohio: Case Western Reserve University, July 1973), pp. 4–5. The reader who wants to see a strong case made for techniques other than fossil fuels and nuclear should start with Wilson Clark, *Energy for Survival* (Garden City, N.Y.: Anchor Books, 1974); Denis Hayes, *Rays of Hope: The Transition to a Post-Petroleum World* (New York: W.W. Norton, 1977); and the works of Amory Lovins.
2. For a list of sample applications, see *Chemical Week*, December 3, 1980, p. 33.
3. Advertisement by Photowatt International, 2414 West 14th Street, Tempe, Arizona 85281, in *National Defense* (January 1983), p. 16a. The ad quotes a price of $9 per peak watt.
4. The photovoltaic cells, however, can be expected to average only 0.33 times peak wattage because of nights and cloudy weather. Nuclear plant factors exceed 50 percent and coal-fired plants are available more than 80 percent of the time.
5. *Wall Street Journal*, March 19, 1982, p. 15; October 29, 1984, p. 28; March 18, 1982, p. 20; June 21, 1982, p. 38; and December 2, 1983, p. 4. For references to the technical literature, see Bent Sørensen, "Turning to the Wind," *American Scientist* 69 (September–October 1981), pp. 500–508.
6. L. J. P. Muffler, "Geothermal Resources," *U.S. Geological Survey Professional Paper 820* (1973), pp. 251–261; Ronald G. Cummings, Glenn E. Morris, Jefferson W. Tester, and Robert L. Bivins, "Mining Earth's Heat: Hot Dry Rock Geothermal Energy," *Technology Review* (February 1979), pp. 58–78.
7. *Wall Street Journal*, March 24, 1982, p. 38.
8. *Wall Street Journal*, March 17, 1982, p. 14.
9. Palmer Cosslett Putnam, *Energy in the Future* (Princeton, N.J.: Van Nostrand, 1953), p. 167. For an account of early attempts to use oil

shale by a participant and critic, see Harry K. Savage, *The Rock That Burns* (Boulder, Colo.: Pruett Press, 1967).

10. William C. Culbertson and Janet K. Pitman, "Oil Shale," *U.S. Geological Survey Professional Paper 820* (1973), p. 498.
11. *Wall Street Journal*, March 24, 1982, p. 5.
12. Ogden H. Hammond and Robert E. Baron, "Synthetic Fuels: Prices, Prospects, and Prior Art," *American Scientist* 64, no. 4 (July–August 1976), p. 412.
13. *Wall Street Journal*, April 1, 1983, p. 13; U.S. General Accounting Office, "Economics of the Great Plains Coal Gasification Project," GAO/RCED-83-210, August 24, 1983.
14. *American Scientist* (January 1979), p. 78.

PART FOUR

Problems and Policies

Health, Safety, and Environment

The energy industries seem to have attracted a disproportionate amount of attention for their impacts on health, safety, and the environment. In part this is simply a reflection of the huge masses of material that are extracted, transported, and processed. The tonnage of coal mined in the United States far exceeds the tonnage of agricultural commodities or any other mineral except stone, sand, and gravel. Consumption of oil involves a similar mass of material. The sheer flows of energy commodities dwarf any of the other commodity flows in the economy.

Combustion of energy commodities to produce useful work can create a large set of environmental problems. It is not possible to have an advanced industrial society without prodigious input of energy, however. It is more useful to think of the costs and benefits of particular changes in techniques, rather than to dream of a pollution-free world.

When economists think of such matters, the analysis soon centers around the concept of externalities. External costs are costs of an activity that are borne by someone other than those directly involved in the immediate transaction. If a steel mill heats the water of a stream enough to make the power plant downstream operate less efficiently, the problem is an external cost of steel production. Some suitable tax and subsidy schemes can be devised to force the steel mill to pay the full costs of its operation and to compensate the power company (which in turn may have to compensate users farther downstream). The externality can be internalized, prices will reflect marginal costs, and the economists will be satisfied.

Suppose, however, that the heat from the plant produces a series of changes in the plants and animals that live in the stream. Among other changes, the trout are replaced by catfish. If anyone kept detailed measurements and records, it might be possible to show that the steel mill and the power plant jointly were responsible for the change. Placing a price on the change would be far more difficult, however. The hours of recreation obtained by fishermen from a body of water seem to be totally insensitive to the quality and quantity—or even the existence—of fish. The change in fish population in any event is just one indication of a set of changes in the quality of the aesthetic experience provided by the stream.

Environmental degradation does not always result in an external cost. Sometimes those who bear the costs are fully compensated even under existing institutional arrangements. Many of the neighborhood effects of a power plant, for example, are soon reflected in property values. If a coal-fired plant drops soot in the neighborhood, the value of the land for residential purposes will decline. It is possible that owners of existing residences at the time the change is made will suffer a loss in the value of their property. Subsequent owners get exactly what they pay for—a dirty location dirt cheap. Recent regulations compelling a cleanup of emissions have provided a windfall for those who bought the housing at low prices from the original owners, who bore the loss.

If political boundaries are drawn properly, even the original owners may benefit if the taxes paid by the new installation decrease the taxes paid by the neighbors of the plant. For a coal-fired plant, such an occurrence requires an extraordinary coincidence between air shed and political boundary. The prospect seems more likely in the case of a nuclear plant, since the tax advantage of the plant far exceeds the cost of increased radiation emitted during the normal operations. The tiny probability of a major disaster inflicts its small cost on such a wide and nebulous area that it probably is not reflected in property values.

Despite the examples of markets or direct negotiations between firms leading to the internalization of apparently external costs, many external costs continue to be imposed by the energy industries. Governments can ignore externalities or use any of a variety of possible approaches to control them. These include (1) assigning property rights to the resource that is being damaged so that someone will have an incentive to recover the cost of the damages, (2) taxing the emissions according to the costs imposed, (3) auctioning permits to discharge a limited total quantity of each pollutant, or (4) regulating the emissions from each source. Although economists generally argue for some version of the first three possibilities, the U.S. government uses direct regulation. The first three techniques make use of the price system to ensure that pollution is decreased in the cheapest possible way. In general, the regulatory approach uses more resources to achieve the same objective.[1]

For a great many environmental questions, the problem is not just the externality but also the publicness of the whole situation; that is, there is no way to find out who are bearing the external costs, let alone to quantify their losses. The factory owner cannot compensate each potential catcher of trout for the fish he thinks he might have caught. Even in those countries where the rights to fish along certain parts of certain streams are regularly bought and sold, the market does not offer a full solution. Many of the aesthetic values are even less measurable than the quantity and quality of fishing. Furthermore, the distribution of wealth and income may be a significant component of the analysis.

We can see this most clearly if we imagine a questionnaire designed to find out the aesthetic costs of converting a freely flowing stream into a reservoir for a hydroelectric project. The first question is, "How much will you pay to stop this project?" The second is, "How much will the power company have to pay you to make you accept the project willingly?" For most people the answers to both questions will be small amounts and may not differ much. But consider the case of the couple who have spent their life savings to acquire their dream: a retirement home overlooking this free-flowing stream beside which they spent many joyous hours of their youth. To the first question they respond that after paying for food, utilities, taxes, and medicine they have left only $10 a month, which they will gladly pay to preserve the view that they worked a lifetime to acquire. To the second question, they state that it would take $100 million to compensate them for the loss of their view. The difference might be narrowed by making the payment a lien against the property to be settled when they are both dead, rather than a cash outlay, and reducing the latter figure to the cost of buying a whole new valley plus some relocation expenses and compensation for the psychic costs of knowing their favorite haunts are flooded. Generally, however, any changes will result in losses to those who much prefer the existing situation, as well as gains to consumers generally (or some other group). Compensation to prevent such losses would be difficult to arrange in practice, as well as very expensive in many cases.

For ordinary marketable goods, such as electrical power, the differing evaluations of the desirability of the good by different consumers poses no problems: Those who like lots of electric power at the prevailing price buy a lot. Those who are not willing to buy much buy little. At the margin each consumer of electric power values the last kilowatt-hour of electricity at its marginal cost to her. If everyone had to pay the same constant price for electricity, regardless of quantity consumed, then each consumer would buy until the pleasure that an additional kilowatt-hour of electricity provided was exactly equal to the pleasure she could get from other uses of the 7 cents that extra kilowatt-hour would cost. Similarly, if prices were correctly set, the true social cost of an additional kilowatt-hour of electricity would also be equal to 7 cents. The benefit to the individuals and the costs to society of the last unit of power would be equal, whereas the earlier units, being cheaper for the firm to produce and valued more highly by consumers, produce a surplus of value over cost for some people in the economy.

For major changes in land use, the analysis is more complex because the change is imposed on all people regardless of their likes or dislikes. If the dam changes the mountain stream into a placid pond, then it is changed for you and for me. If we both like the change but you like it much more, we cannot adjust our purchases until each likes it the same amount at the margin. There is just one quantity with varying evaluations at the margin. This leads to squabbles, not only about the amount of dam building, but also about the sharing of costs. It is even more complex if, as in this situation, some approve of the change while others oppose it. In addition to the standard kinds of negative externalities, therefore, energy projects often involve compulsory consumption and changes that are large enough to make a significant difference to real incomes.

OCCUPATIONAL RISKS

Even when the complications of externalities are not present, the questions of health, safety, and environment often involve intractable issues. Questions of occupational hazards, both to health and to life, furnish a pertinent illustration. In particular, the traditional economic argument found in Adam Smith and subsequent writers is that wages of hazardous occupations include a premium to compensate workers for the greater risk, compared with the average occupation. If we take this argument seriously, then the public need not be concerned with worker safety at all, except possibly to give wide publicity to the hazards of different occupations. According to this view, people who place a high value on life and health will move into safer occupations even though wages are lower. Those who value current income highly will take on the greater risks to get it. If firms find it cheaper to reduce risks to be able to hire qualified people cheaply, then they will do so. Otherwise, it would seem that additional safety is worth less to the employees than it costs.

Public policy is not entirely consistent with this line of reasoning. If the theory of compensating wage differentials is correct, then product prices already reflect the mortality and morbidity rates of the industries in question. Nevertheless, the law has long required companies to pay for workman's compensation insurance, which makes payments to injured workers (or their heirs). The implication is that the wage differentials were insufficient to permit employees to buy their own insurance, or that workers should not be permitted to self-insure, that is, to accept the risk themselves. Since many of the wage differentials persist for long periods of time, the risky occupations may have enjoyed excessive compensation since the imposition of workman's compensation laws. Coal mining presents a very strong illustration of this because of the retroactive addition of "black lung" benefits to compensate people for illness that they knew to be part of the occupational risk of mining coal when they started. According to the market test, coal mining wages were high enough anyway because the industry never had trouble recruiting enough labor.

In addition to compensating wage differentials and workman's compensation, juries in recent years have tended to award extra damages for wrongful death whenever a company permitted any type of safety violation to exist, whether or not the firm condoned the violation. The awards to the heirs are based in part on the expected lifetime earnings of the deceased, which had already incorporated compensation for the risk! This too is reflected in the product price.

The excessive rate of compensation for death and injury is sufficient to induce rational firms to install safer procedures and equipment. Some small firms may be ignorant or unable to afford safety, but in general the pressure to maximize profits should be incentive enough. To this pressure, however, have been added government standards, which sometimes lead to extremely high costs per accident prevented. This too is reflected in the price of the product. If we wished to disregard all occupational injury, mortality, and illness in an analysis of the comparative costs of different sources of power, the logical grounds for such a decision are well prepared. Nevertheless, the constant efforts by society to compensate again for risks that were already reflected in wage rates and then to reduce the risks suggests that the public is not really prepared to disregard the occupational risks as a separate cost.

THE VALUE OF LIFE

If we are not ready to accept the values that individuals implicitly place on their own lives, what is the appropriate price for life? The unpleasant topic arises not only in connection with occupational risks but also with regard to the members of the general public who may be run over by coal trains or oil tankers, blown up by natural gas explosions, burned in fires started by wood stoves, or killed by falling from windmills or solar collectors. On a statistical basis some accidents will occur, and their costs must be calculated – but what is the value of a human life?

This topic arises in connection with many government spending programs including such mundane matters as building highways or locating and equipping ambulances. Despite a vast outpouring of discussion, there is no universally accepted solution, but a few clarifications can be made.[2] First of all, it makes no sense to say that a human life is either valueless or priceless. An explicit decision on a dollar value that is constant throughout all energy activities (unless some reason can be stated for varying it) makes more economic sense than a jumble of standards and implicit or explicit values.

A fixed budget will save more lives if it is allocated so that the marginal life saved in any activity costs the same amount (for example, $500,000). Commonly now, highways are designed as though people were worth relatively little (say, $100,000), whereas airports are designed as though people are worth a lot (perhaps $1 million) and industrial standards implicitly put an even higher value on life (like $10 million). We could save more lives without spending any more on safety by letting working and flying be more dangerous and shifting some resources into making highways safer.

Attempts to put an explicit value on a life saved have sometimes started with the present value of the expected stream of earnings. This measures the value of the person's work to the rest of society. In some states, damages to the family for wrongful death include the expected earnings for the deceased minus the costs of the person's subsistence – the same standard we would apply to a draft horse! For those who enjoy life and do not make others miserable, their value to themselves and the rest of society must be more than the value of their work to society; thus the present value of the stream of earnings expected during a person's life is an underestimate of the appropriate value. This is not the correct answer, but at least it can be estimated and rules out lower values as incorrect.

We can gain some insight by looking at the actions that people will take to reduce the risks of daily life or, conversely, the risks they will take to save time and money. Even assuming that people act with good knowledge of the risks, however, what are we to make of the fact that a businesswoman flies home on a snowy, windy night, heedless of the extra risk that flying under such conditions entails? Is she trying to save the cost of a hotel for her employer? Does she fear missing an important meeting the next day? Does she dread the prospect of spending the next day in the airport lobby if the storm grows worse? Is she eager to see her home and family again after the trip? In fact, we can question whether the slight change in the probability of survival has anything to do with the decision.

CATASTROPHES

Just as the consistency of the individual's decisions about activities with small risks might be questioned, so might the reactions of the public to different patterns of risk. The clearest example is the comparison between generating electric power from coal or nuclear energy. Using coal ensures a steady stream of individually unobtrusive deaths. Using nuclear causes no harm to the public unless a very unlikely set of circumstances produces a major accident. We can estimate that the expected number of deaths per kilowatt-hour of electricity is higher for coal than for nuclear power. Nevertheless, the public is more concerned about the huge disaster (with a tiny probability of occurrence) than about the routine trickle of deaths that adds up to a large number. It is an even more extreme version of the comparison between the safety of automobiles and airplanes. Intellectually, we can all see that plane travel is safer, but the familiarity of the risks and the illusion that we ourselves control them make the auto feel safer to most of us.

Even if we grant that public perceptions of the degree of risk associated with nuclear plants are wrong, what is the appropriate choice of policy? Nuclear plants should obviously be kept very safe, but does that mean that they should be held to a higher standard than coal or other sources of electricity? The incident at Three Mile Island helped to focus the issues more clearly. In particular, if a malfunction is serious enough to hurt people—and even if, like Three Mile Island, it is not that serious—the electric utility will suffer serious financial damage. The firm could easily go bankrupt. We can hope that the officials responsible for any obviously incorrect decisions will find them a barrier to further promotions.

This suggests that individual executives and private utilities have the strongest possible incentive to avoid disaster. Unlike the typical fatalities in other industrial processes that occur one or a few at a time (and thus can be covered by insurance against adverse financial consequences), a nuclear disaster should be expected to wipe out the firm. Standard assumptions of utility-maximizing behavior by risk-averse individuals should be enough to motivate excessive concern for safety, especially when the costs of safe procedures and equipment can be passed on to the consumer of electricity.

Having said this, however, it remains unclear why the utilities did not adopt more stringent procedures in the past. Perhaps serious incidents seemed so remote that no one thought of the implications for the firm and himself of a major accident. Perhaps the limitation on the firm's liability offered by the Price-Anderson Act had led to complacency about the financial implications of individual errors; if so, that was a more serious cost to the country than the few dollars of direct subsidy implied by the act.

Is the impact of infrequent disaster greater on the public than the impact of the same number of deaths more evenly distributed? Certainly, the major accidents receive more attention in the news and seize the popular imagination, but do they cost the economy more in lost output or the people more in private grief? The answer is probably yes to the first question and no to the second. Individual deaths rarely inconvenience the economy because of the redundancy of skills within the individual workplaces. When a fire or plane crash eliminates the entire top management of a firm, however, it can cause serious problems in maintaining continuity of decisions and momentum of growth. The biggest imaginable nuclear disaster might eliminate some small firms, a few of which might have more than local importance.

For the families involved it is hard to see that nuclear casualties are worse than the more routine kind. When a lot of people in one area are killed simultaneously, some families may cease to exist altogether and the resources of the community for dealing with those exposed to radiation and for comforting family members may be strained. This is a bleak picture, but is it worse than the total of grief from numerous independent accidents with conventional sources? That is a conceptually unanswerable question, so we are left to look at the cold statistics on the number of fatalities expected and perhaps not to discount too heavily the occupational fatalities on the grounds that the individual families should have known the risks.

WEIGHTING ILLNESS

There is yet another concern in the interpretation of estimates of the impact of energy sources on human life and health. That is the attempt to summarize the impact on health and mortality in a single number. If one technique produces lung cancer and thyroid nodules, and another produces black lung disease and emphysema, how are the lists of illnesses to be evaluated and compared? There are two types of problems in the endeavor. The first is that despite the years of experience with conventional fuels, we have very little knowledge about effects on health of using them. The second is that there is no uniquely correct way of weighting the various health effects to make an index number that everyone would consider satisfactory. Both these points require some elucidation.

Ignorance of the health effects of energy sources is pervasive. We have extremely good information about the consequences of exposure to high levels of radiation, but only guesses about the extremely low doses that characterize routine operation.[3] The major disasters would involve some large doses, but we have only guesses about the likelihood of occurrence. The traditional fossil fuels certainly can cause health problems.[4] At this stage, however, it is uncertain what actual emissions will turn out to be with the new pollution control equipment, what the effects will be of the remaining pollutants, and what the interactions will be with other emissions also subject to changes.

Even before the complication posed by the tightening of emissions standards, it was hard to disentangle the effects of air pollution from other factors influencing health. In American cities air pollution grew in step with economic advance, but mortality rates declined. The crude correlations obviously yield the incorrect implication that air pollution is healthy, so the statistical studies must allow for the advances in nutrition, medicine, immunization, and sanitation that accompanied prosperity (and pollution) and reduced mortality.

The most promising approach is to look at sickness and mortality in different areas during the same time period. The general assumption is that the overall quality of public health measures and other factors having a major impact on mortality will be similar from area to area in this country, so the differences in mortality and illness can be related to differences in air pollution and other measurable factors. The difficulties are numerous, including (1) poor records of the quantity and types of pollutants and of illnesses, (2) the randomness introduced into the data by the long exposures necessary for most illness to develop, and (3) the movement of people among areas with different pollution levels. Nevertheless, from such studies it has been possible to come up with some estimates of illnesses that might result from conventional power.

Combining the illnesses into some common units and trying to make them additive with deaths has posed conceptual problems. We can simply state some arbitrary rule; for example, four cases of emphysema equal one death. It is difficult enough to do this in a convincing way for any one illness. The problems of extending it to cover all of the significant afflictions from any energy source seem insurmountable and the results thoroughly arbitrary.

An approach that is less arbitrary in the details but more so in the general conception is to ignore illness completely to focus on mortality. An illness will influence the comparison only if it changes life expectancy. The implication of this, of course, is that pain and disability are of no consequence and that any day of life is as highly valued as any other. Aside from the conceptual problem, this is not a technically simple approach. Eliminating 10 percent of the deaths from a particular form of cancer does not give perpetual life to those who would have succumbed; rather it ensures that they will be victims of some other cause of death. Curing cancer (or preventing it) would increase the number of heart attacks and strokes. What is necessary with each change in environmental impacts is to recalculate the entire mortality table.

To account for differences in the quality of life as well as the quantity, attempts have been made to calculate changes in "days free of bed disability," rather than changes in the length of life alone.[5] This is an approach that is more congenial to the young and vigorous than to the old and crippled because it equates spending a day in bed with being dead for that day. Some illnesses that produce pain or other bad effects are not counted at all, whereas some that are relatively mild but leave the patient bedridden are given exaggerated importance.

We could try to derive other weighting schemes for the seriousness of illnesses. They could be weighted according to the cost of cure, but not only would generally fatal illnesses receive almost infinite weight, so also would incurable but mild conditions. Obviously, what is necessary is some sort of index of disutility, but since the market for diseases is poorly developed, it would have to be constructed by survey methods rather than by observation and statistical interpretation of existing data.

Attempts to use information about how much people pay to improve the probability of staying alive and healthy are confused by the smoking issue. The most important step that individuals can take to improve health and increase longevity and especially to reduce the risk of lung cancer is to abstain from cigarettes. The monetary cost of that action is negative, so standard rate-of-return calculations or measures of what a person will pay to increase healthy life or avoid cancer also behave strangely. Many other personal habits that improve health are also cheaper than the unhealthy habits that people enjoy. It is clear that the major costs involved are not monetary, which means that measuring them is difficult.

Chapter 14

The specific health, safety, and environmental issues associated with various energy sources have been sketched in the preceding chapters. Chapter 12 discussed attempts to compare the fatalities associated with the different energy sources. The central message of the studies is that none of the routine energy industries is very hazardous, especially if the new legislation keeps air pollution and hazards to miners under control in the coal industry. The risk to the public is more nearly comparable to the risk of liver cancer from eating peanut butter than to the risk of crashing while hang gliding. Even under the old regulations, the hazards to coal or uranium miners are comparable to those from smoking cigarettes.[6] It seems obvious that concern for safety should not be the central criterion in choosing among any of the usual sources of energy.

The one exception is direct burning of wood within the household because, like other activities undertaken by people for their own account, the safety record is very poor. Wood presents high risks from start to finish: Logging is among the most dangerous occupations and amateur logging is probably worse; axes and other splitting tools could not be approved by the Consumer Product Safety Commission if they were being newly marketed; people are constantly being burned up by wood fires in houses with poor chimneys or careless occupants; and burning wood pollutes the air.

Conservation measures are also far from riskless. A person can fall off a ladder while hanging storm windows or blowing insulation into a house. The asbestos insulation for high temperature use may cause lung cancer, and the urea formaldehyde foam installed in many old houses has been alleged to be so harmful that it is no longer in use. More generally, the archetypal conservation effort at the home or industrial level involves the substitution of some capital (that is, some construction or renovation) for some of the energy input. All construction and manufacture of materials involves risks. Much of that risk is occupational, but some of it falls on third parties as the conservation materials themselves are prepared and transported.

It is also becoming apparent that the successful attempt to save gasoline by making cars smaller and lighter has cost a substantial number of lives. One estimate indicates that in 1978 and 1979 Japanese subcompacts had 40 percent more deaths per registered vehicle than U.S. cars.[7]

That does not mean that conservation efforts are misguided. After all, people do take on these additional risks of their own free will, although in a cloud of ignorance of the consequences and urged on by government regulations, subsidies, and propaganda. People buy smaller cars to save fuel or to maneuver more easily in crowded cities. Nevertheless, the fact remains that conservation too is a risky activity. We can rely on market prices to incorporate that risk. If that seems too callous in the energy-producing industries, then the energy-saving activities must also be accompanied by a tabulation of compensated and uncompensated casualties.

ENVIRONMENTAL COSTS

Questions of risk to human beings and impacts on the environment are overlapping but not identical. Producing wood on mountainous terrain is risky for the logger but is usually considered to be an environmentally sound use of the land. At the other extreme, some of the potential environmental concerns, such as increases in acidity or carbon dioxide content of the atmosphere, could be serious enough to threaten the lives of the people or at least the carrying capacity of some parts of the planet. Since the most dire possibilities will not occur for many years, it is tempting to rule them out of economic analysis completely. Yet if certain effects really turn out to be cumulative and irreversible—if carbon dioxide builds up in the atmosphere forever or if oceans grow dirtier and dirtier until life dies—then it is a trifle harsh to say that the infinite misery of future generations, when discounted by any reasonable interest rate, has a negligible present value to today's population. The relevant question is the seriousness of any impact and the self-correcting features of the environmental system that may prevent problems from getting out of hand. For some of the more speculative possibilities, it seems desirable to continue close surveillance and research but not to go to any great expense until the evidence is stronger.

Most environmental impacts are not so dramatic or long lasting. They are primarily aesthetic effects in the immediate vicinity of the project, perhaps combined with some risks to people (who indeed may be part of the environment). The smog that burns the eyes of Californians is of no concern to me, unless someone can show that it influences my budget (by making vegetables more expensive) or my welfare (by blowing east and dirtying my air). Even local effects, however, can be construed to have a cosmic significance. If a small hydroelectric project floods an area, a sufficiently diligent investigator can undoubtedly find some element of genetic uniqueness among the plant and animal life that will be destroyed.[8] Certainly each piece of scenery is unique, and any development will change it. Does that suggest that the project should be abandoned?

If we could put a price on the environmental impacts, then the market would provide an answer. Losing a unique bullfrog might be a big enough cost to make a marginal project unprofitable. Certainly, the majesty of the Grand Canyon should be valuable enough to prevent our submerging the canyon under a reservoir. Some unique plants, animals, and scenery, however, are just not important enough to stand in the way of good projects. In the absence of an explicit price for the environmental effects, the judgment about which projects to abandon will remain controversial. It is not possible, moreover, to obtain agreement on the cost of the environmental change because, since the quality of the environment is a public good, the valuations of individuals at the margin will differ.

In fact, early support for nuclear power came from environmentalists who wanted to end the environmental assaults of coal mining and to save the few remaining wild rivers from hydroelectric projects. Nuclear power is not totally devoid of environmental impact, of course. In particular, the direct thermal pollution from a nuclear power plant in the United States is somewhat greater than that from a coal-fired plant because the government does not permit the nuclear plant to operate at as high a temperature. Hence it is thermodynamically less efficient and rejects more heat to the environment for a given amount of electricity generated.

Irreversible changes in the environment demand the closest scrutiny. It is necessary to estimate how the valuations of the public for the unique feature will change over time.[9] If the plan is to convert Yellowstone National Park into a geothermal power project, we need estimates of what the future generations would be willing to pay to preserve the geysers and hot springs. Economists often assume that the value of natural wonders will rise relative to the price of manufactured goods as productivity increases.

Aesthetic irreversibilities are not the only concern. Some analysts have become concerned about the possibility of major changes in climate. This prospect would tip the balance toward nuclear power because burning coal produces carbon dioxide, whereas the nuclear plant emits none. (Coal is worse in this respect than the other fossil fuels, since coal is mainly carbon, whereas oil and gas, as hydrocarbons, react with oxygen to form both carbon dioxide and water.) Some fear that the increase in concentration of carbon dioxide in the atmosphere will reduce the loss of infrared energy from the earth, heating the atmosphere by this greenhouse effect. A change of only a few degrees could melt the polar ice caps, turn farmland to desert, flood the coastal plains, and generally wreak havoc.[10] Is it happening? Average temperatures seem to be falling, but perhaps this will be reversed after the dust from the current upsurge of volcanic activity settles. It may be that the effort to control particulate emissions is particularly harmful, since the particulates offset the warming from higher concentrations of carbon dioxide.

More seriously, it is not clear what natural regulatory forces are at work in the entire system. Some of the increased carbon dioxide goes into the ocean. A higher concentration of it should also stimulate the growth of plants. It also appears that the destruction of forests by people accounts for as much of the release of carbon dioxide as does the burning of fossil fuels, but this is a process that cannot continue very long at the present rate before the forests are gone.

Although the buildup of carbon dioxide from fossil fuels may be considered the ultimate catastrophe, some of the allegedly soft techniques of energy production may have similarly harsh environmental effects. Tapping heat differentials in the ocean will bring nutrient-rich bottom water to the top, where biological activity occurs. This upwelling of nutrients is cited as a principal advantage of the plan, but it is not obvious that such a complete environmental disruption is totally benign. Certainly, if the advocate of a nuclear power plant discussed "thermal enrichment" of the source of cooling water, the environmentalists would rush to shut down the project, as they did at the Seabrook, New Hampshire, nuclear power plant. Extracting enough heat from ocean currents might even cause the currents to shift, with catastrophic effects on local climates and economies.

Similarly, it is not clear that large expanses of solar collectors will be benign. They will serve, like buildings and roads, to increase the proportion of energy from the sun that is absorbed. This will produce at least local changes in climate, just as cities and highways and cultivated fields do.

We can go through the list of actual energy sources and recite the pounds of various chemicals released per million BTU. We could also go through the entire list of proposed energy sources citing the worst fears of the most pessimistic observers. Neither listing seems very productive. All sources of energy have environmental consequences. The exact form and degree of such consequences depend on the perceptiveness and diligence of the efforts to limit environmental damage. Even if the officials of the producing firms have good intentions, those intentions must be reinforced by government regulation and enforcement to ensure that the environment will be protected from external costs and that the best-intentioned firms will not suffer a competitive disadvantage.

In strip-mining coal, for example, the most profitable behavior for an unregulated firm is to mine as cheaply as possible and then to abandon the land. The land can be reclaimed by segregating the different layers of soil and rock and carefully replacing them in the order that minimizes acid drainage and enhances fertility. The land can be given the contour (usually nearly level) that maximizes its value or that comes closest to the original contour. Even if the restoration work is done with great care at a cost of several thousand dollars per acre, the land will be worth less than $1,000 per acre in most locations. It is immediately obvious that no firm will carry out such extensive restoration unless the government compels it to.

Any detailed examination of the environmental impacts of energy sources becomes highly specific to the particular industry or even the particular mine or plant. A few general principles can be extracted, however. The first is the importance of diversity in energy sources. Diversity not only decreases the risk of disruption if anything goes wrong but it also makes the best use of the assimilative capacity of the environment and may give a longer lead time before a problem becomes a catastrophe. In particular, if manufacturing photovoltaic cells creates toxic waste, or fossil fuels add too much carbon dioxide to the atmosphere, or any of the nonconventional sources turn out to have harmful effects, the total damages will be restricted by the limited commitment to any one source.

The other great environmental principle is that the least harmful source of energy is likely to be the waste from existing activities. Although it does not always make economic sense to make more intensive use of existing energy, most of the environmental costs have already been borne and so it would seem worthwhile to look more closely at the economic possibilities. The most obvious candidate for closer scrutiny is the process of generating electrical energy from fossil or nuclear fuels, where nearly two-thirds of the energy is wasted as thermal pollution under conventional conditions.

The trend in the construction of modern power plants has been to use higher temperatures and pressures to achieve higher theoretical efficiencies, but limitations of the available materials for operation under such severe temperature conditions have limited the gains that have been achieved, especially during the past two decades.

The three main approaches to improving efficiency are (1) to continue the efforts to raise operating temperatures of conventional equipment, (2) to seek new approaches to very high-temperature operation, and (3) to make better use of the low-grade energy that remains after the steam has powered the turbine generator. The first approach requires no further discussion. Although it paid off handsomely in the early days of the industry, it seems to have reached a plateau at present.

One approach to very high-temperature operation is the magnetohydrodynamic generator (MHD). This research aims to burn coal at such a high temperature that the combustion gases can be ionized. Movement of the electrically conducting gas through a magnetic field would generate the current. An MHD generator could be the first stage in a combined-cycle plant that used the still hot exhaust gases from the MHD stage to generate steam for a conventional plant. The technique of extracting different kinds of work at different temperatures is also the basis for cogeneration, discussed in Chapter 11. Whatever the technique, it seems likely that ways will be found to squeeze more useful work from the fuels that are already being used.

Similarly, from an environmental view the fuels that are by-products of existing economic activities are generally least damaging. These include energy from waste, such as burning municipal refuse to generate electricity, or making use of agricultural and industrial waste products. Such use can also include draining methane from garbage dumps or from coal seams before mining, which reduces the risk of explosion in the mine as well as supplying natural gas.

A closely related activity, the reuse and recycling of waste materials, indicates the pitfalls of attributing environmental superiority to activities that save energy, however. There is no doubt that reuse of manufactured items (such as old automobile parts) uses less energy than manufacturing new ones, although the relationship is not as clear-cut for recycling (for example, melting down scrap into metal that can be used again) and becomes even less so when the associated activities of collection, transportation, segregation, and processing are considered. Despite the presumption in favor of the environmental beneficence of reuse, an inventory of reusable parts (junkyard) often faces opposition on grounds that are proclaimed as environmental. Sometimes the purely aesthetic objections can be overcome by building fences or planting shrubs, but the essential point is the sidetracking of the public discussion by obvious but minor issues, while the significant questions of use of exhaustible resources and irreversible changes in the atmosphere are ignored. Likewise, it is easy to leap to the conclusion that recycling where it saves energy is good for the environment. That may not be so if, for example, the process of preparing waste paper for repulping dirties the water. It is not obvious then that recycling is better for the environment than is growing more trees to provide virgin paper. Tree growing is, after all, environmentally sound.

NOTES

1. For details of the analysis, see any text in environmental economics, such as Edwin S. Mills, *The Economics of Environmental Quality* (New York: W. W. Norton, 1978).
2. T. C. Schelling, "The Life You Save May Be Your Own," in *Problems in Public Expenditure Analysis*, ed. Samuel B. Chase (Washington, D.C.: Brookings, 1968), pp. 127–162. In 1984 the Office of Management and Budget was embroiled in a dispute with the Department of Labor about whether a value of $1 million or $3.5 million should be imputed to the life of a construction worker. See the *Wall Street Journal*, October 24, 1984, p. 3.
3. D. J. Crawford and R. W. Leggett, "Assessing the Risk of Exposure to Radioactivity," *American Scientist* 68, no. 5 (September–October 1980), pp. 524–536; Gilbert W. Beebe, "Ionizing Radiation and Health," *American Scientist* 70, no. 1 (January–February 1982),

pp. 35–44; William Ramsay, *Unpaid Costs of Electrical Energy* (Washington, D.C.: Johns Hopkins University Press/Resources for the Future, 1979), chap. 3.

4. Lester B. Lave and Eugene P. Seskin, "Air Pollution and Human Health," *Science*, August 21, 1970, pp. 723–733.
5. U.S. Department of Health, Education, and Welfare, *Toward a Social Report* (Washington, D.C.: U.S. Government Printing Office, 1969), chap. 1.
6. The increased risk of death because of reactor accidents for a person who lives within five miles of a nuclear power plant for fifty years is about equal to the risk from eating forty tablespoons of peanut butter (liver cancer) or smoking 1.4 cigarettes or spending three hours in a coal mine. See Richard Wilson, "Analyzing the Daily Risks of Life," *Technology Review* (February 1979), p. 45.
7. Dale D. Buss, "Small Cars May Save Fuel but Cost Lives," *Wall Street Journal*, April 27, 1982, p. 1.
8. The snail darter, an apparently endangered fish, helped to stop the Tellico Dam in Tennessee, and the Furbish lousewort, a plant thought to be extinct, halted the Dickey-Lincoln project in Maine. Neither endangered species seems to have been worth saving, but in those cases the hydro projects were not worth saving either. For a careful discussion, see Winston Harrington and Anthony C. Fisher, "Endangered Species," in *Current Issues in Natural Resource Policy*, ed. Paul R. Portney (Washington, D.C.: Resources for the Future, 1982). The International Union for the Conservation of Nature and Natural Resources is even trying to save the pygmy hog sucking louse, which spreads infections among the pygmy hogs, an endangered species living in Assam, India. See James P. Sterba, "Before You Squash That Bug, Be Sure It Isn't One We Need," *Wall Street Journal*, October 5, 1983, p. 1.
9. Anthony C. Fisher, John V. Krutilla, and Charles J. Cicchetti, "The Economics of Environmental Preservation: A Theoretical and Empirical Analysis," *American Economic Review* 62, no. 4 (September 1972), pp. 605–619.
10. John W. Firor and Paul R. Portney, "The Global Climate," in *Current Issues in Natural Resource Policy*; Jae Edmonds and John Reilly, "Global Energy and CO_2 to the Year 2050," *Energy Journal* 4, no. 3 (July 1983), pp. 21–47.

CHAPTER FIFTEEN

Government Policy

To ask whether the United States has an energy policy is to invite a sterile dispute that centers on the definition of *policy*. If we take the broad view of policy as any government action that influences, or is intended to influence, the production or consumption of any form of energy by any person, then it is obvious that the United States has many energy policies. If we restrict policy to a coherent and coordinated set of actions rationally designed to achieve some explicit objective, then the United States does not have an energy policy. The latter can be named a coherent energy policy. This chapter explains why it is impossible and probably undesirable for the United States to implement one. First, however, we need to discuss a number of the areas in which the government has, purposely or inadvertently, influenced energy markets, as well as a number of areas in which it has been suggested that government should have a major role.

Chapter 15

If the federal government has any legitimate functions, ensuring national security must be one of them. Certainly, it is worth examining the connection between energy and national security.[1] This has two aspects. The easier question is whether the United States would have the fuel to carry out military action should that be necessary. The answer is that if we can ensure adequate civilian supplies in normal times, the military requirements can be extracted from the civilian sector during emergencies without extreme sacrifices.

The more serious question is whether the economy is jeopardized by a continuing high level of imports. In the most dramatic case, a foreign nation could just cut off our oil supply, as some OPEC nations did during the embargo of 1973. Nor is this threat restricted to oil-producing nations. The bridge of oil tankers over which our supplies are imported would be very vulnerable to attack by any country. The vulnerability of the supply line is worsened by the necessity of moving such a large proportion of the world's oil supply through the narrow mouth of the Persian Gulf.

Even in the absence of disruptions, the dependence on a few foreign countries for a significant part of our total supply of oil may compromise the independence of our foreign policy. The worst aspect of this is that much of it may just happen as an almost imperceptible slant in the decisions on particular issues as we seek to avoid offending the major oil suppliers. Preserving stability in the Middle East would be difficult even in the absence of oil, but the importance of the oil raises the stakes and may increase the difficulty.

Before discussing remedies for the problems deriving from dependence on unstable sources of supply, let us take a closer look at the various conditions. To what extent are the problems related to our dependence on the Middle Eastern oil producers in particular rather than on imports in general? Would our position be stronger if a larger proportion of our oil came from Mexico and Canada rather than from Saudi Arabia? This simple question raises several types of issues. First is the question of the sources of supply for the United States compared with those of Japan and Western Europe. The United States might contrive to acquire its entire supply from countries other than those in the Middle East. Since the oil reserves of the world are concentrated in the Middle East and production costs are lowest there, however, it is difficult to imagine that the Middle East will cease to be a major exporting region. The implication of making the United States less dependent on that region would be to make Western Europe and Japan more dependent. The improvement in our freedom of diplomatic maneuver would be limited, because in most significant matters the United States would need support from other importing nations. Another way to say the same thing is that oil is fungible. It can flow quickly to the market offering the best terms. It is important therefore to think of the percentage of all exports originating in a particular region in assessing the potential disruption that the region could cause.

Security of supply from military disruption is a somewhat different matter. Presumably, it would be easier to defend pipelines from Mexico and Canada to the United States than tankers from the Middle East, although a large pipeline can be disabled by a single act of sabotage. But the situations under which reliance on this continent would be an important advantage are difficult to foresee. It seems to presuppose a major war among major powers fought with the techniques and weapons of World War II. It also presupposes that Canada and Mexico would be willing to sell so much oil to the United States despite announced policies that forbid it.

The general principle that diversification of supply decreases the risk of major disruptions is correct, but the principle should be applied to the world, not just to the individual country. If the sources of supply are widely diversified, an accident shutting exports from one country will do relatively little damage. Similarly, as the producing countries become more diverse, the need for accommodations in foreign policy dwindles. If the problem is ideological, however, it may span several producing countries, as the 1973 embargo did. Most of the Arab oil exporters did cut back shipments to the United States, but Iran, which is Muslim but not Arab, did not. If the main interest is commercial, then it is in the interests of all producers that output be restricted. Even those who have not joined OPEC have benefited from the organization. Indeed, those who sell at the high prices without having to accept the reductions in output are the largest gainers.

The dwindling share of OPEC in the world oil market is exactly the diversification that is useful for purposes of natural security, but it comes at some cost. From the perspective of world efficiency, it seems unfortunate to devote twenty or thirty dollars of resources to extract oil in difficult, hazardous, and environmentally fragile locations like the North Sea and Alaska when the Middle Eastern oil can still be produced and carried to the refinery for trivial amounts. No one approaches the matter from the perspective of world efficiency, however, and from the view of a consuming nation, the oil that is purchased from another nation at $30 per barrel costs $30 in real resources even if the production cost is negligible.

Nevertheless, the question of where the oil comes from has both short- and long-run implications. Since the Middle East has more than half of the world's reserves, it looks as though the choice is to be dependent now on that volatile area or to be dependent later. It is possible at some cost to develop domestic and other foreign sources to such an extent that Middle Eastern oil is not vital to the world now, but is that wise in view of the prospect that it will become increasingly vital in the future? The correct answer depends largely on forecasts, about which no one can be certain: After all, the Middle East may become a stable, peaceful, reliable supplier; ample reserves may be found in the United States; and oil may be displaced by other sources of energy! If all of these are considered too speculative to count on, then the appropriate solution may not be to end the uncertainty but rather to learn to live with it.

In many of the mineral industries, processors seem to make a practice of maintaining excess capacity in the producing units (mines or wells). The surplus producing capacity keeps the owners of mineral rights in eager competition with one another and prevents cartels of producing countries or firms from lasting long. It is noteworthy that OPEC had its greatest impact at a time when a series of events had combined to eliminate excess capacity for producing oil. Conversely, prices stabilized as excess capacity developed after 1980.

The three remedies for the threat posed by dependence on Middle Eastern oil are control, independence, and preparation for interruptions. Traditionally, strong nations dependent on weaker ones for an essential material have imposed control over the weaker one. It was not always necessary to move in the troops because fear of military action kept the weaker nations in line. Has the world changed so much that the United States could no longer follow such a policy? The embargo was probably the best test, and it showed very little sentiment in the United States for military intervention to keep the oil flowing.

The political response to the embargo of 1973 was a call for energy independence.[2] Whether this was ever a policy is debatable because most of the actions of government delayed the reduction of imports. Nevertheless, it was official rhetoric for a while and many still hold to the goal. To end imports of all energy—or even of all oil—the United States would have to rely more heavily on coal and nuclear power, as well as developing oil shale and other expensive sources of liquid fuels.

The critics of increased domestic production of energy included some environmentalists who wanted more stress on conservation and ridiculed the program as "strength through exhaustion" of our own resources. Opposition also came from economists and others who considered the price too high. The conservation question is examined later, but the question of cost deserves more attention now.

Ordinarily economists think of costs as being reflected in market price. The burden is on the person who alleges that social cost differs from the market price. In this case, energy independence seems expensive from the viewpoint of consumers; the price of domestic alternatives would exceed the price of the imported oil. Do imports impose such large external costs that the market prices must be rejected? Such arguments were reviewed in Chapter 9, but if national security is the sole externality, then it is useful to ask if a cheaper means than energy independence will achieve an acceptable level of national security.

The United States has long maintained stockpiles of other strategic imports such as alloying metals that can be tapped in national emergencies. It has also maintained the Elk Hills naval petroleum reserve—a field of proved reserves that was preserved for military use in emergencies. It was natural therefore to think of expanding such emergency capacity as a cushion against interruptions.

By the end of 1984 the Strategic Petroleum Reserve (SPR) amounted to almost 450 million barrels of crude oil stored mainly in salt caverns.[3] It is not easy and cheap to dump oil into a hole in the ground in a way that ensures it can be retrieved and moved quickly to market when it is needed. Nevertheless, the SPR has much to recommend it because a large enough stockpile gives the advantage of energy independence without the high costs of domestic production at high levels and without running down our reserves.

It seems ironic, however, to import oil to dump into the ground at the same moment that domestic producers are pumping oil out. Holding natural fields in reserve sounds superficially more attractive: Develop fields but cap them and mothball the pumps until an emergency. The chief objection to this approach is that to ensure that oil can be delivered from the field at a rapid enough rate, huge reserves would have to be set aside and huge development expenditures would have to be borne. The main problem is that oil cannot be extracted from a natural field at a very rapid rate without reducing the amount that is ultimately recovered.

Since the SPR amounted to almost 450 million barrels by the end of 1984, a year when imports of crude and products averaged 5 million barrels per day, it was a ninety-day supply. It is certainly not impossible to imagine protection against an interruption of Middle Eastern production of even a year by a combination of further increases in the SPR and moderate reductions in domestic consumption.[4]

What does such a level of protection cost? The explicit costs of building and maintaining facilities are considerable. In addition, the cost of carrying oil in inventory must be considered. Adelman estimates that a stockpile of 750 million barrels would cost about $6 billion per year to hold. This could be financed by a "disruption tax" of $3 per barrel of imports.[5]

Does the interest charge on the oil inventory measure the cost to the economy of holding it? The answer depends on the behavior of the price of oil. If the interest rate is 15 percent and the price of the crude rises 15 percent per year, it costs nothing for the government to hold it. If oil prices are rising faster than the interest rate, the government can profit by holding oil, but that would seem to be an unstable situation since anyone else could profit by holding oil as well. If oil prices rise only slowly, there is some cost in holding inventories, and the cost becomes heavy if the price of oil is actually declining.

In equilibrium the price of oil is supposed to rise at such a rate that economic rent (for the marginal barrel) rises at the rate of interest. This implies that prices rise less rapidly than the rate of interest, but how much less rapidly depends on the proportion of rents to the price of oil. The difficulty is that the market currently is not in full equilibrium. It is hard to see which well or producing area is marginal and thus hard to know what rate of economic rent is governing the price trend. In the ideal textbook case, the resources that can be brought to market cheapest are exploited first. This corresponds with Middle Eastern oil. Only if the lowest-cost producers are unable to expand without encountering high marginal costs will inferior deposits be drawn into production.

Before the success of the OPEC combination in the 1970s, the world seemed to be moving slowly in the direction implied by the theory. The low-cost sources of the Middle East were accounting for an increasing share of production. Exploration had slowed in the high-cost areas. Many producing wells were capped when the rate of output declined enough, rather than being reworked.

Once the cartel succeeded in raising prices, the situation became more complex. The increase in prices made a great variety of energy sources profitable, and development proceeded quickly. It may be assumed that individual firms controlling resources ranked various properties according to expected profitability, but it is also clear that in addition to the errors in guessing about profits, some of the projects undertaken by firms that had no better prospects were worse than those that were postponed by the better endowed.

Under these conditions, can anything be said about the expected path of future prices? If prices were to decline further, many of the drilling efforts now under way would not appear profitable. It is conceivable that the cartel could break down so completely that Middle Eastern oil would displace new drilling in more expensive locations. Much depends on the willingness of Saudi Arabia and other low-cost producers to withhold potential production. As long as the low-cost reserves in the Middle East are underutilized, the economic pressures are toward lower prices and increased reliance on that unstable region. This is a long way around to the conclusion that both the interest costs and the national security advantage of holding an oil reserve are real.

A tank of oil does not have to be labeled a strategic petroleum reserve to serve that function. Normal working inventories can be maintained at different levels depending on current and expected prices, interest rates, and other economic variables. Even with the growth of the SPR in recent years, crude oil stocks of those in the business were nearly as large at the end of 1983 and stocks of finished products of refiners were even larger. It is very difficult to operate without adequate inventories, but an alternative to a designated SPR could be tax or other inducements to firms to hold a larger inventory.

Greater commercial inventories would also solve the problem of deciding when to release reserves. A reserve does not protect against disruption unless it is drawn down when a shortage occurs. During the embargo of 1973, for example, the government was so slow to allocate supplies to relieve distress that inventories were actually larger when the embargo ended than when it began. More recently, the government has announced a policy of immediate and aggressive use of the reserve, but it remains to be seen whether the government is capable of stabilizing action in view of the political and other complexities.[6]

ECONOMIC GROWTH

The energy problems of the 1970s provoked a reexamination of the entire issue of economic growth. Some initial confusion about the relationship among economic growth, energy growth, and population growth has been clarified.[7] As noted in Chapter 1, the number of BTUs per dollar of GNP has declined. The ratio can certainly continue to decline as technological change makes the use of energy more efficient. Indeed, the efficiency with which energy is used improved so rapidly that total energy consumption in the United States in 1983 was less than that in 1972.

In contrast to energy, economic growth as measured by GNP per capita must continue. This does not mean the accumulation of an ever-increasing tonnage of goods. It is easily possible to imagine the tonnage of consumption declining as automobiles, electronic equipment, and housing become more compact. Those who are worried about the environmental impact of mass consumption usually think of economic growth in such gross physical terms. There is no reason why economic growth cannot be consistent with decreased environmental impact if it occurs as the further refinement of the same quantity of materials.

"Zero economic growth" would be difficult to live with.[8] We can think of traditional societies in which, with stagnant populations, the children simply marched in their parents' footsteps forever. Modern industrial economies cannot revert to such a condition, even if the order and predictability of such societies may occasionally seem attractive in our harried moments!

A modern industrial economy is never in equilibrium. The normal business fluctuations, with periods of growth creating inconsistencies among various sectors, require adjustments that are brought about during the decline. Again, the decline is self-limiting as the advance of technology proceeds and the consumers and businesses accumulate lists of items that they will purchase as soon as they are able to obtain financing.

Even if politicians should ever succeed in fine-tuning the economy to produce steady, sustainable growth in such aggregates as GNP and employment, cycles will continue among the components of the economy as technology, accumulating wealth, and changes in consumer preferences cause some industries to grow while others decline.

International trade reinforces the necessity of growth. The strong and complacent industry serving the nation with apparent success can find itself displaced from the highest-valued markets by imports that provide greater precision, closer tolerances, a different range of sizes, superior service, or any of dozens of other qualities. As the imports strip off one market after another, the domestic industry suffers from lowered revenues and operates with excess capacity. Under those conditions it is difficult to justify (or finance) the major investments in new productive facilities that may be necessary to capture the full benefits of technological change. If the industry tries to make do with old facilities, however, it falls farther and farther behind the competition. It may eventually die unless some major infusion of new capital, knowledge, and entrepreneurial spirit reverses the decline.

This sketch of the technical reasons why zero economic growth is impossible as a stable equilibrium does not exhaust the topic. The social consequences of stagnation are more serious than the economic. If people cannot look forward to growth in their own level of well-being from general economic growth, they turn their attention to extracting a "fair share" of the finite level of resources. The social and political atmosphere soon becomes venomous with bickering about the division of a fixed national income that appears increasingly meager by international standards. The economic decline of Great Britain during the 1960s and 1970s illustrates the consequences of giving priority to distribution of income and quality of life rather than economic growth and efficiency.

Growth must go on in the economic sense, but what of energy? In part this will be a matter of conscious policy, because it is possible to devise growth-oriented policies, or conservation-oriented policies, or to adopt a neutral course and let the market determine the energy intensity of economic activity. The rate of population growth will also have some influence on energy, both directly because of the demands of additional people for additional housing and other energy-intensive goods and indirectly through whatever influence population might have on economic growth. Population, however, is too large a subject to consider here. It is also not clear that government can influence it by normal instruments of policy.

A growth strategy must aim at economic growth, rather than energy growth, because production of energy for its own sake does not provide benefits. We can always find ways to use more income or wealth, but once the room is comfortably warm, more heat does not add to utility. A growth strategy therefore means a strategy of economic growth unconstrained by fears about the adequacy of energy resources. Still, the question of the objectives of economic growth must be confronted because the goal will determine the types of sacrifices we are willing to make in order to grow. If, for example, we are inspired by the goal of improving the welfare of future generations, it would be ironic to exhaust their minerals and destroy their environment in the name of improving their welfare.

Whereas irreversible impacts on the environment are serious concerns to an advocate of growth, the transitory environmental impacts need not be. Air pollution from an electric power plant would not be controlled until it threatened to reduce productivity. Similarly, the ugly appearance of strip mining could be ignored. More to the point, if economic growth were the overriding objective, the barriers that have stood in the way of using oil shale, high-sulfur coal, and nuclear power would be eliminated. After all, in any field of endeavor, if we spent our time looking for every danger that might lurk around the corner, we could never accomplish anything.

Any of the plausible reasons to pursue economic growth imply the necessity of taking steps to ensure national invulnerability to interruptions of imports. That does not mean energy independence if that is more expensive than imports, but it does imply research to develop domestic means to supply energy if imports are restricted or become more expensive. In concrete terms this means funding demonstration plants for producing liquid fuel from oil shale and coal so that technology and operating experience are available when needed. It also suggests that research on the breeder reactor and especially nuclear fusion be pushed ahead so that a backstop is available before conventional fuels become too expensive.

The easily available fossil fuels would last longer if environmental standards were lowered. The power-consuming precipitators and dust collectors could be eliminated, standards for water quality could be lowered, and the auto emissions standards could be dropped. Furthermore, some additional tolerance about the areas judged suitable for hydroelectric development would also promote growth.

Naturally, the best thing that could be done for efficiency—and hence growth—would be to drop all the regulations that influence the pricing and use of fuels. If prices are set free, the market will decide such questions without the waste that inevitably characterizes government allocations and the maintenance of low prices for favored classes of users. Similarly, in the interests of efficiency and growth, the distortion of effort and use of capital implied by subsidies for particular activities like insulating houses or installing solar collectors and windmills should be ended.

Outside the energy area the government has bigger tasks, but still a limited number, if it wants to encourage growth. The basic message is the same as it is for energy: Eliminate the controls and regulations that distort the market allocation of capital and labor. The most important task is to encourage capital formation, risk taking, and entrepreneurial spirit. There is not much that government can do directly in these areas, but it can remove the taxes that inhibit economic effort (shifting from income taxes to consumption taxes would help). It can also prune away the complex of regulations in areas like labor relations, marketing techniques, financial reporting requirements, and product liability that make it difficult for a small firm to operate.

Capital accumulation is vital for economic growth, but it is particularly important for energy supply. In case after case in the preceding chapters, particular sources of energy were found to be technically satisfactory but not economically feasible because of the high capital charges. This is particularly a problem with the nonconventional sources. Solar power requires the use of a great deal of capital to collect an input that comes freely from the sun. Similarly, windmills, hydroelectric projects, tidal power, nuclear power, and even heavy insulation all use massive amounts of capital for the energy they produce or conserve. Any reduction in the interest rate makes it profitable to adopt techniques that have the effect of conserving the stocks of fossil fuel and postponing the difficult and costly transition to such other sources of energy as oil shale, breeder reactors, and nuclear fusion.

The only way to supply massive amounts of capital to the energy industries without harming the economy is to encourage savings. If people can be persuaded to save by the tax laws, propaganda, financial institutions that make saving easy, safe, and convenient, the example of prominent people, and religious or moral precepts, then firms can invest without inflation and the economy can grow efficiently. Government also can save directly by cutting expenditures enough to run a surplus. By retiring some of the government debt, the federal government can supply funds directly to the financial markets where they will act to drive down the interest rates.

CONSERVATION OF RESOURCES

Many people do not share the enthusiasm for economic growth that is the premise of the preceding section. Although most people have a mix of objectives, we can usefully consider a policy based on the objective of conserving as large a portion of the reserves of fossil fuels as it is feasible to do. It is common to think of conservationists and environmentalists as advocating similar policies; in the case of fossil fuel conservation, that may not be so.

People might be concerned to conserve fossil fuels because they are so convenient as portable sources of energy and as chemical feedstocks; that is, they might consider them to be a unique legacy that this generation is obligated to conserve for the future. They might also be concerned about the damage that coal mining or oil spills can do to the environment or by the pollution from using fossil fuels and the possible consequences of high levels of carbon dioxide in the atmosphere. Then, too, they may just advocate conservation today to minimize the adjustments they will have to make during their own lifetimes.

The possible roles of government in pursuing conservation may be classified into four levels: First, the government may restrict itself to repealing the laws that inhibit conservation. On a second level, government may take some steps to speed the adjustment of the economy to the increased level of conservation that has become profitable with higher energy prices. Third, the government may provide tax or other incentives for particular measures to conserve energy. Finally, the government may enforce conservation by requiring or forbidding particular behavior.

It would seem sensible for the government to rescind immediately all laws that discourage conservation of energy. Perhaps the federal government will begin by exempting automobiles from emission controls, while local governments will start by eliminating minimum lot sizes and minimum house sizes! In most communities if you wanted to live in a fur-lined cave to conserve fossil fuels for future generations, you would not be able to obtain a building permit. Nearly all zoning laws, land use controls, and building codes are aimed at decreasing density and increasing the cost of housing. One result is the reduction of density to the point that mass transit is not feasible. Furthermore, the large, solid houses specified by the codes are so expensive that drafty old houses cannot be replaced by tight new construction. In many places houses that appear bizarre because of solar heating or extraordinary insulation are not permitted.

A diverse array of laws and regulations inhibits many attempts to save energy. In many cities, for example, landlords are required to provide heat at particular levels, times, and circumstances. Often the tenant is better heated than the homeowner, now that fuel prices have risen. Responding to similar pressures, politicians have often kept energy prices low through controls, attempting to benefit favored groups. One result is to discourage efforts at conservation. For the regulated utilities, marginal cost pricing would offer a substantial incentive to conserve if it were extended to all customers.

In general, all emission controls—not just those on automobiles—require energy to operate and inhibit the use of waste as fuel. Similarly, nearly all attempts to increase safety add to energy requirements by adding weight and complexity to machinery and other products and inhibiting experimentation with lightweight solutions.

It can readily be seen from this brief list of suggestions that even the first level of government encouragement of conservation has some controversial aspects, because even conservationists have other goals. The second level of intervention—speeding the adjustment of the economy to the new prices—is less controversial, although it involves more detailed intervention.

The basic means by which the adjustments can be speeded are research, propaganda, devising new institutional arrangements, and legal coercion. Research could include such matters as rating automobiles for mileage, which the government purports to do, or testing the safety and effectiveness of different kinds of home insulation, which the government never got around to. It could include a wide range of applied research on the most efficient ways to build homes, operate appliances and automobiles, carry out standard industrial processes, and so on. Some bits and pieces of this sort of work were done, but much of the effort of the Department of Energy has been directed at pure propaganda.

Much of the propaganda is innocuous, and some of it may even be useful to a few people: To save fuel close doors and windows; to save electricity turn off the light when no one is using it, and so on.[9] More specific information about the costs and benefits of particular conservation measures seems more likely to be convincing. The other extreme from the specific information about savings from caulking around windows is the generalized assertion that less is better. Propaganda can be used for the purpose of convincing people that they would really rather live a totally Spartan existence; for example, standard advertising techniques can be used to persuade people that they prefer shivering in a cold room to traveling, buying more appliances, and wallowing in creature comforts. As an intermediate stage, the government can exhort people to sacrifice for patriotic reasons. All such techniques are likely to work best in dealing with temporary national emergencies when people can see that they are postponing pleasures to survive a crisis, rather than giving up good living forever. In the interest of nudging people toward the equilibrium that is to be expected after higher energy prices have had their full impact, it is more useful to suggest how people can enjoy life to the fullest using less energy.

Devising new institutional arrangements to deal with the new realities of energy is an area where government leadership could cut adjustment times substantially. The classic example is the case of rental housing where heating is paid by the renter. The owner has no immediate incentive to insulate because he does not pay the heating bills. The renter has no incentive to insulate because the term of her lease is shorter than the payback period. In the long run, the market may solve the problem by permitting higher rents in insulated houses or by the development of various specific arrangements between individual landlords and tenants. Are there ways to get the work done now with the payment to come from the savings in heating costs? This problem has proved to be too difficult to solve.

As another example, all sorts of devices by which plants might make use of otherwise wasted energy (cogeneration, waste heat boilers) are easier to arrange if firms can sell surplus energy easily and use backup power from a utility. Recent changes in federal legislation have been helpful, but there are still problems with firms becoming subject to regulation as public utilities if they sell energy. Another institutional problem relates to lending practices of banks, reinforced by FHA guidelines. If banks traditionally limit mortgage payments to a set percentage of family income, then solar power (or heavy insulation) is at a particular disadvantage. Such techniques substitute capital for operating cost and thus permit the family to pay larger mortgage bills. Traditional banking rules are biased toward techniques with low capital cost and high operating cost.

The institutional questions span a wide range. For example, many private groups influence energy consumption. The lighting standards for schools and offices widely used by architects and engineers were developed and published by a lighting industry trade association. This example is commonly cited, but undoubtedly there are numerous other instances in which heavy energy costs are embedded in standards and customary practices. Some have been written into building codes, local ordinances, professional standards, trade union agreements, and government regulations. Such institutional problems are very difficult to root out, particularly since no single one seems worth the effort it would require to eliminate it. Furthermore, they were adopted to serve some useful objective, and both the origin and the authority to eliminate them are often unclear. Even when engineers succeed in changing the design standards found in the current textbooks, the engineer on the job may be using a handbook that is twenty years old.

At a third level of involvement, the government can establish positive incentives for conservation. This is politically popular, especially when the incentives take the form of tax breaks. Nevertheless, such measures can accumulate to have substantial consequences for economic efficiency. If they are successful, they will draw capital and labor into the favored activities—often without a clear and public accounting of the total distortion of economic activity. The favored activities will not be the ones to which the resources of the economy should be devoted unless the judgment of politicians about efficiency is better than the judgment of the market.

The United States has used the income tax as a vehicle for subsidizing insulation and the development of certain renewable resources. In the interest of encouraging experimentation, the tax breaks for some of the more innovative activities can certainly be defended. The subsidies for such routine and predictable activities as insulation and caulking are harder to justify except as a technique for speeding the adjustment to higher prices.

The government can subsidize specific activities through expenditures as well as taxation. There have been some programs to subsidize energy conservation measures for low-income households. These seem like unpromising activities for a national government to undertake. A subsidy program involving more manageable numbers of recipients has been that for urban mass transit. The justification is that since private passenger cars use a lot of gasoline, mass transit must be more efficient. A bus carrying one passenger uses much more fuel than an ordinary car carrying one passenger, so the superiority of mass transit is not self-evident. It depends on ridership. We can doubt that the subsidies for new capital equipment were the most efficient way to increase ridership. The implication is that the design of subsidy programs is a tricky business, so there is no guarantee that subsidies will be more efficient uses of federal revenues than will the notoriously inefficient tax deductions and credits.

Taxes can be used to penalize specified activities as well as to reward others. The main existing example of a penalty tax for energy use is the special assessment on automobiles that do not meet the fuel efficiency standards. This tax is not well targeted, however, because it is levied on all cars produced by a manufacturer if the average fuel mileage of cars made by the firm does not attain the standard. A more rational approach is to tax each car according to fuel consumption per mile. More rational still, if the objective is to discourage fuel consumption, is to levy a stiff tax on automobile fuel; that is, the existing federal fuel tax could be raised substantially. This would let individuals decide whether they wanted to pay it and continue their former use, to buy cars that used less gasoline, or to use their cars less. The problem with levying the tax on the fuel inefficiency of the car is that it does not encourage the response of driving less, which is surely the preferable one from the view of reducing the hazards and congestion and the aesthetic and urban problems arising from the automobile.

One certain effect of a very heavy tax on gasoline and diesel fuel would be an increase in the feasible application of electric and other alternatively fueled vehicles. If the objective is to reduce demand for petroleum, that is enough. Some have argued, however, that a general tax on energy consumption ("the BTU tax") would be useful to encourage conservation. If the objective is to conserve fossil fuels only, then the tax could be eliminated on other sources of energy including nuclear power, hydro power, and other renewables.

Although direct subsidies via government spending and indirect subsidies via taxation are the main incentives that government uses, there are other possibilities. For example, there can be special encouragement for resource-conserving practices in the exploitation of minerals by exemptions from some of the more onerous environmental requirements. The detailed requirements for restoration of original contour in coal mining might be waived for firms that extract a high enough proportion of all coal. Likewise, the law could encourage unitization of oil and gas fields to reduce the risk that fragmented ownership will bring about excessively rapid rates of exploitation.

Finally, the government can put its own house in order. That means not only tending to buildings and vehicles used by government but also examining various programs to see whether they are consistent with conservation. A program to provide special supplementary welfare payments to poor people with large heating bills, for example, is not consistent with a conservation policy. At a more subtle level, various government agencies have continued to encourage urban sprawl and long commutes, even while the stated objective is conservation. If energy conservation were considered a sufficiently important objective, the government might be able to shift various policies toward energy conservation. This is a difficult task, however, because the policies were adopted to serve some purpose or satisfy some interest. Unless the policy was a mistake in its own terms or the interests have changed, the policy will not easily be modified. Thus the Farmers Home Administration insures mortgages in rural areas, even for those who commute to the city to work, precisely because strong interest groups want a large population in the rural areas.[10]

A fourth level of government support for conservation is direct coercion. The borderline between specific taxes and coercion is sometimes hard to draw because failing to obey a regulation or law usually results in a fine (not imprisonment or execution) that can be treated as a tax. The penalty for failing to comply with automobile fuel efficiency standards can be treated as either a tax or a fine. More generally, however, the government can prohibit certain activities, such as the use of gaslights in the yards of private homes or the sale of gas appliances that require a pilot light to burn continuously. The Natural Gas Policy Act was filled with direct prohibitions: Large new boilers are to burn coal rather than oil or gas; existing large boilers must be converted to coal. It has also been suggested that insulation standards for houses be extended to the existing stock by requiring that every house be upgraded at the time of sale.

There is precedent for taking such heavy-handed action in the name of conservation. At the beginning of World War II, manufacturers of tin plate were required to switch to the then-experimental method of electrolytic tinning under penalty of not obtaining any allocation of that scarce metal. The electrolytic method decreased tin consumption substantially.

Such techniques seem more appropriate during wartime than during the routine functioning of the economy. In addition to any qualms a civil libertarian might have about federal bureaucrats checking to see whether I have snuffed out the gaslight in my garden, the efficiency of conservation by regulation is open to question. If a firm uses an expensive technique because a cheaper one is forbidden, then it is using expensive labor, capital, and other materials to replace cheaper energy. This can be justified only by the assumption that the market price of energy is less than its cost to society. People who believe that should advocate a corrective tax rather than direct regulation.

Conservation has been used to defend a wide range of specific actions by government, but a few general points emerge from the preceding pages. The first is that the keystone of a conservation strategy is the same as the keystone of a growth strategy; the government should end price controls and other behavior that leads to excessive use of energy resources. Moving a step beyond that, there is enough work in weeding out the institutional pressures toward excessive energy use to keep the government's energy specialists occupied for years. Although the assignment may seem negative, it is important work and much of it could not be accomplished except by government. If the public has some abiding taste for inflicting even more conservation on the country as a whole, the most reasonable approach is a BTU tax on all fossil fuels or perhaps all nonrenewable energy sources. The intellectual underpinnings of such a special conservation effort are shaky, but the old message from public finance still applies to any effort to promote conservation: The more general the tax, the fewer the distortions. In particular, a general tax is much less damaging than a series of specific prohibitions.

RESEARCH AND DEVELOPMENT

The easiest way to evade tough questions is to call for more research and development. Why should government subsidize R&D, and especially why should it do so in the energy area?[11] The explanations from economic theory include the external benefits of R&D. If a firm succeeded in developing fusion power to the point where it became commercially successful, it would probably have difficulty in earning a very large share of the returns. It would be able to patent some of the equipment used, but once the basic approach that works is well known, it is often possible to invent around the patent. Furthermore, fusion would be a long-term solution to energy problems but patents expire in seventeen years. The standard economic rationale for government support of R&D is that a private party would not be able to capture all the benefits and hence would not do enough research. This argument is strongest for basic research and grows progressively weaker as we move toward the actual hardware of pilot plants, where much of the knowledge may be embodied either in patented equipment or in unwritten operating practice. The external benefits expected from energy research can therefore be used to justify some funding by government, especially of the early stages.

A related justification is the fear of possible external costs from continuing in the traditional way. If burning fossil fuels adds enough carbon dioxide to the atmosphere to do some damage, then this is a cost to everyone. By making fusion feasible, that cost of fossil fuels could be eliminated. Yet it is not in the interest of any individual to make that kind of investment because the private cost of the fossil fuels does not reflect the possibility that they may harm everyone. If we were certain of the adverse consequences, the government could levy a tax large enough to make the marginal private cost of fossil fuels equal to the marginal social cost.

It is sometimes argued that government should undertake investments that are too big for private firms. Nuclear fission power (and now fusion) may have been too large for any private firm to risk. Often the argument that private firms lack the capacity to undertake the research is just another way of saying that the expected returns are too low to justify an investment. Why should government devote resources to projects that appear to have low payoffs (or none at all)? The justification has to be given in terms of the familiar categories of externalities, national security, or the need for the government to look further ahead. Government research on fusion power began in the 1950s and will certainly not bear fruit for at least half a century from its starting point. It would be easier under such conditions to justify the whole enterprise by sheer curiosity than to pretend that we could calculate some acceptable rate of return.

If we assign government some grander role than petty calculator, such as ultimate insurer against catastrophe, then the government does have a responsibility to support research of two kinds. The first is to ensure that some source of energy—even if it is an expensive backstop—will always be available. The second is to probe deeply into health, safety, and environmental effects. The former would seem to be the way to justify funding research on fusion (in addition to pure curiosity), but it offers no guide for answering the political question of whether that funding should be $400 million or $800 million per year.

For a politician the most hazardous time bomb is not the environmental catastrophe of next century but the unforeseen change in income distribution of next year. The large price changes of the 1970s had a magnified impact on economic rents from oil and gas reserves, which produced a corresponding change in the value of the reserves. At the time when consumers had to contend with enormous increases in energy costs, producers were enjoying large increases in income and wealth. This inherently painful change was even more politically charged because of the regional distribution of gainers and losers. The old industrial cities of the North saw the increase in fuel prices as yet another nail in their coffins, whereas the already booming Southwest made further spectacular gains.[12] Experience diverged even for colleges and universities; schools in the snowbelt adopted desperate measures to reduce the impact of high heating bills, while the University of Texas saw its endowment soar as oil royalties flowed in.

Whereas imports had once served to decrease the economic advantage of the energy-rich states, in the 1970s they began to increase it. In the 1950s, as the cheap Middle Eastern oil began to enter the country, the producing regions sought protection. They were then able to supply any oil the country would buy, but not as cheaply as the imports. The energy squabbles began then with the East Coast pressing for more imports and the producing regions calling for a greater degree of independence. The political solution was to impose quotas. Moreover, the right to import specified amounts was given by government to certain firms, rather than being auctioned off to benefit the taxpayers.

At the time of the embargo and later, it was the consuming interests who pressed for controls on prices.[13] Although producers would have received higher prices in the absence of controls, it is not clear that retail prices would have been different, since the international oil companies responded to the controls by shipping their most expensive oil to the United States; moreover, some products were being imported at world prices. There are grounds for arguing that the government control transferred profits from (small) producers to (large) refiners in the name of equity. But logical analysis is not the strong suit of politicians. The controls stayed on to satisfy consumers that Congress was doing something for them—even if that something was to increase imports, prices, and inconvenience! When the controls finally were removed, it was only because a windfall profits tax had been imposed with the objective of capturing economic rent.

Some of the energy-producing areas also feel aggrieved. In particular, some of the newly strip-mined areas are inhabited by people who valued the peace and quiet of an earlier time. If such people do not block mining completely, they may seek some compensation. The most natural approach is to try to extract a large share of the economic rent for the state or locality. Often the effort is discussed in terms of natural heritage or leaving some capital endowment to future generations as compensation for the endowment of natural resources that is used up.[14] Whatever the verbiage, implementation of such a tax is conditioned on the existence of economic rents. (In the short run, the returns to capital investments can be taxed away, but that will inhibit new development.)

Economic rent is a slippery object of taxation. In a recent example, Montana imposed a heavy severance tax on coal. This issue reached the U.S. Supreme Court on the grounds that the tax was an unconstitutional state tax on interstate commerce. Since the tax was levied as a severance tax on all coal rather than a tariff on exports, Montana won the legal case, but that just obscured the economic issue. If there is economic rent, someone will get it. If firms could really pass the tax on to their customers, it indicates that the firms or landowners had misjudged the market. It appears most likely that in the long run a severance tax will be paid by the firm or the owner of the mineral rights (whichever had managed to obtain the economic rent) and that it will reduce output somewhat by raising the cost of the marginal coal. It is not a tax on interstate commerce, or on all consumers, but on the recipients of economic rents.

Interstate differences in the degree of energy self-sufficiency, especially in gas and oil, have raised the most troublesome political questions. Traditionally, these have involved differences of opinion about restricting imports, where the lines are always drawn between producing and consuming regions of the country. Once the government moved into controlling prices (and the inevitable controls on allocation that followed), the stakes in the political controversy became higher. The political discussion suggests that the sole objective of energy policy in the United States during the 1970s was to prevent large oil companies from making profits! The government failed to achieve that objective, but in the process of pursuing it managed to destroy many of the incentives for increased production and more efficient consumption.

PROSPECTS FOR A COHERENT ENERGY POLICY

The United States has not been able to devise a coherent energy policy of any description, and elementary political analysis indicates that it is impossible. The reasons for this pessimistic conclusion center around the complexity of the issues relative to congressional capacities and the variety of diverse interests involved. It is important to bear in mind that the political process produces particular laws, not policy. Each law reflects the pressures of the moment of passage when it seemed expedient for a majority of Congress to endorse it and when the president was willing to go along. But circumstances and the lineup of pressure groups change, so it is to be expected that laws of one moment will be inconsistent with those of another. Only rarely, however, will a reversal in the preferences of the majority be strong enough to provoke repeal of existing legislation.

I mentioned some of the sources of diverging interests in discussing regional differences. The main problems are the differences between consumers and producers and the bewildering variety of interests of different firms. A smaller country with no domestic oil production has a much simpler task in forging a unified national policy. The policy might turn out to be wise or stupid, but it is likely to be coherent if everyone in the country wants the same thing.

In the United States, however, some oil companies want favors for domestic production, whereas others produce in various foreign countries. Integrated firms have different interests from separate producers, refiners, or marketers. Some gas producers want the price set free immediately; distributors are concerned that they will lose their customers. Railroads, as well as coal mining companies, want to encourage coal production, but railroads do not want coal slurry pipelines to have the right of eminent domain. There has also been a great deal of maneuvering to give an edge to either eastern or western coal. A variety of firms and research groups are interested in nuclear fission, fusion, oil shale, solar, and other sources of energy.

Moreover, the lineup of firms on particular issues is highly specific to the exact proposals. Some have decried the monopolization of energy by a few large firms, but the energy industries are unconcentrated compared with much of manufacturing. More important, even the firms that participate in several parts of the industry (for example, gas, oil, pipelines, and coal) usually have such different mixes of imports, domestic purchases, refining capacity, and retail sales that the degree of similarity of interest differs from issue to issue.

Thus neither industries nor regions speak to the politicians with one voice on energy issues. Other groups are divided as well. Even the self-professed public interest groups are divided internally among consumerist advocates of lower prices, conservationist advocates of higher prices, environmentalist enemies of coal, and opponents of nuclear power. The only policy that would satisfy all of the constraints that the public interest groups would impose would be to provide all Americans with large flat rocks on which they could bask in the sun like lizards to absorb their quota of energy. Even this would not satisfy the Consumer Product Safety Commission if the rocks had sharp edges.

The diversity of interests in the private sector of the economy is represented also in the various federal agencies. It is not easy to define a coherent energy policy that does not encounter resistance from some agency because of inadequate attention to other legislated objectives. The Environmental Protection Agency is especially likely to be in the middle of any attempt to increase output of coal, oil, oil shale, or hydroelectric power. Any attempt to increase the amount of mineral extraction will undoubtedly run into conflicts with the way that the Department of the Interior administers public lands. Remember that each agency has employees who have chosen to pursue certain goals, have been trained to do so, and are constantly in contact with outside groups and particular political leaders who approve of those goals. It is not easy to change the direction of the agency, even by order of the president or department secretary.

In addition to these diverging objectives, energy policy is difficult to formulate because the topic is complex. The political process, like other forms of show business, deals with audience reactions to appearances. Politicians are successful practitioners of such arts but frequently have no understanding of science, engineering, or economics. The analytical frame of mind may even be inconsistent with the thinking of someone who is inclined to consider how the public will react to particular words rather than how the world will behave.

The inherent complexity of energy issues means that our representatives will not understand what they are doing and also that they can be confident that most voters will not understand either. As a result, the laws that are passed will include inconsistent wish lists that Congress instructs the agency to achieve: decrease price, increase supply, ensure fair allocation, prevent windfall profits, encourage domestic production, conserve resources, and so on. The agency is put in a position where it cannot comply with all of its congressional directives.[15]

Under such conditions we might expect an aggressive manager of the agency to pick from the list that Congress gave him or her the objectives that the manager personally prefers and then to point out that the remainder were impossible with the limited resources available. This sort of thing does occur, but in the highly charged political atmosphere that surrounds energy issues (because of the strong regional interests and large redistributions of income involved), another outcome is possible. The agency may devote its efforts to creating the appearance of action and subsidizing studies that forecast crises in the absence of further resources for the agency. This is a reasonably good description of the early behavior of the Department of Energy and its predecessors. It is difficult to find any concrete accomplishments except for the routine reporting of data relating to energy. Nor is this surprising; if the politicians refuse to make the difficult choices between competing objectives, why should the bureaucrats be expected to do so?

Since the good that the government of the United States can do in the energy area is limited, it would seem wise to restrict the range of activities that government undertakes. The market, if it is left uncontrolled, can help to promote conservation and the development of domestic energy sources. Indeed, it was the response of individual firms to the higher prices for oil, gas, and electricity that brought about the declines in the energy intensiveness of GNP at the end of the 1970s. We can only speculate about how much earlier the change would have occurred had government not intervened to "protect" the consumer from high prices. Certainly, however, we can count as a cost of price controls not only the excessive consumption and the time people spent waiting in line for supplies but also the oil and gas wells drilled in locations chosen because of the way the price control laws were structured, rather than because of geology and transportation systems.

Conceivably, the government has the competence to play some role. But the individual tasks of ensuring safety, preventing extreme environmental disruption, subsidizing basic research into backstop technologies, and administering a strategic petroleum reserve are so immense that they constitute a full agenda for government. Fortunately, the items on this list can be carried out without mutual coordination, although in questions of safety and environmental impact it would be helpful if one agency could survey the entire range of energy sources.

The more detailed planning—who should be allowed to use natural gas, who must be forced to convert to coal, what mix of energy sources will prevail two decades hence—are beyond the capacity of the United States government. The results of the attempt to exert this level of control include delay and reversals of policy, inconsistent decisions, and political uproar. Most important, by assuming that decisions could be made administratively or politically, the government delayed making the politically difficult decision to leave such matters to the market.

Nor should we suppose that the market is an inferior mechanism to which the United States must resort only because of its large size and diverse interests. The alternative to market decision making (other than pure whim or intuition of the decision makers) must be reliance on some sort of formal model of the supply of and demand for energy. The formal models of energy have not worked out well for a variety of obvious reasons. The supply of energy is determined not only by geology and exploration but also by political conditions relating to the stability of supply from foreign countries and to the decisions in the United States on environmental and safety issues. The modeling process is not good at dealing with such matters or with instabilities on the demand side that have an equivalent impact.

Can markets forecast geological discoveries and political cataclysms better than a formal model? Obviously, we could build into a formal model the best possible forecasts of such events—but only if we know which forecasts are best. Models are constructed by people with no special expertise in such matters; if they were very good at forecasting matters of such momentous importance, their time would be too valuable to waste on modeling! In short, they could be speculating in the private market, making their own fortunes while also making the market anticipate the future better than does the model.

The government can use the model to test various contingencies—this indeed is the most valuable use of models. It might be possible for a clever technician to construct a model that predicts accurately the result of a revolution in Saudi Arabia. Could the Department of Energy really base its decisions on a plan incorporating that contingency without arousing the wrath of both the State Department and the House of Saud? The problem is that a forecast used by government is a political document. This limits the flexibility of an open government in planning for political contingencies.

The fact that planning does not work well and cannot really be used anyway is an advantage. Regardless of the quality of a comprehensive plan, it is a device of inherently limited imagination because it represents an ossified part of the thinking of a few people. As such it can incorporate new facts and analysis only with great difficulty and at infrequent intervals. Ideas that are not popular enough to belong in the consensus drop out of play completely. The market permits new and minority views to influence the outcome. It is less likely to lead to overreliance on a few expensive or obsolete techniques. The United States should be able to survive very well in the absence of any central energy planning. All it requires is the realism and political will to give up price and allocation controls and to leave our destiny in our own individual hands.

NOTES

1. See David A. Deese and Joseph S. Nye, eds., *Energy and Security* (New York: Ballinger, 1981); Douglas R. Bohi and W. David Montgomery, *Oil Prices, Energy Security, and Import Policy* (Baltimore, Md.: Johns Hopkins University Press/Resources for the Future, 1982).
2. See Henry S. Rowen, "Comprehensive Energy Policy: King Stork?" in *Options for U.S. Energy Policy* (San Francisco: Institute for Contemporary Studies, 1977).
3. *Monthly Energy Review*, December 1984, p. 45.

4. S. Fred Singer, "Our Strategic (for Others) Oil Reserve," *Wall Street Journal*, August 3, 1983, p. 18 argues that since the market will direct oil to the highest bidder, our SPR is a reserve for the entire world. Japan will benefit from it more than will the United States, and Saudi Arabia will benefit even more! For a detailed analysis, see William W. Hogan, "Oil Stockpiling: Help Thy Neighbor," *Energy Journal* 4, no. 3 (July 1983), pp. 49–71.
5. M. A. Adelman, "Coping with Supply Insecurity," *Energy Journal* 3, no. 2 (April 1982), p. 15.
6. Richard B. Mancke, *Squeaking By: U.S. Energy Policy since the Embargo* (New York: Columbia University Press, 1976); *Wall Street Journal*, March 6, 1984, p. 32.
7. Sam H. Schurr, "Energy Efficiency and Productive Efficiency: Some Thoughts Based on American Experience," *Energy Journal* 3, no. 3 (July 1982), pp. 3–14; Sam H. Schurr, Sidney Sonenblum, and David O. Wood, *Energy, Productivity, and Economic Growth* (Cambridge, Mass.: Oelgeschlager, Gunn & Hain, 1983).
8. For a thoughtful analysis, see *The No-Growth Society*, ed. Mancur Olson and Hans H. Landsberg (New York: W. W. Norton, 1973).
9. See, for example, "Energy Activities with Energy Ant," Federal Energy Administration (Washington, D.C.: U.S. Government Printing Office, May 1975).
10. William S. Peirce, *Bureaucratic Failure and Public Expenditure* (New York: Academic Press, 1981), pp. 42–47.
11. John E. Tilton, "The Public Role in Energy Research and Development," in *Energy Supply and Government Policy*, ed. Robert J. Kalter and William A. Vogely (Ithaca, N.Y.: Cornell University Press, 1976).
12. *National Journal*, March 22, 1980, pp. 468–474.
13. Paul W. MacAvoy, *Energy Policy: An Economic Analysis* (New York: W. W. Norton, 1983), chap. 2 provides an analysis of policy relating to oil prices; the political economy is analyzed by Joseph P. Kalt, "Oil and Ideology in the United States Senate," *Energy Journal* 3, no. 2 (April 1982), pp. 141–166.
14. Warren Aldrich Roberts, *State Taxation of Metallic Deposits* (Cambridge, Mass.: Harvard University Press, 1944).
15. For an example, see "Controlling Petroleum Prices," chap. 11 in Peirce, *Bureaucratic Failure and Public Expenditure*. For an example of the use of a formal model to analyze the impact of government policies, see Paul F. Dickens III, David L. M. Nicol, Frederic H. Murphy, and Julie H. Zalkind, "Net Effects of Government Intervention in Energy Markets," *Energy Journal* 4, no. 2 (April 1983), pp. 135–149.

Index